FIVE KINGDOMS

A phylogeny of life on Earth based on the Whittaker five–kingdom system
and the symbiotic theory of the origin of eukaryotic cells.

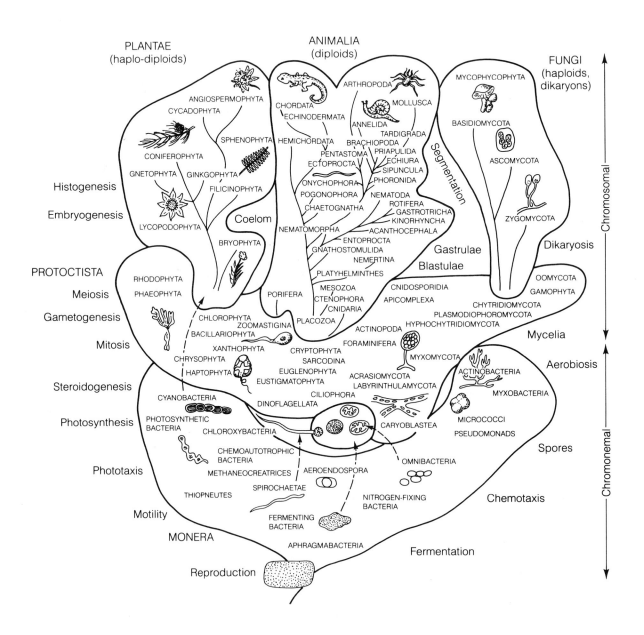

FIVE KINGDOMS

An Illustrated Guide to the Phyla of Life on Earth

Lynn Margulis
BOSTON UNIVERSITY

Karlene V. Schwartz
UNIVERSITY OF MASSACHUSETTS, BOSTON

W. H. FREEMAN AND COMPANY
San Francisco

Project Editor: Patricia Brewer
Manuscript Editor: Andrew Kudlacik
Designer: Robert Ishi
Production Coordinator: Bill Murdock
Illustration Coordinator: Cheryl Nufer
Supplemental artist: Donna Salmon
Compositor: York Graphic Services
Printer and Binder: Arcata Book Group

Library of Congress Cataloging in Publication Data
Margulis, Lynn, 1938–
    Five kingdoms.

    Includes bibliographies and index.
      1. Biology—Classification. I. Schwartz, Karlene V.,
1936–   . II. Title.
QH83.M36   574'.012   81–7845
ISBN 0-7167-1212-1   AACR2
ISBN 0-7167-1213-X (pbk.)

Printed in the United States of America

2 3 4 5 6 7 8 9 0 KP 8 9 8 7 6 5 4 3 2

CONTENTS

*Genera illustrated in each phylum are given in italics below the title of the phylum.

CHAPTER 5 PLANTAE 245

FOREWORD

Like bureaucracy, knowledge has an inexorable tendency to ramify as it grows. In the early nineteenth century, the great French zoologist Georges Cuvier classified all "animals" —moving beings, both microscopic and visible—into just four great groups, or phyla. A century earlier, Linnaeus himself, the father of modern taxonomy, had lumped all "simple" animals into the single category "Vermes" —or worms.

Cuvier's four animal phyla have expanded to more than forty, distributed in two kingdoms, the Protoctista (for microscopic forms and their descendants) and the Animalia (for those that develop from embryos)—and remember, we have said nothing of plants, fungi, other protoctists, and bacteria as yet! The very names of these groups are imposing enough—kinorhynchs, priapulids, onychophorans, and the recently discovered gnathostomulids. Some biologists can spit out these names with a certain virtuosity; but like my seven-year-old son who knows both Babe Ruth's and Ty Cobb's lifetime batting averages but doesn't know what a batting average is, most of us know rather few of the animals behind the names. This ignorance arises for two primary reasons: the names are simply now too many, and modern training in zoology is now so full of abstract theory that old-fashioned knowledge of organic diversity has, unfortunately, taken a back seat.

Margulis and Schwartz have generated here that rarest of intellectual treasures—something truly original and useful. If the originality comes before us largely as a "picture book," it should not be downgraded for that reason—for primates are visual animals, and the surest instruction in a myriad of unknown creatures must be a set of figures with concise instruction about their meaning—all done so admirably in this volume. It is remarkable that no one had previously thought of producing such a comprehensive, obvious, and valuable document.

My comments thus far have been disgracefully zoocentric. I have spoken only of animals, almost as if life were a ladder with animals on the top rungs and everything else inconspicuously and unimportantly below. The old taxonomies included two kingdoms (plants and animals, with unicells placed, in Procrustean fashion, into one or the other camp), or at most three kingdoms (animals, plants, and unicells). With this work, and its 89 phyla distributed among five kingdoms, we place animals (including ourselves) into proper perspective on the tree of life—we are a branch (albeit a large one) of a massive and ramifying tree. The greatest division is not even between plants and animals, but *within* the once-ignored microorganisms—the prokaryotic Monera and the eukaryotic Protoctista. The five kingdoms are arrayed as three great levels of life: the prokaryotes (bacteria of the kingdom Monera), the eukaryotic microorganisms and their derivatives (Protoctista), and the eukaryotic larger forms (Plantae, Animalia, and Fungi). These last three familiar kingdoms represent the three great ecological strategies for larger organisms: production (plants), absorption (fungi), and consumption (animals).

Some people dismiss taxonomies and their revisions as mere exercises in abstract ordering—a kind of glorified stamp collecting of no scientific merit and fit only for small minds who need to categorize their results. No view could be more false and more inappropriately arrogant. Taxonomies are reflections of human thought; they express our most fundamental concepts about the objects of our universe. Each taxonomy is a theory about the creatures that it classifies.

The five-kingdom system and the expansion of phyla to nearly 100 represent a set of new and exciting ways of thinking about organisms and their evolution. I greatly value the opportunity to express my thoughts on these important issues, but must close by stating that I write this foreword only by a cruel accident of fate. Robert H. Whittaker, the great Cornell ecologist and general biologist, who developed the five-kingdom system and would have written this foreword, died of cancer in 1980, long before his time. The virtually universal adoption of his system by major biology textbooks (usually the slowest medium to accept change of any kind) is a testimony to the power of his ideas and to the importance of this work and its subject.

May 1981

Stephen Jay Gould
HARVARD UNIVERSITY

This book is a catalog of the world's living diversity. We wrote it for science students, their teachers, and anyone else who is curious about the extraordinary variety of living things that inhabit this planet. We have described and illustrated all major groups of organisms, reckoned as almost one hundred phyla belonging to the five kingdoms proposed more than two decades ago by R. H. Whittaker. Much of this knowledge, particularly about the microbial world, has been gathered only in the last decade and represents the frontier of research. Botanists and zoologists agree with few exceptions on the classification of plants and animals into phyla. Microbiologists, however, have not yet had the opportunity to erect phyla that can satisfy bacteriologists, phycologists, protozoologists, and others. Thus, we have been compelled to construct some new phyla for microbes, on principles consistent with the accepted classification of larger organisms.

The Introduction broadly describes life based on cells and explains why viruses are not considered cells. It discusses the profound differences between the two great styles of cellular organization, the prokaryotic (nonnucleated) style of Kingdom Monera and the eukaryotic (nucleated) style of the other four kingdoms. A chronology of the last four billion years of geological history is described and summarized as a chart on pages 16–17.

The introductory sections of the next five chapters describe the general features of each kingdom: Monera, Protoctista, Fungi, Animalia, and Plantae. Each introduction is accompanied by an illustrated phylogeny, a family tree showing the evolutionary relationships among the groups of the kingdom described. All the phylogenies, especially those of the microorganisms, are tentative, because the evolutionary history of no group is known for certain. At the end of each chapter is a bibliography of suggested readings on the members of each phylum described in the chapter. We have selected one or two species to illustrate each phylum, represented by one or more photographs, an anatomical drawing, and a brief essay. These presentations, 89 in all, constitute the main part of the book.

An Appendix classifies about a thousand genera, including all those mentioned in the book, to phylum level; it also includes the common names of many organisms. Finally, there is an extensive Glossary.

Some arthropod specialists tell us that there may be ten million species of organisms, most of them insects, on the Earth. No one can hope to know them all. Because of different traditions among the biological disciplines, there is not yet a consensus on the nomenclature and systematics of every taxon. Thus, not all of the taxa in this book will be uniformly acceptable. Nevertheless, we believe that our scheme is generally acceptable, consistent, and useful. We welcome corrections, suggestions for finer photographs and drawings, and any information that can improve the next edition.

June 1981   *Lynn Margulis*
                                                              Karlene V. Schwartz

ACKNOWLEDGMENTS

The artists' way of looking at organisms sharpened our own understanding immensely; for this we thank Laszlo Meszoly, Michael Lowe, Illyse Atema, Dorion Sagan, Emily Hoffman, Linda Reeves, Peter Brady, and Robert Golder. Were it not for Michael Gorczyca, the Glossary would not exist. Natasha Villamia, Matthew Zavitkovsky, Rosaline Glicksman, Betsey Grosovsky, Gregory Vogel, Jeremy Sagan, Geraldine Kline, Janet Williams, and Joan Howard have given talented assistance during manuscript preparation. We are grateful to William Ormerod and James G. Schaadt for their aid in identifying organisms in the field and for taking and selecting photographs.

Thanks are due the hundreds of specialists in the United States and abroad who expressed enthusiasm by sending theses, manuscripts, photographs, and drawings, read portions of our manuscript, and clarified our illustrations. We are grateful to R. H. Whittaker for the initial impetus to illustrate his five-kingdom concept. We are only sorry that he did not live long enough to enjoy the finished book. We thank the librarians at the University of Massachusetts in Boston, the University of Hawaii, and the Marine Biology Laboratory at Woods Hole for their unstinting efforts in securing books and journals. We wish to acknowledge the critical comments and continuing encouragement of our colleagues at Boston University and the University of Massachusetts, Boston.

Our biggest debt is to the Life Sciences Division at NASA, which has supported our research since 1973 under grant NGR-004-025, Microbial Contributions to the Precambrian Earth. During the pre-Viking days, the possibility of Martian life and the form that extraterrestrial life might take were widely discussed. In conversation with NASA scientists, primarily

Richard S. Young, Donald L. DeVincenzi, and Gerald A. Soffen, we became aware of the general lack of appreciation of the diversity of life on Earth, especially the microbial world. Many of the world's organisms might not have been identified as living by scientists in search of life on the red planet. The lack of an illustrated guide to life on Earth was felt not only in Viking scientific circles but among biology students and teachers who had noticed that no single comprehensive book introduced taxa of both large and microscopic organisms. Our efforts have been to remedy this, and we depended in these efforts on dedicated work-study and graduate students at Boston University who helped us to collect information. Only NASA grant support made this possible.

The preparation of the index depended on Jeremy Sagan's skillful programming and hours of aid from B. D. D. Grosovsky, K. E. Nealson, L. Read, W. Solomon, D. Sagan, and I. Taylor.

One of us (L. Margulis) acknowledges the support of the Guggenheim Foundation (Fellow, 1979), and the other (K. V. Schwartz) the support of the Biology Department, University of Massachusetts, Boston, for sabbatical in Hawaii (1979–1980). We are both of course grateful to our families (Nick, Dorion, Jeremy, Zachary, Jenny, Lowell, and Jonathan) for urging us onward.

L.M.
K.V.S.

INTRODUCTION

Great contributors to our concepts of kingdoms and phyla

Carolus Linnaeus (Carl von Linné, Swedish) originated the concept of binomial nomenclature
  and a comprehensive scheme for all nature.
Antoine Laurent de Jussieu (French) established the major subdivisions of Kingdom Plantae.
Georges Léopold Cuvier (French) established the major "embranchements" (phyla) of Kingdom Animalia.
Ernst Haeckel (German) was the innovator of Kingdom Monera for bacteria.
Herbert F. Copeland (American) reclassified all the microorganisms, recognizing
  Kingdom Protoctista for the nucleated microorganisms.
Robert H. Whittaker (American) founded the five kingdom system, recognizing Kingdom Fungi.

Antoine Laurent de Jussieu
1748–1836
[The Bettmann Archive]

Georges Léopold Cuvier
1769–1832
[Bibl. Museum Hist. Nat. Paris]

Carolus Linnaeus
1707–1778
[The Bettmann Archive]

Herbert F. Copeland
1902–1968

R. H. Whittaker
1924–1980
[Courtesy of R. Geyer]

Ernst Haeckel
1834–1919
[The Bettmann Archive]

CLASSIFICATION SYSTEMS

At least three million and perhaps ten million species of living organisms are now alive, and an even greater number has become extinct. The effort to discern order in this incredible variety has given rise to systematics, the classification of the living world. Modern systematists group closely related species into *genera* (singular, *genus*), genera into *families*, families into *orders,* orders into *classes,* classes into *phyla* (singular, *phylum*), and phyla into *kingdoms* (see Table 1). This conceptual hierarchy grew gradu-

Table 1   The Classification of Two Organisms

Taxonomic level	Man	Garlic
Kingdom	Animalia	Plantae
Phylum (Division)*	Chordata	Angiospermophyta
Subphylum†	Vertebrata	
Class	Mammalia	Monocotyledoneae
Order	Primates	Liliales
Family	Hominoidea	Liliaceae
Genus	*Homo*	*Allium*
Species	*sapiens*	*sativum*

*Botanists use the term *division* instead of *phylum.*
†Intermediate taxonomic levels can be created by adding the prefixes *sub* and *super* to the name of any taxonomic level.

ally, in the course of about a century, from a solid base established by the Swede Carolus Linnaeus (1707–1778), who began the modern practice of binomial nomenclature. Every known organism is given a unique two-part name, Latin in form. The first part of the name is the same for all organisms in the same genus; the second part distinguishes them from each other. For example, *Acer saccharum, Acer nigrum,* and *Acer rubrum* are the scientific names of the sugar maple, the black maple, and the red maple.

Groups of all sizes, from species on up, are called *taxa* (singular, *taxon*), and *taxonomy* is the analysis of an organism's characteristics for the purpose of assigning the organism to a taxon. Since the time of Linnaeus, the growth of biological knowledge has greatly extended the range of characteristics used in taxonomy. Linnaeus based his classification on the visible structures of living organisms. Later, extinct organisms came into the picture. In the nineteenth century, the discoveries of paleontologists and Darwin's revelation of evolution by natural selection encouraged systematists to believe that their classifications reflected the history of life—classifications were converted into *phylogenies,* family trees of species or higher taxa. To this day, very few lineages from fossil organisms to living ones have actually been traced, yet the truest classification is still held to be the one that best reflects the evidence for relationship by common ancestry.

In the twentieth century, advances in embryology and biochemistry have given the taxonomist new tools. For example, phylogenies can now be based on patterns of larval development or on the linear order of amino acids that compose proteins. In recent years, there have been great improvements in the techniques of electron and optical microscopy. The internal structure of the smallest life forms and of the constituent cells of large forms can now be studied in unprecedented detail.

From Aristotle's time to the middle of the twentieth century, most biologists were content to divide the living world into two kingdoms, plants and animals. Since the middle of the nineteenth century, however, many systematists have seen that certain organisms, such as bacteria and fungi, differ from plants and animals more than plants and animals differ from each other. Third and fourth kingdoms to accommodate these anomalous organisms were proposed from time to time. However, most biologists either ignored these proposals or considered them unimportant curiosities, the special pleading of eccentrics.

The climate of opinion began to change in the 1960s, largely because of the knowledge gained by new biochemical and electron-microscopical techniques. These techniques revealed fundamental affinities and differences on the subcellular level that encouraged a spate of new proposals for multiple-kingdom systems—one included thirteen kingdoms! Among them, a system of five kingdoms first proposed by R. H. Whittaker in 1959 has distinguished itself by gathering support steadily during the past two decades. With a few modifications, it is the one used in this book. Briefly, the five kingdoms are Monera (bacteria), Protoctista (algae, protozoans, slime molds, and other less known aquatic and parasitic organisms), Fungi (mushrooms, molds, and lichens), Animalia (animals with or without back-bones), and Plantae (mosses, ferns, cone-bearing plants, and flowering plants).

CELLS OF THE FIVE KINGDOMS

Antony van Leeuwenhoek (1632–1723), the discoverer of the microbial world, called the microorganisms that he found everywhere in vast num-bers "little animals." For more than a century after his discoveries, it was commonly held that these little animals arose spontaneously from inani-mate matter. Louis Pasteur (1822–1895) and John Tyndall (1820–1893) showed conclusively that, like macroscopic organisms, microbes are pro-duced only by other microbes.

Ernst Haeckel (1834–1919), the German champion and popularizer of Darwin's theory of evolution, made several proposals for a third kingdom of organisms. The boundaries of Haeckel's new kingdom, Protista, fluctuated in the course of his long career, but his consistent aim was to set the most primitive and ambiguous organisms apart from the plants and animals, with the implication that the higher organisms evolved from protist ancestors. Within the protist kingdom, Haeckel recognized the bacteria and blue-green algae as a major group, the Monera, distinguished by their lack of a cell nucleus.

The nucleus is a structure found in the cells of all large, familiar orga-nisms. Typically spherical and conspicuous under the microscope, the nucleus is separated by a *nuclear membrane* from the rest of the cell, the *cytoplasm.* If a cell is stained with a standard dye at certain stages of its life cycle, the nucleus can be seen to contain several rod-shaped bodies, the *chromosomes,* which preferentially take up the dye. The chromosomes are composed of several proteins and *deoxyribonucleic acid* (DNA). DNA is a very long skinny molecule that is capable of duplicating itself. It bears the information that the cell uses to synthesize the specific proteins it needs to live and reproduce. The number of chromosomes in a cell nucleus is a characteristic of each species—leaf cells of the wild aster *Haplopappus gracilis* contain only four, whereas those of some species of *Kalanchoë* (also flowering plants) contain about 500; each human tissue cell contains 46 chromosomes.

Mitosis, the kind of cell division that results in two identical daughter cells, begins when the chromosomes duplicate themselves, forming two identical sets. A *spindle* forms, consisting of long, thin *microtubules* of protein that radiate from poles at opposite ends of the cell and attach themselves to the chromosomes. In some way, the microtubules separate the two chromosome sets. Thus, when the cell has divided, each daughter cell is assured of a complete set of chromosomes identical to those in the parent cell.

In 1937, the French marine biologist Edouard Chatton wrote a little paper for an obscure journal published in Egypt suggesting that the term *procariotique* (from Greek *pro,* meaning *before,* and *karyon,* meaning *kernel* or *nucleus*) be used to describe bacteria and blue-green algae, and the term *eucariotique* (from Greek *eu,* meaning *true*) to describe animal and plant cells. During the past four decades, Chatton's insight into the nature of cells has been abundantly verified. Virtually all biologists now agree that

> this basic divergence in cellular structure which separates the bacteria and the blue green algae from all other cellular organisms, probably represents the greatest single evolutionary discontinuity to be found in the present-day world.*

This fundamental distinction is represented in the five-kingdom system used in this book—Kingdom Monera contains *prokaryotes* and only prokaryotes; the organisms of the other four kingdoms are all *eukaryotes.*

Both in structure and in biochemistry, eukaryotes and prokaryotes differ by far more than the presence and absence of a cell nucleus (see Table 2). As you can see by comparing the illustrations in Chapter 1 with those in Chapter 2, prokaryotic cells are generally simpler in structure (but not necessarily in chemistry) and smaller than eukaryotic cells. Their DNA is not organized into chromosomes; because it is not combined with protein,

---

*R. Y. Stanier, E. A. Adelberg, and M. Doudoroff, *The Microbial World,* 3rd ed., 1963, Prentice-Hall, Englewood Cliffs, N.J.

Table 2   Major Differences Between Prokaryotes and Eukaryotes

Prokaryotes	Eukaryotes
Mostly small cells (1–10 μm). All are microbes.	Mostly large cells (10–100 μm). Some are microbes; most are large organisms.
DNA in nucleoid, not membrane-bounded. No chromosomes.	Membrane-bounded nucleus containing chromosomes made of DNA, RNA, and proteins.
Cell division direct, mostly by binary fission. No centrioles, mitotic spindle, or microtubules. Sexual systems rare; when sex does take place, genetic material is transferred from donor to recipient.	Cell division by various forms of mitosis; mitotic spindles (or at least some arrangement of microtubules). Sexual systems common; equal participation of both partners (male and female) in fertilization. Alternation of diploid and haploid forms by meiosis and fertilization.
Multicellular forms rare. No tissue development.	Multicellular organisms show extensive development of tissues.
Many strict anaerobes (which are killed by oxygen), facultatively anaerobic, microaerophilic, and aerobic forms.	Almost all are aerobic (they need oxygen to live); exceptions are clearly secondary modifications.
Enormous variations in the metabolic patterns of the group as a whole.	Same metabolic patterns of oxidation within the group (Embden-Meyerhof glucose metabolism, Krebs-cycle oxidations, cytochrome electron transport chains).
Mitochondria absent; enzymes for oxidation of organic molecules bound to cell membranes (not packaged separately).	Enzymes for oxidation of 3-carbon organic acids are packaged within mitochondria.
Simple bacterial flagella, composed of flagellin protein.	Complex 9+2 undulipodia composed of tubulin and many other proteins.
In photosynthetic species, enzymes for photosynthesis are bound as chromatophores to cell membrane, not packaged separately. Various patterns of anaerobic and aerobic photosynthesis, including the formation of end products such as sulfur, sulfate and oxygen.	In photosynthetic species, enzymes for photosynthesis are packaged in membrane-bounded plastids. All photosynthetic species have oxygen-eliminating photosynthesis.

prokaryotic DNA is not as easy to see in a stained cell as eukaryotic chromosomes usually are.

In an electron micrograph of a bacterium (see Chapter 1), the DNA can be seen as a light, ill-defined area called the *nucleoid,* which is not separated by a membrane from the rest of the cell. Lacking chromosomes, prokaryotes do not divide by mitosis. As a prokaryote grows, the nucleoid divides into two masses by attaching itself to parts of the cell membrane that are growing apart from each other. Eventually, the cell may form a new wall that divides the cell into two (see Phyla M-2 and M-11, Fermenting bacteria and Aeroendospora), or it may simply constrict in the middle until the two halves separate (see Phylum M-9, Nitrogen-fixing aerobic bacteria).

Eukaryotic cells contain certain *organelles,* fairly large bodies, some of them separated by their own membranes from the cytoplasm (see Figure 1). *Mitochondria* (singular, *mitochondrion*), ovoid organelles that specialize in producing energy by oxidizing simple organic compounds, are found in nearly all eukaryotes, but in no prokaryote. Green plant cells and algal cells contain one or several *plastids,* membrane-enclosed bodies in which complex structures made of membranes, chlorophyll, and other biochemicals perform photosynthesis. In photosynthetic bacteria and blue-green algae, on the other hand, chlorophyll and other photosynthetic chemicals are found as small granules on membranes that are not contained in plastids. In this characteristic as in many others, such as their method of cell division, the blue-green algae clearly resemble bacteria. In fact, many biologists now call them blue-green bacteria or *cyanobacteria* (as we do) to emphasize their bacterial nature and their lack of relationship with the nucleated algae and green plants, with which they traditionally have been classified.

The cells of most eukaryotes—many plants, most protoctists, and most animals—at some stage of their life cycle have flexible intracellular whiplike extensions called *undulipodia* (singular, *undulipodium;* traditionally called *cilium* or *eukaryotic flagellum*). All undulipodia are composed of microtubules in bundles that, in cross section, show a characteristic nine-fold symmetry (see Figure 2). They are enclosed in a ciliary (or undulipodial) membrane, which is simply an extension of the cell membrane. This nine-fold

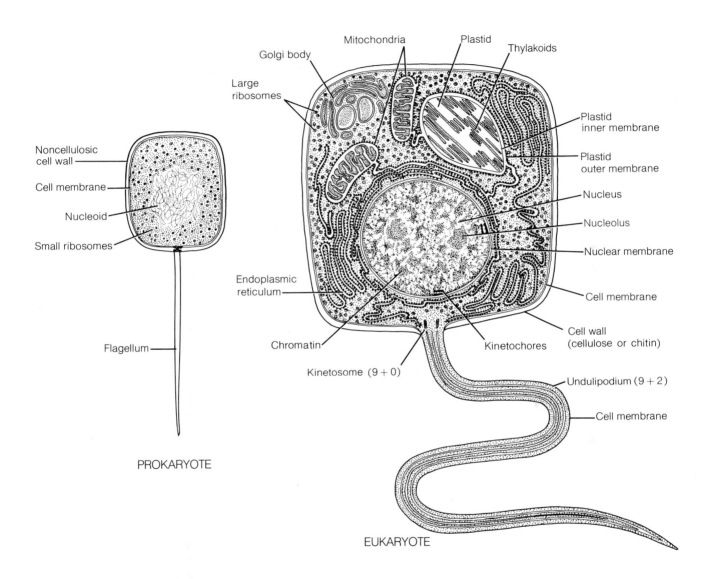

Figure 1
Typical prokaryotic and eukaryotic cells, based on electron microscopy. Not every prokaryote or eukaryote has every feature shown here. ''9+0'' and ''9+2'' refer to the cross sections of kinetosomes and undulipodia, shown in Figure 2.

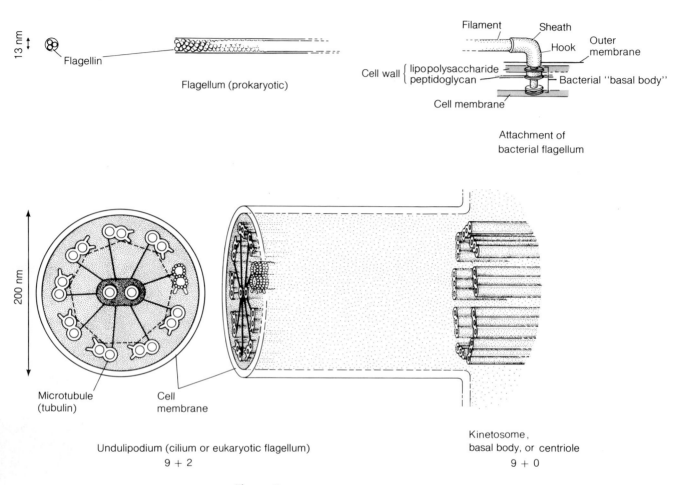

13 nm

Flagellin

Flagellum (prokaryotic)

Filament   Sheath
                    Outer
              Hook   membrane

Cell wall { lipopolysaccharide
            peptidoglycan

Cell membrane

Bacterial "basal body"

Attachment of
bacterial flagellum

200 nm

Microtubule   Cell
(tubulin)   membrane

Undulipodium (cilium or eukaryotic flagellum)
9 + 2

Kinetosome,
basal body, or centriole
9 + 0

Figure 2
The structures of an undulipodium and a bacterial flagellum.

symmetry is also found in the *kinetosomes* or *basal bodies* from which the undulipodia appear to grow, and in *centrioles,* small barrel-shaped bodies that, in many cells, appear at the poles of the spindle during mitosis. Undulipodia are composed of more than 40 different proteins. The main one, which makes up the walls of the microtubules, is called microtubule protein, or *tubulin.*

Many prokaryotic cells also bear long thin moveable extensions, called *flagella* (singular, *flagellum*). However, these are not undulipodia—they are not composed of microtubules and they do not have nine-fold symmetry.

Flagellar shafts are extracellular structures that protrude through the cell wall; they are composed of a single globular protein called *flagellin.* The beating motion of an undulipodium is caused by the transformation of chemical into mechanical energy along the full length of the organelle; a flagellum is moved by the rotation of its basal attachment to the cell.

LIFE CYCLES

In *sexual reproduction,* which is peculiar to eukaryotes, two nuclei, called *gamete nuclei,* from parent organisms merge to form a new nucleus. In many organisms, the gamete nuclei are carried by individual cells called *gametes.* In species that have distinguishable male and female sexual types, the gametes are called the *sperm* and the *ovum* (plural, *ova*), or *egg.* By definition, the male gamete is the one that travels, while the female is more sedentary. The fusion of the gamete nuclei produces the *zygote nucleus,* which contains two of each kind of chromosome, one from each parent. Such a nucleus is called *diploid;* a cell or nucleus, such as a gamete nucleus, having only one set of chromosomes, or an organism made of such cells, is called *haploid.*

In animals and in flowering plants, the zygote grows and divides repeatedly by mitosis; the resulting diploid cells differentiate, eventually to form an adult organism. To complete the cycle, the organism must produce gametes—that is, from diploid cells it must produce haploid cells. This is accomplished by *meiosis,* or *reduction division.* In most meioses, a diploid cell duplicates its chromosomes once (as in mitosis) but then divides twice in succession, producing four haploid cells. In some organisms, these cells serve immediately as gametes; in others, the haploid cells grow and divide by mitosis for a considerable time before some of them differentiate into gametes or gamete nuclei.

In animals, meiosis is called *gametic meiosis* because it immediately precedes the production of gametes. After fusion of the gametes, the zygote eventually develops into an adult animal—composed of diploid cells. Some plants, such as ferns (Phylum Pl-4, Filicinophyta), release haploid spores produced by meiosis from a diploid parent, called a *sporophyte.* The spores develop into plants that look quite different from their parent. These haploid plants are called *gametophytes,* because they eventually (without meiosis) produce gametes, which are of two types, motile and nonmotile. When a motile gamete (a sperm) finds a nonmotile one (an egg), they fuse into a zygote that develops into a sporophyte fern, completing the cycle. In flowering plants (Phylum Pl-9, Angiospermophyta), the gametophyte generation consists of only a few cells, embryo sac and pollen grain, contained in the flower, which is itself part of the sporophyte parent. The gamete nuclei fuse within the flower before seed is set.

Nearly all fungi can reproduce asexually by releasing spores produced by a single parent. Many of them are also capable of sexual reproduction, although it is different from sexual reproduction in animals and plants. Fungal spores are haploid and are generally produced in great numbers—a single giant puffball, *Calvatia gigantea* (Phylum F-3, Basidiomycota), may contain several trillion spores. The spores are released into the air. If they alight at a suitably moist place, they start to grow and divide by mitosis to produce long threads, only one cell wide, called *hyphae* (singular, *hypha*). The cells within a hypha may contain one, two, or many nuclei, depending on the species of fungus.

In sexually mature fungi, two cells from adjacent hyphae may *conjugate*—they combine to form a single *heterokaryotic* cell that contains at least one haploid nucleus from each "parent" hypha (the Greek word *heteros* means *different*). Conjugation, which is not restricted to fungi, is the fusion of growing extensions of cells, or of cells themselves, followed by migration of nuclei between the cells. By definition, it does not involve undulipodiated gametes. In many cases, the two cells, called *conjugants,* are indistinguishable in appearance.

Most fungi display opposite mating types—they cannot conjugate with themselves or with another individual of the same mating type, although often one cannot discern the mating type of a particular individual except by its willingness or unwillingness to conjugate with another. In any case, the two kinds of haploid nuclei typically do not fuse immediately, but the new cell and its nuclei divide mitotically to produce hyphae composed of heterokaryotic cells. It is as if the "male" and "female" nuclei are willing to live together but are reluctant to take the final step. Eventually, however, in some of the heterokaryotic cells, two nuclei descended from different parents finally fuse to form a diploid zygote nucleus. This stage is extremely transient. The zygote nucleus immediately undergoes meiosis (called *zygotic meiosis*) and forms haploid spores, completing the life cycle.

The full sexual life cycle, in which the number of chromosomes per nucleus is doubled by nuclear fusion, or fertilization, and then halved by meiosis, distinguishes the three most complex kingoms, the plants, animals, and fungi, from the other two, the monerans and protoctists. The life cycle of monerans, which lack nuclei and chromosomes, naturally does not include fertilization and meiosis. However, some monerans do have sex in a very broad sense. When two bacteria of the same species happen to come into contact, a thin, temporary bridge between them may form. Through this bridge, one of the bacteria donates a copy of some of its DNA—the amount varies widely, from almost none to all of the donor's DNA. The recipient bacterium incorporates some or all of the donated DNA into its own DNA. Like sex in eukaryotes, this exchange, called *bacterial conjugation,* results in an ''offspring'' (the receiving bacterium) having more than one parent. However, the genetic contribution from the two ''parents'' is very rarely equal, as it is generally in fungi, animals, and plants.

New genetically distinct bacteria can also be produced by *transduction.* Certain viruses, called *transducing bacteriophages,* pick up small amounts of DNA from host bacteria; subsequent host bacteria receive and incorporate this DNA when the viruses penetrate them. In nature, transduction may be more important than conjugation for the transfer of DNA between bacteria.

Unfortunately for those who wish nature would behave consistently, the protoctist life cycle cannot be typified as easily as those of the other four kingdoms. Protoctists are all the eukaryotes that are not animals, plants, or fungi. They include all the single-celled eukaryotes, called *protists,* and their immediate multicellular descendants. Because they are grouped together as leftovers of a classification system, it is not too surprising that their life cycles are extremely various.

In some protoctist phyla, new individuals are formed by sexual processes; in others, only by asexual ones. The life cycle of some brown algae

Figure 3
Top, Botulinum ϕ, a virus that attacks *Clostridium botulinum;* bar = 0.1 μm. Bottom, tobacco mosaic virus, which causes the mosaic blight of tobacco plants; bar = 1 μm. [Photographs courtesy of E. Boatman; botulinum drawing by R. Golder; TMV drawing by M. Lowe.]

(Phylum Pr-12, Phaeophyta) is very much like that of ferns: alternation of full-grown diploid and haploid individuals. The life cycle of some red algae (Phylum Pr-13, Rhodophyta) and some green algae (Phylum Pr-14, Gamophyta) is more like that of fungi: sex by conjugation, nuclear fusion followed immediately by meiosis, a very short diploid stage. The diatoms (Phylum Pr-11, Bacillariophyta) and many chytrids (Phylum Pr-26, Chytridiomycota) spend most of their life cycle in the diploid state, as animals do. The naked amoebae (Phylum Pr-3, Rhizopoda), the golden-yellow algae (Phylum Pr-4, Chrysophyta), and the amoeboflagellates (Phylum Pr-8, Zoomastigina, Class Schizopyrenida) have no sex at all. Nearly any pattern that works can be found among the members of this kingdom.

VIRUSES

All organisms in the five kingdoms either are cells or are composed of cells. Some arguably living forms that do not fit this description are the viruses (Figure 3). Composed of DNA (or the related RNA, ribonucleic acid) enclosed in a coat of protein, viruses are much smaller than cells. Although they reproduce, they can do so only by entering a host cell and using its living machinery. Outside the host cell, they neither reproduce, feed, nor grow. Some of them can even be crystallized, like minerals. In this state, they can survive for years unchanged—until they are wetted and placed into contact with their particular hosts.

Viruses are probably related more closely to their hosts than to each other. They may have originated as nucleic acids that escaped from cells and began replicating on their own—always, of course, by returning to use the complex chemicals and structures in their former home cells. Thus, the polio and flu viruses are probably more closely related to people, and the tobacco mosaic virus (TMV) to tobacco, than polio and TMV are to each other.

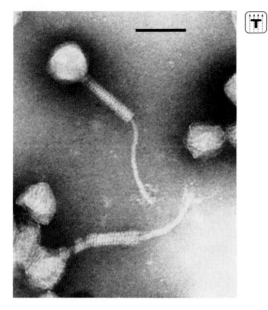

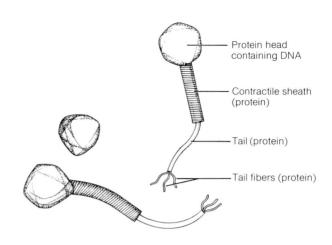

Protein head
containing DNA

Contractile sheath
(protein)

Tail (protein)

Tail fibers (protein)

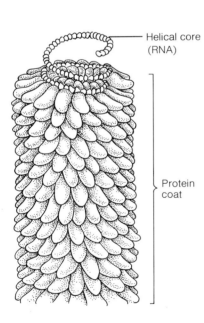

Helical core
(RNA)

Protein
coat

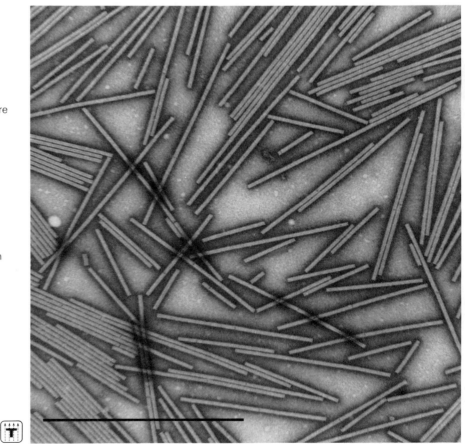

EARTH HISTORY

The Earth is almost five billion years old. The oldest fossils yet discovered, bacterium-like filaments from rocks of the Pilbara gold fields near North Pole, Western Australia, are three and a half billion years old. Bacterium-like spheroids have also been found in rocks of that age from the Swaziland rock system of southern Africa. Thus, for most of its history, the Earth has supported life. Almost everything we know about the history of the planet and of the life on it has been gleaned from evidence in its rocks. Geologists have worked out a chronology of this history from the composition of rocks, the order of their formation, and the fossils they contain (Figure 4). The upper layers of an undisturbed sequence of sediments are assumed to be younger than the lower ones; rock layers at different places on the Earth's surface are compared and matched by examining the fossils in them; absolute ages are determined by radioactive dating methods.

The longest time divisions are called *aeons.* The sequence of rocks from the latest, the Phanerozoic, is known in such detail that this aeon has been divided into *eras,* these into *periods,* and the periods into *epochs* (not shown in the figure). Of the other aeons, only the Proterozoic is known well enough to allow a tentative division into eras. The Phanerozoic is so well mapped because of its abundance of fossils. In fact, except for a few late Proterozoic impressions of invertebrate animals found at Ediacara (South

Figure 4
A chronology of Earth history.

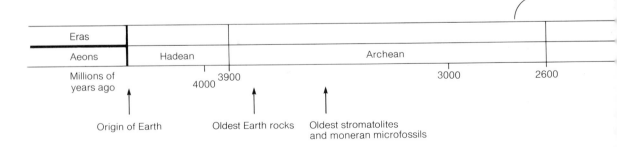

Australia) and at a few other places, and some billion-year-old fossil algae found in rocks of the Little Dal Group in western Canada, the entire fossil record of eukaryotic organisms is Phanerozoic.

The most abundant fossils from before the Phanerozoic are the rock formations called *stromatolites.* A typical stromatolite is a column or dome of rock a few inches wide made of thin horizontal layers. The layers are apparently the remains of sediment trapped or precipitated and bound by growing communities of microorganisms—bacteria and blue-green algae (cyanobacteria). Such communities exist today, but only in a few isolated, extreme environments, such as salt ponds. In the Archean and Proterozoic aeons, they were the dominant form of life. For at least two and a half billion years, more than half of the Earth's existence, the planet was the uncontested territory of Kingdom Monera. The stromatolite communities began to withdraw to their modern places of refuge only after the rise of protists and animals—some of the new organisms must have grazed voraciously in the lush bacterial pastures.

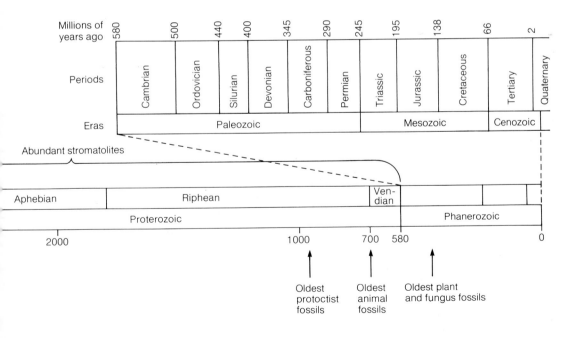

READING THIS BOOK

We have recognized and described 89 phyla—16 moneran, 27 protoctist, 5 fungal, 32 animal, and 9 plant. In each kingdom, we have tried to arrange them in an order from the simplest, and presumably the earliest to have evolved, to the more complex, presumably more recent forms. However, there is nothing to prevent you from browsing—the introduction to each chapter describes the general features of an entire kingdom.

As a rule, the highest taxa within a kingdom, the phyla and classes, represent the most ancient evolutionary divergences; the lowest taxa, the genera and species, represent the most recent. However, this cannot be an absolute rule because the evolutionary relationships of many groups are not known. In many cases, organisms are grouped in the same taxon only because they have some clear distinguishable trait in common (for example, the thread of Cnidosporidia, Phylum Pr-20), whether or not their common ancestry has been documented. With these cautions in mind, we have opened each chapter with a phylogeny showing the possible evolutionary relationships among the phyla of that kingdom.

The phyla differ enormously in size; some have only a few species and others have millions. Each phylum description begins with a list of well-known genera that we have classified in that phylum. Not all of these are mentioned in the text. Even so, we hope that the lists will give experienced students a firmer grasp of the concept of each phylum, and that all students will use the lists as clues to further reading. (There is a list of suggested readings at the end of each chapter.) In the Appendix, these lists are consolidated into a single alphabetical list that gives the phylum and common names (if any) of each genus.

At the top of the right-hand page of each phylum is a scene with an arrow pointing to the typical habitat of the members of the phylum. Figure 5 shows the five different scenes: temperate seashore, rocky, sandy, and muddy; temperate forests, lakes, and rivers; deserts and high mountains; tropical forest; and tropical seas, including reefs, continental shelves and slopes, the abyss, and the tropical seashore (the last two are not shown but are indicated by arrows pointing toward the bottom and the top of the scene). In many cases, members of a phylum are found in habitats not

Figure 5
The five scenes used to designate typical habitat.

Temperate seashore—rocky, sandy, and muddy

Temperate forests, lakes, and rivers

Deserts and high mountains

Tropical forest

Tropical seashores, continental shelves, abyss

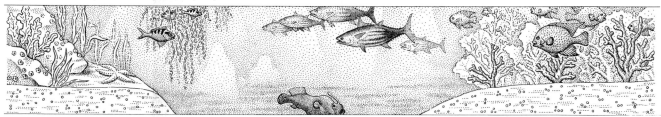

illustrated—this is certainly not a complete list of the world's habitats. However, the scenes include those in which life is frequently and abundantly found.

The text is accompanied by photographs and drawings of representative species. (The illustrated species are listed with each phylum in the Contents.) In most phyla, the main photograph is duplicated as a labeled anatomical drawing. Unfamiliar words are listed in the Glossary at the end of the book.

Most of the organisms were photographed alive. The chief exceptions are those photographed in a vacuum with an electron microscope. Transmission electron microscopy must be done with samples that have been chemically fixed, embedded in a transparent matrix, and cut extremely thin. Even scanning electron microscopy, which allows a more three-dimensional view of a sample, usually requires deadly fixation techniques.

The legend of each photograph states the type of microscopy (if any) that was used to take the photograph: LM stands for light (optical) microscopy, TEM and SEM for transmission and scanning electron microscopy. The legend also gives the length represented by the scale bar in the photograph. A colophon with each photograph indicates the kind of optical equipment needed to see the subject of the photograph (Figure 6).

Bibliography

Copeland, H. F., *The classification of lower organisms.* Pacific Books; Palo Alto, California; 1956.

Hogg, J., "On the distinctions of a plant and an animal, and on a fourth kingdom of nature." *The Edinburgh New Philosophical Journal* (New Series) 12:216–225; 1861.

Whittaker, R. H., "On the broad classification of organisms." *Quarterly Review of Biology* 34:210–226; 1959.

Whittaker, R. H., and L. Margulis, "Protist classification and the kingdoms of organisms." *BioSystems* 10:3–18; 1978.

COLOPHON					
OPTICAL EQUIPMENT	Naked eye	Hand lens	Light microscope	Scanning electron microscope	Transmission electron microscope
SIZE OF SUBJECT IN METERS (approximate)	10^{-3}–10	10^{-4}–10^{-2}	10^{-6}—10^{-4}	10^{-8}–10^{-2}	10^{-9}–10^{-5}

Figure 6
Key to photograph colophons.

MONERA

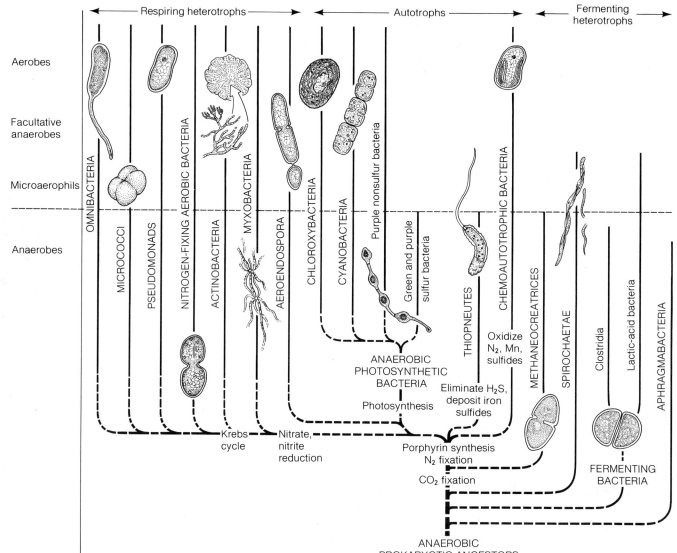

Respiring heterotrophs — Autotrophs — Fermenting heterotrophs

Aerobes

Facultative anaerobes

Microaerophils

Anaerobes

OMNIBACTERIA
MICROCOCCI
PSEUDOMONADS
NITROGEN-FIXING AEROBIC BACTERIA
ACTINOBACTERIA
MYXOBACTERIA
AEROENDOSPORA
CHLOROXYBACTERIA
CYANOBACTERIA
Purple nonsulfur bacteria
Green and purple sulfur bacteria
THIOPNEUTES
CHEMOAUTOTROPHIC BACTERIA
METHANEOCREATRICES
SPIROCHAETAE
Clostridia
Lactic-acid bacteria
APHRAGMABACTERIA

Oxidize N₂, Mn, sulfides

ANAEROBIC PHOTOSYNTHETIC BACTERIA

Eliminate H₂S, deposit iron sulfides

Photosynthesis

Krebs cycle

Nitrate, nitrite reduction

Porphyrin synthesis N₂ fixation

CO₂ fixation

FERMENTING BACTERIA

ANAEROBIC PROKARYOTIC ANCESTORS

MONERA

Greek *moneres,* single, solitary

Small though they are, bacteria are crucial to health, agriculture, forestry, and the very existence of the air we breathe. The modern food-processing industry began with the awareness of the nature of bacteria. Canning, preserving, drying, and pasteurization are all sterilizing techniques that prevent contamination by even a single bacterium. The success of these techniques is truly remarkable in view of the ubiquity of bacteria and their unbelievable numbers. Every spoonful of garden soil contains some 10^{10} bacteria; a small scraping of film from your gums might reveal about 10^9 bacteria per square centimeter of film—the total number in your mouth is greater than the number of people who have ever lived. Bacteria make up a significant percentage of the dry weight of all animals. They cover the skin; they line nasal and mouth passages, and live in the gums and between the teeth; they pack the digestive tract, especially the colon.

Germs, or pathogens, are simply bacteria (occasionally, fungi) capable of causing infectious diseases in animals or plants. The word *germ,* like the word *microbe,* has no definite technical meaning. A germ is simply a small living organism capable of growth at the expense of another organism; a microbe is simply a small organism of interest to someone. Bacteria cure as well as cause disease. Many of our most useful antibiotics—streptomycin, erythromycin, chloromycetin, and kanamycin—come from bacteria (penicillin comes from fungi).

Although much of the nomenclature is in dispute, more than 5000 species have been named and described in the bacteriological literature. Many more are still unidentified. Few bacteriologists would deny that the vast majority of bacterial species have never been carefully studied and described.

Bacteria are morphologically rather simple. The most complex do undergo developmental changes in form: simple bacteria may metamorphose into stalked structures, grow long branched filaments, or form tall fruiting bodies that release resistant sporelike microcysts. Some produce highly motile colonies. However, detailed knowledge of bacterial structure seldom affords insight into function. In this respect, bacteria are very different from animals and plants. Because their differences lie chiefly in their metabolism, or internal chemistry, many species of bacteria can be distinguished only by the chemical transformations they cause. A universally applied diagnostic test of bacteria is whether they stain purple with the *Gram test,* a staining method developed by the Danish physician Hans Christian Gram (1853–1938). *Gram-positive* organisms (which stain deep purple) differ from *Gram-negative* ones (which stain light pink) in the chemistry of their cell walls.

Bacteria can effect a large number of different chemical transformations and are metabolically far more diverse than all of the eukaryotes. Although some very complicated molecules are made only by certain plants and fungi, the biosynthetic and degradative patterns—the chemistry of reproduction and energy generation—in all plant and fungal cells is remarkably similar. Animals and protoctists exhibit even less variation in their chemical repertoires. In short, the metabolism of eukaryotes is rather uniform; it follows the same fundamental patterns of photosynthesis, respiration, glucose breakdown, and synthesis of nucleic acids and proteins. Certain plants and fungi add flourishes of metabolic virtuosity, but the basic patterns are not altered. Bacteria, on the other hand, are not only very different from eukaryotes but also from each other.

The work of most microbiologists is closely concerned with the role bacteria play in health and disease. The role of bacteria in the environment has been studied much less, but it is equally significant. Bacteria produce and remove all the major reactive gases in the Earth's atmosphere: nitrogen, nitrous oxide, oxygen, carbon dioxide, carbon monoxide, several sulfur-containing gases, hydrogen, methane, and ammonia, among others. Protoctists and plants also make substantial contributions to atmospheric gases, but none that is different from those of prokaryotes. On the other hand, many important reactions are catalyzed only by bacteria.

There is a profound difference between the soil of the Earth and the regolith—loose particles of rock—on the surface of Mars and of the Moon. Of course, Mars and the Moon are very dry and have much less atmosphere than the Earth; but soil and regolith differ in much more than just moisture content. The surface of the Earth—its soil and sediments—is rich in complex organic compounds; less tractable ones, such as tannic acids, lignin, and cellulose, abound as well as much more actively metabolized organics, such as sugars, starches, organic phosphorus compounds, and proteins. All of these organic compounds are the products of chemosynthesis or of photosynthesis, processes that use chemical energy or sunlight to convert the carbon dioxide of the air to the organic compounds of the living biosphere and, ultimately, of organic-rich sediments from which we obtain oil and coal. In fact, the soil and rocks of the Earth contain about one hundred thousand times as much organic matter as living forms do.

Chemosynthesis, the production of organic matter from inorganic by means of chemical energy, is limited to certain groups of bacteria. Photosynthesis is often attributed only to plants, but many bacteria also photosynthesize. Chemosynthesis and photosynthesis are forms of autotrophic nutrition: deriving food and energy from inorganic sources. Heterotrophy, the alternative mode of nutrition, is deriving food and energy from preformed organic compounds—from either live or dead sources. Like plants, most photosynthetic bacteria convert atmospheric carbon dioxide and water to organic matter and oxygen; unlike plants, other bacteria are capable of very different modes of photosynthesis—for example, using hydrogen sulfide instead of water. Bacterial photosynthesis and chemosynthesis are absolutely necessary for cycling the elements and compounds upon which the entire biosphere depends, including, of course, ourselves.

The notion that the food chain starts with the plants, followed by the herbivores, and ends with carnivorous animals is very shortsighted. Zooplankton of the seas feed on protists, which feed on bacteria; bacteria break down the carcasses of animals and algae, releasing back into solution such elements as nitrogen and phosphorus required by the phytoplankton. Bacteria facilitate entire food chains, or rather food webs, by transforming inorganic materials into complex organic compounds and being eaten by other organisms. Life on Earth would die out far faster if bacteria became extinct than if the animals, plants, and fungi disappeared. In fact, we have good reason to believe that life on our planet thrived long before the three more complex kingdoms appeared. Some of the ways we depend on bacteria will be explained in the descriptions of moneran phyla.

Bacteria have an ancient and noble history. They were probably the first living organisms and, in respect to everything but size, have dominated life on Earth throughout the ages. The oldest fossil evidence for bacteria goes back some 3400 million years, whereas the oldest evidence for animals dates back about 700 million years and the oldest fungi and land plants date back about 470 million years. The earliest evidence for protoctists is more equivocal, but they are probably not more than 1000 million years old. It is widely believed that 2000 million years ago the cyanobacteria—oxygen-eliminating photosynthetic prokaryotes that used to be called blue-green algae (Phylum M-7)—effected one of the greatest changes this planet has ever known: the increase in concentration of atmospheric oxygen from far less than 1 percent to about 20 percent. Without this concentration of oxygen, people and other animals never would have evolved.

As a group, bacteria are the most hardy of living beings. They can survive very low temperatures, even total freezing, for years. Some species thrive in boiling hot springs and others survive even in very hot acids. By forming spores, particles of life containing at least one copy of all the genes of a bacterium, they can tolerate total desiccation. Bacteria are the first to invade and populate new habitats: land that has been burned or newly emerged islands. They are capable of survival at great oceanic depths and atmospheric heights, although no organism—not even the hardiest of bacteria—is known to complete its life cycle suspended in the atmosphere.

Some activities of bacteria are still only dimly perceived. The incorporation of metals such as manganese and iron into nodules on lake and ocean floors may possibly be accelerated by bacterial action. Layered chalk deposits called stromatolites are thought to have been produced by the trapping and binding of calcium-carbonate-rich sediment by growing communities of bacteria, especially by cyanobacteria. Even the gold in South African mines is found with rocks rich in organic carbon. In Witwatersrand, the miners find the gold, deposited apparently more than 2500 million years ago, by following the "carbon leader." The carbon is probably of microbial origin. Copper, zinc, lead, iron, silver, manganese, and sulfur seem to have been concentrated into ore deposits by biogeochemical processes involving bacterial growth and metabolism.

All bacteria reproduce asexually. Although a few bacteria are known to be products of unequal sexual donations from two "parents," the extent of bacterial sexuality in nature is not really known. This is partly because the extent of bacterial diversity is not well known, and partly because, with bacteria as with most organisms, sex life is a most elusive object of study. Genes may also be transferred between bacteria by viruses. Some bacteria even excrete DNA. Whether this behavior is a part of bacterial sexuality in nature is not known; in the laboratory, however, DNA excreted by one bacterium is taken up and used by another in the formation of genetically new individuals.

No communities of living organisms anywhere on Earth lack bacteria, but few places are dominated by them. Some exclusive bacterial habitats, most often found in intemperate climates, are the bare rocks of cliffs, the interior of certain carbonate rocks, and muds lacking oxygen. Perhaps the most spectacular are the hot boiling muds of Yellowstone Park, in Wyoming, and the salt flats and shallow embayments of tropical and subtropical areas. Many such flats and bays are dominated by microbial mats, cohesive, more or less flat structures on soil or in shallow water that are caused by the growth and metabolism of microbes, primarily filamentous cyanobacteria. By entrapping bits of sand, carbonate, and other sediment, such growing communities of microbes can grow to be quite conspicuous manifestations of biological activity.

Except for the rather extreme environments where microbial mats or thermal springs abound, eukaryotes dominate our landscape, or so it seems to the naked eye. However, a microscopic examination of any forest, tide pool, riverbed, chaparral, or other habitat apparently dominated by eukaryotes will reveal prokaryotes in abundance. In activity and potential for rapid unchecked growth, they are unexcelled among living organisms. When environmentalists mourn the destruction of habitats by pollution, they are usually thinking of the loss of fish, fowl, and fellow mammals. If their sympathies were with the cyanobacteria and other bacteria instead, they would perceive eutrophication of lakes, for example, as a sign that life is flourishing.

Bacteriologists have not conformed their nomenclatural and taxonomic practices to those of other biologists. Thus, inevitably, several of our groupings differ from those found in the standard reference works such as *Bergey's Manual of Determinative Bacteriology.* Our innovations aim to make the taxonomic level of *phylum* conceptually comparable throughout the five kingdoms. To our knowledge, this has never before been attempted. We recognize sixteen phyla—fewer than in the animals or protoctists but more than in the plants or fungi. These phyla group the bacteria by clearly distinguishable traits, both morphological and metabolic. Where evolutionary information is available, we have tried to keep natural groups together. However, so little is known about the past history of bacteria that our phyla are for the most part frankly pragmatic. We have retained familiar English names, where they were concise and appropriate, for some groupings here given phylum status. New names include Aphragmabacteria (M-1), Thiopneutes (M-4), Methaneocreatrices (M-5), Chloroxybacteria (M-8), Aeroendospora (M-11), and Omnibacteria (M-14).

The habitat scenes are notably arbitrary in this chapter because so many bacteria can be found in both animal and plant hosts, and in soil and water samples, from vastly different habitats and locations. The natural history and ecology of bacteria have been so little explored that little can be said about the distribution and quantity of bacteria in the world's environments.

M-1 Aphragmabacteria

Greek α, without; *phragma*, fence

Acholeplasma
Anaplasma
Bartonella
Cowdria
Ehrlichia
Mycoplasma
Spiroplasma
Thermoplasma
Wolbachia

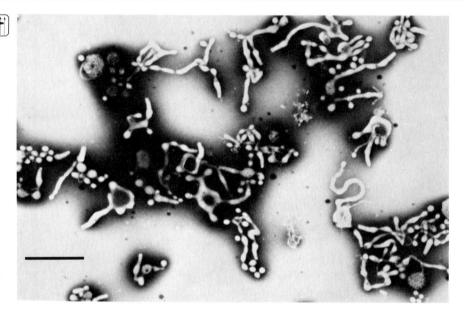

A *Mycoplasma pneumoniae*, which lives in human cells and causes a type of pneumonia. TEM, negative stain, bar = 1 μm. [Courtesy of E. Boatman.]

The cell walls of most bacteria, although various in composition, contain polysaccharides. In the cell wall, these long sugarlike molecules are attached to short polypeptide molecules, which—like full-fledged proteins, their chemical cousins—contain nitrogen in addition to carbon, hydrogen, and oxygen. In contrast, aphragmabacteria are bounded by a single triple-layered membrane composed of lipids—hydrocarbon compounds that dissolve only in such organic media as alcohol, carbon tetrachloride, and acetone. Aphragmabacteria are incapable of synthesizing the polysaccharide compounds muramic acid and diaminopimelic acid, which form the finished walls of all other bacteria. Lacking cell walls, they are resistant to penicillin and related drugs that inhibit wall growth.

Some aphragmabacteria are less than 0.2 μm in diameter and invisible even with the best light microscopes. Their shapes vary: irregular blobs, filaments, and even branched structures reminiscent of tiny fungal hyphae.

How they reproduce is not clear. In some, tiny coccoid structures appear to form inside the cells, emerging when the "parent" organism breaks down. Others seem to form buds that become new organisms. Some apparently reproduce by binary fission—that is, by division of the cell into roughly equal halves.

On agar plates, aphragmabacteria typically form tiny colonies, with a dark center and lighter periphery, like a fried egg.

Most well-studied species cause diseases in mammals and birds. Few have been cultivated outside their hosts; of those that have, many require a very complicated growth medium that includes steroids, such as cholesterol. These lipid compounds, produced by most and required by all eukaryotes, are seldom, if ever, found in prokaryotes. However, in aphragmabacteria of the genus *Mycoplasma*, cholesterol constitutes more than 35% of the membrane's lipid content. For a bacterium, this is an extremely large fraction and may represent the legacy of a long biological association between *Mycoplasma* and animal tissue rich in lipids. All strains so far cultured require long-chain fatty acids (lipids) for growth, and most ferment either glucose (a sugar) or arginine (an amino acid). The fermentation products are usually lactic acid and some pyruvic acid.

Mycoplasmas are of economic and social importance because they are the cause of certain types of pneumonia in humans and domestic animals. They are responsible for the death of cells in laboratory tissue cultures and probably are widespread in insect, animal, and plant cells but are usually too small to identify, even with electron microscopes.

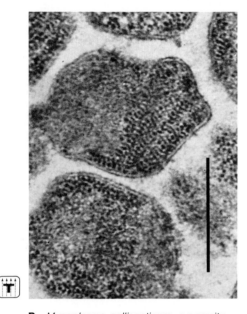

B *Mycoplasma gallisepticum,* a parasite of chickens. TEM, bar = 0.5 μm. [Courtesy of J. Maniloff. *Journal of Cell Biology* 25:139–150; 1965.]

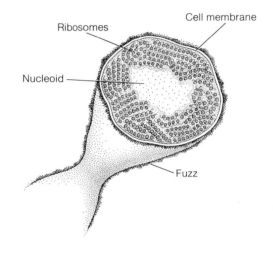

C A generalized mycoplasma. [Drawing by L. Meszoly.]

Bacteriologists have often used the word *mycoplasma* to refer to a class or a family, rather than a phylum, consisting of the genus *Mycoplasma* and similar organisms. In any case, all recognize several other genera: *Acholeplasma,* which do not require steroids; *Thermoplasma;* and *Spiroplasma.*

Thermoplasma is a genus with a single species: *T. acidophilum.* The only studied members come from a hot coal refuse pile and from the famous hot springs in Yellowstone Park. First described in the 1970s, thermoplasmas are unique because of the extremely hot and acidic conditions under which they live, thriving at nearly 60°C and pH values of from 1 to 2—the pH of concentrated sulfuric acid! Having no competition under such conditions, thermoplasmas can easily be grown in pure culture. However, observation of live cells is very difficult because, at 37°C (human body temperature) or cooler, and at pH 3 or greater, thermoplasmas die. They are the only prokaryotes known to contain DNA coated with basic proteins similar to histones, the chromosomal proteins of most eukaryotes. It is believed that the protein coating protects their DNA from destruction in hot acid.

Spiroplasmas were isolated from the leaves of citrus plants affected with a disease called Stubborn. Whether the spiroplasmas actually cause the disease has not been determined with certainty. Like other aphragmabacteria, the spiroplasmas are variable in form, lack cell walls, and form colonies that look like fried eggs. Some are helical in shape and motile, showing a rapid screwing motion or a slower waving movement, yet they lack flagella or any other obvious organelles of motility. How do they move? Nothing is known.

It is certainly possible that the various genera of aphragmabacteria are not directly related to each other but evolved separately by loss of walls. On the other hand, some may be truly primitive, having never had walls. Evidence that at least some aphragmabacteria are primitive is the observation that some strains have only about 4.5×10^8 daltons of DNA—ten times less than most bacteria.

M-2 Fermenting Bacteria

Acidaminococcus
Arthromitus
Bacteroides
Clostridium
Diplococcus
Eubacterium
Fusobacterium
Lactobacillus
Leptotrichia

Leuconostoc
Megasphaera
Peptococcus
Peptostreptococcus
Ruminobacter
Ruminococcus
Streptococcus
Veillonella

Fermenting bacteria are obligate anaerobes; they cannot tolerate oxygen, which inhibits their growth or quickly kills them. Nutritionally, they are heterotrophs of the kind called chemoorganotrophs—they require a mixed set of organic compounds to grow and reproduce. Fermentation is a metabolic process that uses organic compounds to produce energy; the end products are a different set of organic compounds, from which some chemical energy has been extracted.

Fermenting bacteria are distinguished by their inability to synthesize metals and nitrogen-containing ring compounds called porphyrins. Organisms other than members of this phylum are capable of synthesizing some sort of porphyrin.

We group fermenting bacteria into four classes, often referred to as families or other taxa in the microbiological literature: Gram-negative cocci, lactic-acid bacteria, clostridia, and peptococcaceae.

Gram-negative cocci are nonmotile spherical bacteria with complex nutritional requirements. Some ferment carbohydrates, lactate, pyruvate, oxaloacetate, succinate, or sulfur-containing organic compounds, generally producing gases such as hydrogen (H_2), carbon dioxide (CO_2), hydrogen sulfide (H_2S), and ammonia (NH_3); others require CO_2 for growth; some make volatile fatty acids containing from two to six carbon atoms. Many Gram-negative cocci have been isolated from the intestines of cows. Three genera are recognized: *Veillonella, Acidaminococcus,* and *Megasphaera.*

Lactic-acid bacteria, such as *Lactobacillus, Streptococcus,* and *Leuconostoc,* are rod-shaped organisms famous for their ability to ferment sugar, in particular, that in milk, and to produce lactic acid as well as acetate, formate, succinate, CO_2, and ethanol. Their complex nutritional requirements include amino acids, vitamins, fatty acids, and compounds depending on species and strain. A 5% to 10% solution of CO_2 enhances their growth. They stain Gram positive. They do not form spores. Although some may tolerate oxygen, none use it for metabolism.

Clostridia are ubiquitous, versatile, anaerobic flagellated microorganisms that generally form spores. As a group, they will ferment almost anything organic except plastics—sugars, amino acids and proteins, polyalcohols, organic acids, purines, collagen, and cellulose. Some sixty species, all placed in the genus *Clostridium* or, if filamentous, in the genus *Arthromitus* (= *Coleomitus*), are known from soils, freshwater and marine sediments, animal alimentary tracts, and every other anaerobic environment. *Clostridium denitrificans* and several others can fix atmospheric nitrogen into organic compounds. They produce many different end products, chiefly acetic and butyric acids and

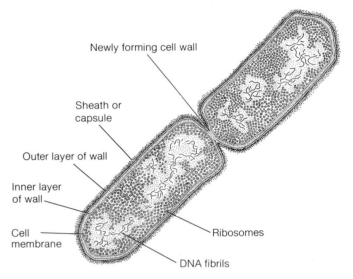

Newly forming cell wall

Sheath or capsule

Outer layer of wall

Inner layer of wall

Cell membrane

Ribosomes

DNA fibrils

alcohols. All clostridia are able to form heat- and desiccation-resistant spores when conditions are unfavorable for growth, thus defending themselves against the hostile aerobic world about them. *Clostridium botulinum* and several others produce powerful poisons. Some will grow in animal tissues, causing fearsome diseases such as gas gangrene and botulism. Some give off gases such as H_2S, NH_3, and CO_2 as fermentation end products in addition to organic compounds.

Peptococcaceae are spherical Gram-positive organisms that lack flagella and do not form spores. Ranging in diameter from 0.5 μm to 2.5 μm, they occur singly, in pairs, in irregular masses, three-dimensional packets, and long and short chains of cocci. Many produce gases such as CO_2, H_2S, and H_2, and they ferment carbohydrates, amino acids, and other organics. They have complex growth requirements; many have been isolated from the mouth or intestine of animals. Three genera are recognized: *Peptococcus, Peptostreptococcus,* which uses protein or its breakdown products for energy, and *Ruminococcus,* which can break down cellulose.

A number of genera of Gram-negative rods, some having tens of species, belong in this phylum but do not fall easily into any of the five classes. These include *Eubacterium, Bacteroides, Fusobacterium,* and *Leptotrichia.*

Bacteroides fragilis, an obligate anaerobe found in animal gut tissue, is seen here just prior to cell division. TEM, bar = 1 μm.

[Photograph courtesy of D. Chase; drawing by L. Meszoly.]

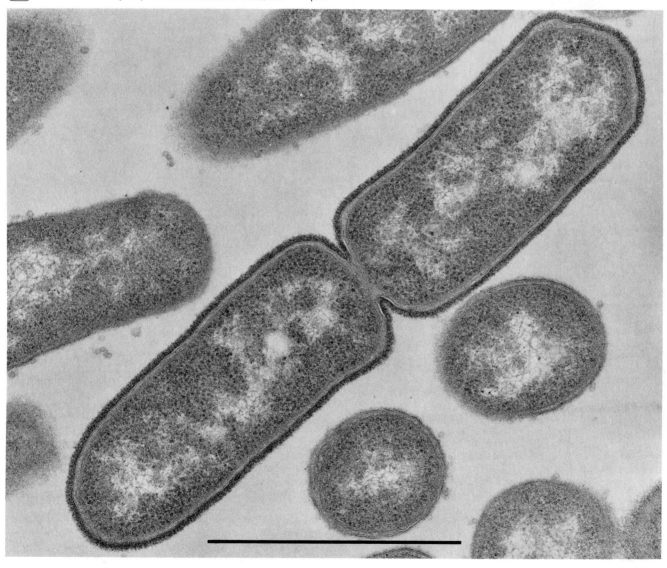

M-3   Spirochaetae

Latin *spira,* coil; Greek *khaite,* long hair

Borrelia
Clevelandina
Cristispira
Diplocalyx
Hollandina
Leptonema
Leptospira
Pillotina
Spirochaeta
Treponema

Spirochetes look like tightly coiled snakes. Unlike other motile bacteria, they have from two to more than a hundred *internal* flagella (the number depends on the species) in the space between the different layers of the cell wall. (These have been called axial filaments or endoflagella in the literature.) Spirochetes are found in marine and fresh waters, deep muddy sediments, and in the gastrointestinal tracts of several different kinds of animals. Their long, slender, corkscrew shape enables them to move flexibly through thick, viscous liquids with great speed and ease. In more dilute environments, many swim quickly with complex movements—rotation, torsion, flexion, and quivering.

An unusual flagellar arrangement distinguishes spirochetes from all other bacteria. Their elaborate cell wall is probably responsible for their corkscrew shape and characteristic movements. Each flagellum originates near an end of the cell and extends about two-thirds of the body length. Thus, flagella anchored at opposite ends of the cell often overlap, like the fingers of loosely folded hands. Presumably, internal rotation of these flagella is responsible for motility of spirochetes, just as external flagellar rotation is for other motile bacteria.

There are three major groups of spirochetes: leptospires (small), spirochaetales, and cristispires (large). Leptospires are the only spirochetes that require gaseous oxygen; most others are quickly poisoned by the slightest trace. Certain leptospires live in the kidney tubules of mammals. Often they are carried with the urine into water supplies and can enter the human bloodstream through cuts in the skin, causing leptospirosis. Because the leptospires are so tiny, the disease is often misdiagnosed.

In the spirochaetales there are three genera. *Borrelia* and *Treponema* include some species that are internal parasites of animals. Infamous as the cause of syphilis, *Treponema pallidum* also takes the blame for yaws, a debilitating and unsightly tropical eye disease. These spirochetes have from one to four flagella at each end of the cell. The genus *Spirochaeta* contains free-living marine and freshwater spirochetes less than a micrometer wide having body plans that resemble those of treponemes.

The cristispires are much larger; some may be three micrometers wide and hundreds of micrometers long. All known species have been found in symbiotic relationships with invertebrate animals. Cristispires inhabit the crystalline style of clams and oysters; the style is an organ that helps these mollusks grind their algal food. Included in this relatively unstudied group are the pillotina spirochetes, which are found associated with other bacteria and protoctists in the hindgut of dry-wood and subterranean termites as well as of wood-eating cockroaches. Four genera of pillotinas are known: *Pillotina, Hollandina, Diplocalyx,*

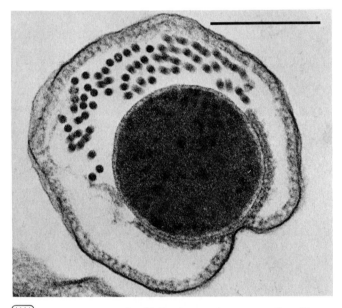

A *Diplocalyx* sp., in cross section. These large spirochetes, which belong to the family Pillotaceae (the pillotinas), have many flagella. The several genera all live in the hindguts of wood-eating cockroaches and termites. This specimen was found in the common North American subterranean termite *Reticulitermes flavipes* (Phylum A-27, Arthropoda). TEM, bar = 1 μm. [Courtesy of H. S. Pankratz and J. Breznak.]

and *Clevelandina.* The insect hosts ingest wood, but the microbes inhabiting their intestines digest it. Not only pillotinas but also the other genera of cristispire spirochetes are found in environments where active breakdown of algal or plant cellulose is taking place. The spirochetes themselves probably lack cellulases, enzymes that initiate the breakdown of wood, but do have enzymes for digesting the products of the initial breakdown. In fact, the termite spirochetes are usually found in intimate contact with protists (Phylum Pr-8, Zoomastigina) that contain cellulases.

Most spirochetes are difficult to study in the laboratory; they have complicated growth requirements—many require large, complex fatty acids. Very few have been cultured, and none of these are cristispires.

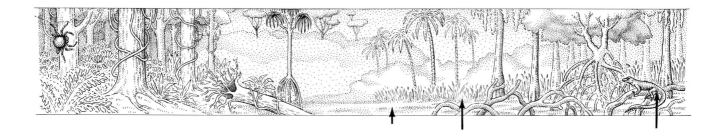

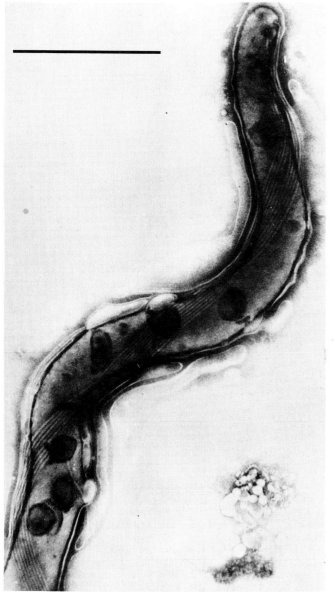

B Unidentified spirochete from the hindgut of a subterranean termite *Reticulitermes hesperus* (Phylum A-27, Arthropoda), found in the far western United States. TEM, bar = 0.5 μm. [Courtesy of D. Chase.]

C Parameters of a spirochete.

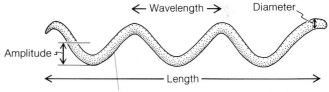

D A generalized pillotina spirochete. Not all members of the family have all of these features. [Drawing by L. Meszoly.]

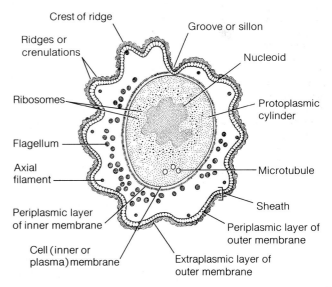

Greek *theion*, sulfur; *pneutes*, breather

The thiopneutes are obligate anaerobes; they are quickly poisoned by exposure to oxygen. Because they do not retain the purple color of Gram stains, they are called Gram negative. Thiopneutes are distinguished by their metabolic behavior: they require sulfate (SO_4^{2-}), just as we require oxygen, for respiration. In this energy-yielding process, electrons from food molecules are transferred to inorganic compounds; by this transfer, the food molecules are oxidized and the inorganic compounds are reduced in oxidation state. All the thiopneutes reduce SO_4^{2-} to some other sulfur compound, such as elemental sulfur or hydrogen sulfide (H_2S). To reduce SO_4^{2-}, they must synthesize cytochromes, electron-transporting proteins whose activity is due to their inclusion of a porphyrin molecule. The thiopneutes contain distinctive porphyrin proteins, which vary slightly with species. Although they obtain energy by "breathing" SO_4^{2-}, thiopneutes also take in some kind of organic compounds, usually the three-carbon compounds lactate or pyruvate, as a source of electrons. Thus, they are not autotrophs.

Three genera of sulfate reducers are now recognized: *Desulfovibrio* (about five species); *Desulfotomaculum* (three species); and *Desulfuromonas* (one species). The last two were not characterized until the 1960s and 1970s.

Desulfovibrios are unicellular bacteria widely distributed in marine muds, estuarine brines, and freshwater muds. They are incapable of forming spores. Either single polar flagella or bundles of them (lophotrichous flagella) provide motility. Several species require sodium chloride for growth, and hence are considered marine bacteria. Desulfovibrios contain c_3 cytochromes and a pigment called desulfoviridin, which gives them a characteristic red fluorescence. Many also synthesize hydrogenases, enzymes that generate hydrogen, which protects the organisms from the hostile aerobic world.

Desulfovibrios and their relatives release gaseous sulfur compounds, including H_2S, into the sediments, thus playing a crucial role in cycling sulfur—a constituent of all proteins—throughout the world. In iron-rich water, the H_2S formed by these bacteria reacts with iron, leading to the deposition of pyrite (iron sulfide, also known as fool's gold). It is thought that Archean and Proterozoic iron deposits may be due, at least partly, to the activity of sulfate-reducing bacteria. No symbiotic or pathogenic forms of sulfate-reducing bacteria have been reported; the group is free living. Some species may be capable of fixing nitrogen.

Desulfotomaculum is a genus of unicellular bacteria, motile by means of many flagella distributed over the surface of each cell (peritrichous flagella). They reduce SO_4^{2-} or sulfite (SO_3^{2-}) to H_2S. Thus, in many environments, *Desulfotomaculum* causes the

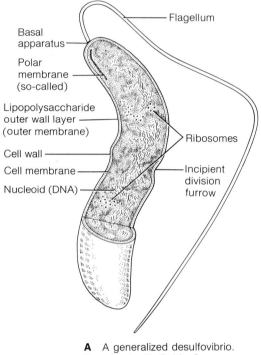

A A generalized desulfovibrio.
[Drawing by R. Golder.]

Labels: Flagellum; Basal apparatus; Polar membrane (so-called); Lipopolysaccharide outer wall layer (outer membrane); Cell wall; Cell membrane; Nucleoid (DNA); Ribosomes; Incipient division furrow

precipitation of iron, just as *Desulfovibrio* does. The genus is distinguished by the formation of resistant endospores and is commonly found in marine and freshwater muds, in the soil of geothermal regions, in the intestines of insects and bovine animals, and in certain spoiled foods. Some require SO_4^{2-} to grow, and all reduce it to H_2S. Some can tolerate temperatures as high as 70°C. The oxidation of organic substrates by these organisms is always incomplete, leading to the production of small organic compounds such as acetate.

Desulfuromonas is a new bacterial genus, first described in 1977. Its members are motile and do not form spores. They can reduce sulfate to elemental sulfur, which they excrete. They are all free living, primarily in muds.

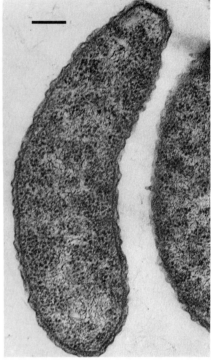

B *Desulfovibrio desulfuricans.* TEM, bar = 1 μm. [Courtesy of T. M. Hammill and G. J. Germano.]

C *Desulfovibrio desulfuricans.* These comma- or vibrio-shaped bacteria inhabit anaerobic sulfur-rich environments. TEM, negative stain, bar = 1 μm. [Courtesy of T. M. Hammill and G. J. Germano, *Canadian Journal of Microbiology* 19: 753–756, 1973.]

M-5   Methaneocreatrices
(Methanogenic bacteria)

Latin *creatrix,* creator

Methanobacillus
Methanobacterium
Methanococcus
Methanosarcina

Methanogenic bacteria are very strange indeed. They cannot use sugars, proteins, or carbohydrates as sources of carbon and energy. In fact, they can use only three compounds as food: formate, methanol, and acetate. Methanogens are characterized by their extraordinary way of gaining energy: they form methane (CH_4) by reducing carbon dioxide (CO_2) and oxidizing hydrogen (H_2)—a scarce commodity. They obtain both these gases from the air, although they cannot tolerate oxygen. Some may also use formate, methanol, or acetate as a source of electrons for reducing CO_2. The characteristic overall metabolic reaction is

$$CO_2 + 4H_2 \longrightarrow CH_4 + 2H_2O$$

Methanogens can be Gram-positive or -negative, motile (by means of flagella) or immotile. They come in all three classic bacterial shapes: rod, spirillum, and coccus. Some species can tolerate moderate or high temperatures. They are found worldwide in sewage, in marine and freshwater sediments, and in the intestinal tracts of animals. Three genera are formally recognized: *Methanosarcina, Methanobacterium* and *Methanococcus.*

Methanobacillus omelianski, which was thought for many years to be a methane-producing bacterium requiring ethanol, is now recognized as a symbiosis between two similarly shaped but metabolically distinct bacteria. One, known simply as "organism S," is an anaerobic, fermenting bacterium (Phylum M-2) that produces hydrogen from ethanol. It forms no CH_4 whatsoever. At a certain concentration, the H_2 produced by *Methanobacillus omelianski* is toxic to the fermenting bacterium. The other bacterium, called "strain MOH," is a methanogen that combines the H_2 provided by its colleague with atmospheric CO_2 to form CH_4 in the autotrophic reaction given above.

Methanogenic bacteria are the source of marsh gas, which is produced in swamps and in sewage-treatment plants, and indeed of much natural gas. If methanogens did not move organic carbon from the sediments to the atmosphere in the form of CH_4, much of the carbon produced by photosynthesis would remain buried in the ground.

For every atom of carbon buried as carbohydrate or other reduced compound produced by oxygenic photosynthesis, a molecule of oxygen (O_2) has been released (see Phylum M-7, Cyanobacteria). Without the intervention of methanogens, the amount of buried carbon and the amount of O_2 in the atmosphere would increase. An excess of atmospheric O_2 could lead to spontaneous fires that would threaten the entire biosphere. However, most of the CH_4 released by methanogens reacts with atmospheric O_2 to form CO_2 and maintain the balance.

Methanogenic bacteria produce some two billion tons of methane per year—a quantity equivalent to several percent of the total annual production of photosynthesis on the entire Earth! The entire quantity of methane in the Earth's atmosphere, amounting to more than one part per million, is produced by methanogenic bacteria. Although some bacteria give off methane as an end product of carbohydrate fermentation, no eukaryotes or any other prokaryotes are capable of formation of methane from CO_2 and H_2. A good part of the world's methanogenesis, perhaps some 30%, comes from those fermentation tanks on four legs we recognize as cows, elephants, and other cellulose-eating animals. Their rumen stomachs could never function, just as our sewage-treatment plants could not function, without the anaerobic methanogenesis of these peculiar bacteria. Even the methane gas released by people is made by the bacteria in their guts!

Methanogens have certain distinctive features suggesting that they evolved separately from other bacteria and have been evolving separately, even from other anaerobes, for several billion years. They differ from other prokaryotes in the sequences of nucleotides that make up the RNAs in their ribosomes. In fact, the 16S RNAs of methanogens are so different from those of other organisms that some biologists would give them kingdom status as the Archaebacteria (old bacteria). The only other microbes having this unusual sort of ribosomal RNA are salt-requiring aerobic bacteria of the genus *Halobacterium.* (see Phylum M-10, Pseudomonads). Halobacters probably evolved from methanogens, even though they have very few obvious traits in common. The suggestion that methanogens and halobacters be placed together in a new kingdom deserves scrutiny.

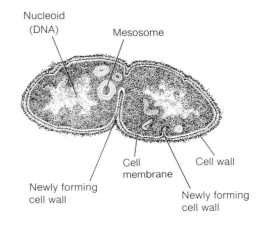

Nucleoid (DNA)

Mesosome

Cell membrane

Cell wall

Newly forming cell wall

Newly forming cell wall

Methanobacterium ruminantium, a methanogenic bacterium taken from a cow rumen. The bacterium has nearly finished dividing: a new cell wall has nearly formed. Notice that a second new cell wall is beginning to form in the right-hand cell. TEM, bar = 1 μm. [Photograph from J. G. Zeikus and V. G. Bowen, "Comparative ultrastructure of methanogenic bacteria." Reproduced by permission of J. G. Zeikus and the National Research Council of Canada from the *Canadian Journal of Microbiology* 21: 121–129 (1975). Drawing by I. Atema.]

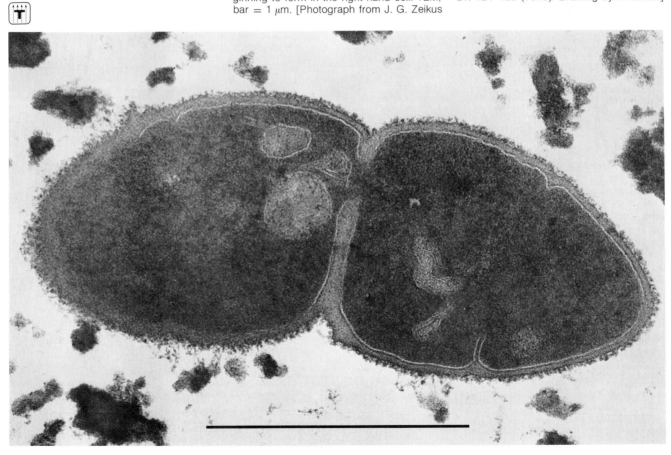

Chloroflexus
Chlorobium
Chloropseudomonas
Chromatium
Pelodictyon
Rhodomicrobium
Rhodopseudomonas
Rhodospirillum
Thiocapsa

Thiocystis
Thiodictyon
Thiopedia
Thiosarcina
Thiothece

Probably the most important evolutionary innovation on the Earth, if not in the solar system and the galaxy, was photosynthesis, the transformation of the energy of sunlight into the chemical energy of food molecules. Photosynthesis began in anaerobic bacteria billions of years ago.

There are three major classes of bacterial photosynthesizers: green sulfur bacteria, purple sulfur bacteria, and purple nonsulfur bacteria. Photosynthesis is an essentially anaerobic process, and none of these bacteria can carry it out when they are exposed to oxygen. Except for *Chloroflexus,* the green sulfur bacteria are hypersensitive to oxygen; they can grow only photosynthetically and in the total absence of oxygen. Some species of purple nonsulfur bacteria can grow microaerophilically, that is, under oxygen concentrations less than the modern norm, or even aerobically, but only in total darkness. In that case, they derive their energy from the breakdown of food, as many other respiring bacteria and nearly all eukaryotes do.

Oxygen release is not at all necessary to photosynthesis, even though it is characteristic of plant and algal photosynthesis. What is essential is the incorporation of carbon dioxide (CO_2) from the air into organic compounds needed for growth of the photosynthesizer, and the conversion of the energy of visible light into chemical energy in a form useful to cells. The conversion of light energy requires chlorophyll and other pigment molecules, although not always precisely the same chlorophyll molecule as that found in plants. The chemical energy currency produced is adenosine triphosphate (ATP), a nucleotide used in transformation reactions of all cells. Although the details of the enzymatic pathways of photosynthesis are still being worked out, it is already clear that they are different in the three groups of photosynthetic bacteria, and in plant and cyanobacterial (Phylum M-7) CO_2 fixation as well.

To reduce the CO_2 in the air to organic compounds, cells need a source of electrons, which, as a rule, are carried by hydrogen atoms. The source of these electrons varies with the organism. In the green sulfur bacteria, the electrons generally come from hydrogen sulfide (H_2S), although they may also come from hydro-

A *Rhodomicrobium vanielii,* a photosynthetic purple nonsulfur bacterium that lives in ponds and grows by budding. A new bud is forming on bottom. TEM, bar = 10 μm. [Courtesy of E. Boatman.]

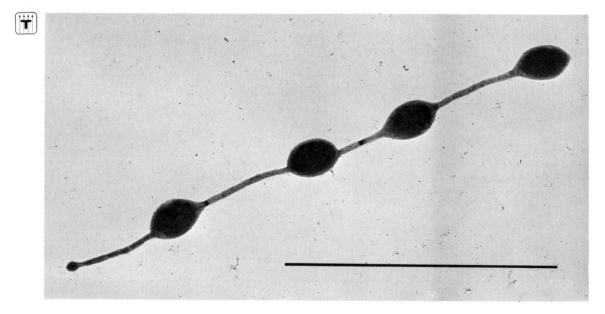

gen gas (H_2). The purple sulfur bacteria also use H_2S as the hydrogen donor. In the purple nonsulfur bacteria the hydrogen donor is a small organic molecule such as lactate, pyruvate, or ethanol. Thus, the general photosynthetic equation can be written as

$$2nH_2X + nCO_2 \xrightarrow{light} nH_2O + nCH_2O + 2nX$$

where X varies according to species. The molecule H_2X is the hydrogen donor. In these bacteria, it is never water; thus, oxygen is not a by-product of their photosynthesis. In algae and plants, on the other hand, water is the hydrogen donor and oxygen is released. If H_2S is the hydrogen donor the by-product of photosynthesis is sulfur, which may be stored or excreted by cells as elemental sulfur or as more oxidized sulfur compounds.

The photosynthetic bacteria are a morphologically very diverse group. They are found as single cells, motile and immotile, in packets, in stalked budding structures, and in extensive sheets of cells in which the spaces between the cells are filled with coverings, called sheaths, composed of mucous material. Some contain gas vacuoles, giving them buoyancy and a sparkling appearance. Some are filaments. Under anaerobic conditions, most of the purple sulfur bacteria convert H_2S to elemental sulfur, which they deposit inside their cells in tiny but visible granules; the presence, distribution, and shape of these granules can be used to distinguish species.

The photosynthetic bacteria are delightfully colored in an astonishing range of pinks and greens, although in the bright sunlight in the top layers of anaerobic muds they become very dark, nearly black. Because each species has an optimum growth at a given acidity, oxygen tension, sulfide and salt concentration, moisture content, and so on, they grow in layers—each in its appropriate niche—which can be seen in sediments inhabited by layered communities of photosynthetic and other bacteria.

Among the many different species of photosynthetic bacteria some are tolerant of extremely high or extremely low temperatures or salinities. In each group, some species are capable of fixing atmospheric nitrogen.

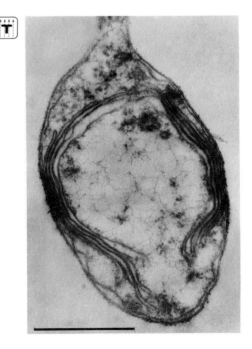

C *Rhodomicrobium vanielii,* showing layers of thylakoids (photosynthetic membranes) around the periphery of the cell. TEM, bar = 0.5 μm. [Courtesy of E. Boatman.]

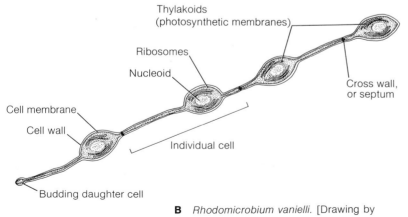

B *Rhodomicrobium vanielli.* [Drawing by I. Atema.]

M-7 Cyanobacteria

(Blue-green algae or blue-green bacteria)

Greek *kyanos*, dark blue

Anabaena
Anacystis
Chamaesiphon
Chroococcus
Dermocarpa
Entophysalis
Gloeocapsa
Hyella
Lyngbya

Microcystis
Nostoc
Oscillatoria
Spirulina
Synechococcus
Synechocystis

Until less than two decades ago, cyanobacteria were always called *blue-green algae* or *cyanophyta* and were considered part of the plant kingdom. Indeed, in their physiology, cyanobacteria are remarkably similar to algae and plants. Together with the members of Phylum M-6 (Anaerobic photosynthetic bacteria) and the plants they sustain all other life on Earth by converting solar energy and carbon dioxide (CO_2) into food and oxygen. Algae, plants, and cyanobacteria photosynthesize according to the same rules:

$$12H_2O + 6CO_2 \xrightarrow[\text{chlorophyll}]{\text{visible light}} C_6H_{12}O_6 + 6O_2 + 6H_2O$$

Like algae and plants, cyanobacteria respire the oxygen they produce by photosynthesis. Like plants and algae, they may both photosynthesize and respire; unlike algae and plants, they may not do both simultaneously—they respire only in the dark.

Investigations with the electron microscope have shown that the blue-green algae, although physiologically like plants, are morphological counterparts of photosynthetic bacteria. The similarity between cyanobacteria on the one hand and nucleated algae and plants on the other is now understood to apply only to the plastid portions of algae and plants.

Like other photosynthetic organisms, cyanobacteria contain membranes called thylakoids in which are embedded photosynthetic pigments. Chlorophyll *a* (their only chlorophyll) and two similar lipid-soluble pigments, phycocyanin and allophycocyanin, give them their bluish tinge. Many have an additional bluish pigment called phycoerythrin. These pigments (called phycobilins) are biosynthesized largely by the same reactions that produce porphyrins. Chemically, they resemble nitrogen-containing porphyrins, with the ring opened up to form a chain. Bile pigments of animals also have this structure. In the cyanobacterial cell, phycobilins are attached to proteins; the pigment-protein complex is called phycobiliprotein.

Unobtrusive though they are, there are thousands of living species of cyanobacteria; in the ancient past, they dominated the landscape. The Proterozoic Aeon, from about 2500 million until about 600 million years ago, was the golden age of cyanobacteria. Remains of their ancient communities include trace fossils called stromatolites, layered sedimentary rocks produced by the metabolic activities of microorganisms, especially of filamentous cyanobacteria. Stromatolites are still being produced in salt flats and shallow embayments of the Persian Gulf, the Bahamas, western Australia, and the west coast of Mexico. In the Proterozoic, however, such communities were rampant on all the continents and even in fairly deep open waters; they occupied the kinds of environments that coral reefs do today.

There are two great classes of cyanobacteria:

Class Coccogoneae: coccoid cyanobacteria
  Order Chroococcales: coccoids that reproduce by binary fission. Some genera are *Gloeocapsa, Chroococcus, Anacystis, Synechocystis,* and the stromatolite-building coccoid *Entophysalis.*
  Order Chamaesiphonales: coccoids that reproduce by releasing exospores. Unlike other microbial spores, these are not necessarily resistant to heat and desiccation. *Chamaesiphon* and *Dermocarpa* are two genera in this group.
  Order Pleurocapsales: coccoids that reproduce by forming endospores; the parent organism disintegrates when the spores are released. Unlike the endospores of other microbes, these are not resistant.

Class Hormogoneae: filamentous cyanobacteria.
  In many members of this class, filament fragments containing as many as several dozens of cells break off, glide away, and begin new growth. These fragments are called hormogonia.
  Order Nostocales: filaments that either do not branch or exhibit false branching, that is, a branch formed not by only a single growing cell but by slippage of a row of cells.
  Order Stigonematales: filaments that exhibit true branching. A single cell may have two places of growth on it, leading to the formation of two new filaments from the cell. These cyanobacteria are among the morphologically most complex of the prokaryotes.

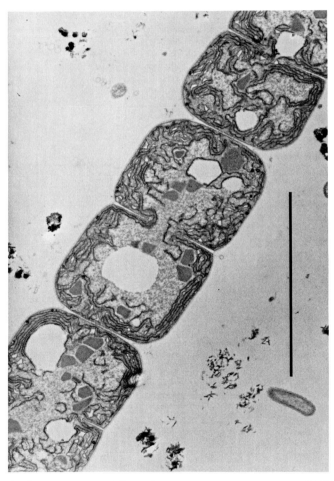

Anabaena, a common filamentous cyano-bacterium, grows in freshwater ponds and lakes. Within the sheath, the cells divide by forming crosswalls. TEM, bar = 5 μm. [Photograph courtesy of N. J. Lang; drawing by R. Golder.]

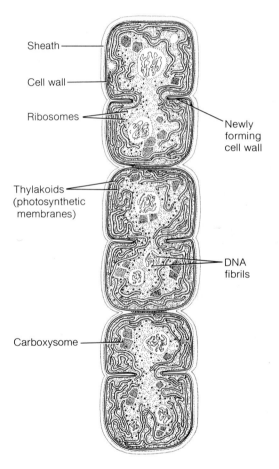

Sheath

Cell wall

Ribosomes

Newly forming cell wall

Thylakoids (photosynthetic membranes)

DNA fibrils

Carboxysome

M-8 Chloroxybacteria
(Prochlorophyta)

Greek *chloros*, yellow green

The chloroxybacteria were first described in the late 1960s. Although they had been seen before, it had always been assumed that they were protists of the green algal sort. When they were carefully studied with the electron microscope, however, it became clear that these coccoids are prokaryotes. Physiologically, they are algae: their photosynthesis produces oxygen according to the green-plant photosynthetic equation (see Phylum M-7). Morphologically, they are photosynthetic bacteria like the cyanobacteria. They have been named prochlorophytes to indicate their relationship to green plants (Greek *phyton* means plant). Indeed, they do resemble the chloroplast, the photosynthetic organelle of plants and green algae. However, because these organisms are bacterial in structure, not plantlike, and because they are green oxygen producers, we suggest the name *Chloroxybacteria*.

Unlike all other photosynthetic bacteria, but like green algae and plants, chloroxybacteria contain chlorophyll *b* in addition to chlorophyll *a*. Seven strains are known, primarily from the South Pacific. All are members of the same genus, *Prochloron;* they are simple nonmotile coccoids. All have been found in extracellular associations with tropical and subtropical invertebrates, primarily with marine chordates of the subphylum Tunicata (Phylum A-32, Chordata). *Prochloron* has been found on two species of *Didemnum*—as gray or white surface colonies on *D. carneolentum* and as internal but not intracellular cloacal colonies in *D. ternatanum.* They also reside as colonies in the cloaca of *Diplosoma virens, Lissoclinum molle, Lissoclinum patella,* and *Trididemnum cyclops.* Exactly why they grow only on the surface or in the cloaca of these tunicate hosts is a mystery. Because no techniques have yet been developed to cultivate *Prochloron* in the absence of the animal host, the nature of the symbiosis is not understood at all.

All members of *Prochloron* are unicellular and spherical, having diameters of from 6 μm to 25 μm, depending on the strain. Their cell walls are typical of prokaryotes in that they are sensitive to the enzyme lysozyme and contain muramic acid. Although they lack the phycobiliproteins of cyanobacteria, they contain carotenoid pigments. A large percentage of the total carotenoid is beta carotene, the pigment in carrots, green algae, plants, and cyanobacteria. Unlike most prokaryotes, *Prochloron* stores α-1-4 glucan, a highly branched carbohydrate typical of prokaryotes. On the other hand, *Prochloron* also contains a linear unbranched amylose like the starch of green algae and plants.

Apparently, *Prochloron* lacks sucrose, the sugar most commonly synthesized by green algae and plants, but does contain small amounts of glucose and smaller four- and five-carbon sugars, as well as alcohol derivatives of sugars. *Prochloron* lacks β-hydroxybutyrate, a substance that may compose more than half the dry weight of many Gram-negative bacteria; it also seems to lack cyanophycin, a copolymer of arginine and aspartic acid that is quite commonly found as granules in cyanobacteria.

All in all, one can think of *Prochloron* as a free-living prokaryotic counterpart of the chloroplasts of green algae and plants.

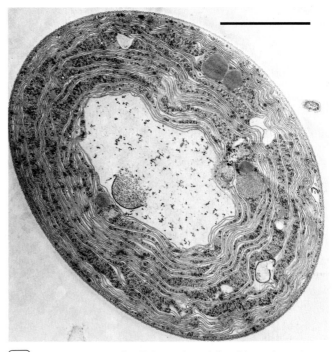

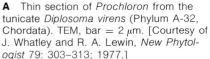

A Thin section of *Prochloron* from the tunicate *Diplosoma virens* (Phylum A-32, Chordata). TEM, bar = 2 μm. [Courtesy of J. Whatley and R. A. Lewin, *New Phytologist* 79: 303–313; 1977.]

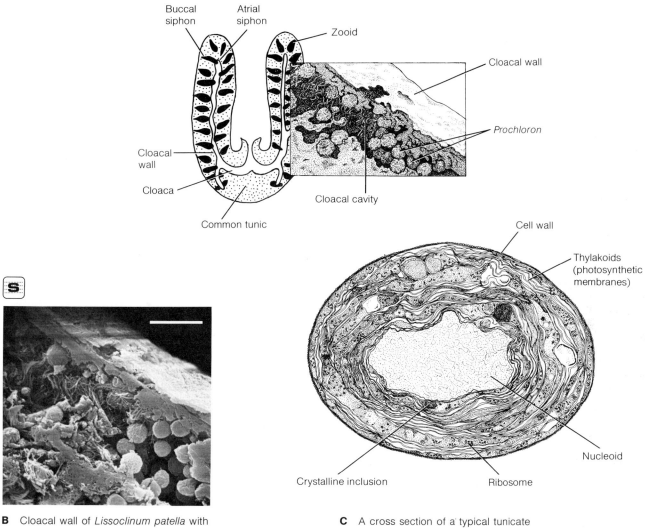

B Cloacal wall of *Lissoclinum patella* with embedded small green spheres of *Prochloron.* The tunicate *Lissoclinum patella* is native to the South Pacific. SEM, bar = 20 μm. [Courtesy of J. Whatley.]

C A cross section of a typical tunicate (top left), prochlorons embedded in the tunicate cloacal wall (top right), and a cross section of a single *Prochloron* (bottom). [Drawings by E. Hoffman.]

M-9 Nitrogen-Fixing Aerobic Bacteria

Azomonas
Azotobacter
Beijerinckia
Derxia
Rhizobium

Most bacteriologists consider the nitrogen-fixing bacteria to be a family called Azotobacteriaceae or Rhizobiaceae. All members are Gram negative and most are flagellated. Although strict aerobes, many are capable of growth under oxygen tensions lower than atmospheric, and some prefer highly aerated waters supersaturated with oxygen. They are common in soil and water everywhere in the world, and often are isolated from the surface of leaves. Five genera are recognized: *Azotobacter, Azomonas, Beijerinckia, Derxia,* and *Rhizobium.*

Azotobacters are large ovoid or rod-shaped cells that produce copious extracellular slime, grow rapidly, and form cysts. The thick-walled cysts are rounded desiccation-resistant structures that enclose the cells, which themselves tend to be more rod shaped. Azotobacters respire carbohydrates for growth. Many of them form black pigments that are insoluble in water.

The azomonads are very similar to azotobacters, but form neither cysts nor spores. They are large cells some 2 μm in diameter that occur singly, in pairs, or in clumps. They, too, form slime. Typical *Azomonas* cells are motile by means of polar or peritrichous flagella.

Named for the great Dutch microbiologist M. W. Beijerinck (1851–1931), the beijerinckias are small rods that embed themselves in a very tenacious extracellular gum or slime of their own making. Within the cells are produced conspicuous internal bodies made of lipid material that are found at both ends of the cell. Beijerinckias tend to grow more slowly than azotobacters or azomonads.

The genus *Derxia* is quite like *Beijerinckia* except that the many lipid bodies in each cell, unlike the lipid bodies of all the other genera of the group, are catalase negative; that is, they are unable to synthesize the iron porphyrin enzyme catalase, which breaks down toxic hydrogen peroxide to water and oxygen. *Derxia* cells can be variously shaped, but are generally rodlike with rounded ends; some grow to be rather large. Neither cysts or spores are known. For growth, the cells oxidize a wide range of alcohols and organic acids. In the laboratory, they grow in colonies that eventually become highly wrinkled and take on a mahogany brown color. They have been found in tropical soils in Asia, Africa, and South America. Derxias grow optimally at temperatures between 25°C and 35°C.

All rhizobia are motile and form neither cysts nor spores. The spectacular success of the Leguminosae, the pea family, is due in large measure to the regular nitrogen-fixing symbioses formed between plant and *Rhizobium* bacteria. Soybeans, alfalfa, clover, common beans, lentils, and many other economically important crops owe their growth on nitrogen-poor soils to their bacterial symbionts. Some *Rhizobium* species are quite specialized for symbiosis with only certain species of host plants.

Rhizobia penetrate the tiny root hairs of plants; eventually, they induce the plant roots to form nodules composed of cells infested with rhizobia. Within the nodules, the rhizobia have been transformed—now called bacteroids, they are unable to move or reproduce. The nodules become the sites of nitrogen fixation.

The plant itself, like all eukaryotes, is incapable of nitrogen fixation. Rhizobia in soil are also incapable of it. Recent experiments have shown that nitrogen-fixing capacity exists in soil rhizobia, but is repressed, and that it can be derepressed by the plant after the bacteria become bacteroids in the nodules. Derepression can also be accomplished in cultured bacteria under certain conditions. All nitrogen-fixing prokaryotes are capable of transforming atmospheric nitrogen (N_2) into organic nitrogen ($R—NH_2$). To accomplish this feat, they require an enzyme complex called nitrogenase, which is composed of two interacting proteins, molybdoferredoxin and azoferredoxin. Both of these proteins contain iron and sulfur; molybdoferredoxin also contains molybdenum. Nitrogen-fixing bacteria thus require the element molybdenum to grow, although in some media it can be replaced by vanadium.

Without the nitrogen-fixing capability of this bacterial phylum, we would all starve of protein deficiency. Extracting nitrogen from the air is an extremely costly process, as manufacturers of fertilizer know. Although these bacteria pay the price—they depend on the carbohydrates produced by their plant hosts—it is well worth it because they grow with very little competition in environments that lack nitrogen compounds.

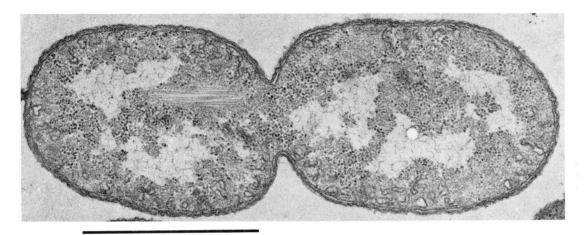

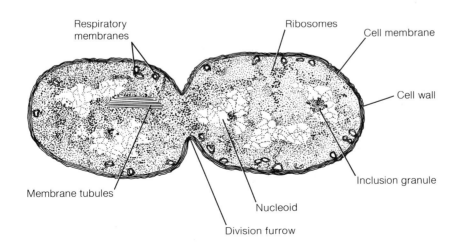

Azotobacter vinelandii, commonly found in garden soils. In this photograph, division into two cells is nearly complete. TEM, bar = 1 μm. [Photograph courtesy of W. J. Brill; drawing by I. Atema.]

Respiratory membranes

Ribosomes

Cell membrane

Cell wall

Inclusion granule

Membrane tubules

Nucleoid

Division furrow

M-10 Pseudomonads

Greek *pseudes,* false; *monos,* single

Bdellovibrio
Halobacterium
Hydrogenomonas
Pseudomonas
Xanthomonas
Zoogloea

Pseudomonads are the weeds of the bacterial world. As a group, they have an amazing ability to break down organic carbon compounds of all kinds. They metabolize not only the usual sugars, organic acids, and alcohols, but also organic ring compounds, such as those found in petroleum, complex tough polysaccharides in natural fibers, and a wide variety of plant metabolic products.

Pseudomonads are typically straight or curved Gram-negative rods, singly or multiply flagellated at one end. They all contain catalase and use oxygen as the terminal electron acceptor in respiration. In addition, some can respire anaerobically using nitrate as the terminal electron acceptor, reducing it to nitrite, nitrous oxide, or molecular nitrogen. Most can grow on a simple mineral medium with one- or two-carbon compounds, such as methanol, ethanol, and acetate, as their sole source of energy. However, they cannot ferment carbon compounds (break them down anaerobically by using them as electron acceptors). Pseudomonads must respire their food sources; thus, they give off carbon dioxide.

The phylum is named after the enormous genus *Pseudomonas.* The most metabolically distinguished of the pseudomonads, *Hydrogenomonas,* grows on purely inorganic substances; it oxidizes molecular hydrogen with oxygen from the air to derive energy, and uses carbon dioxide as a source of carbon. *Hydrogenomonas* and all other pseudomonad autotrophs are facultative autotrophs—they also feed on energetic organic carbon sources, which they prefer when given the option.

Xanthomonas, yellow straight rods that form yellow colonies on agar, are found in plants. Xanthomonads have far more complex nutritional requirements than many species of *Pseudomonas.* Many xanthomonads are serious plant pathogens, causing a variety of diseases, including leaf spots and cankers.

A member of the genus *Zoogloea* begins its life cycle as a single cell motile by means of single polar flagellum. Having produced a large population in pond or lake waters, the cells aggregate into flocs or globs that may be visible to the unaided eye. From these flocs grow out finger-like projections of a firm consistency. The cohesiveness of the colony is due to the intertwining of fibrils produced by the cells.

Zoogloeas cannot break down long-chain carbohydrates. However, they can oxidize all sorts of sugars, ethanol, and certain amino acids. They require vitamin B_{12} for growth; other vitamins may be stimulatory. There are only two accepted species: *Z. ramigera,* originally isolated from sewage sludge, and *Z. filipendula,* originally isolated from pistons and other submerged objects in waterworks near Berlin.

Members of the genus *Bdellovibrio* are small polarly flagellated curved rods, some marine and some freshwater. These amazing predatory bacteria reproduce only in the periplasm of their victims, usually omnibacteria (M-14).

The two species of *Halobacterium, H. halobium* and *H. salinarium,* are included in the pseudomonads because of their obligate aerobic metabolism and their morphology: Gram-negative rods or cocci. Also, they are motile. However, they are quite different from other pseudomonads in their strange adaptation to salt—they live in saturated solutions of sodium chloride in salt works and brine all over the world. They produce bright pink carotenoids and even can be spotted from airplanes and orbiting satellites as pink scum on salt flats. Many of their proteins have been modified so that they simply do not function except at high salt concentrations. Their cell walls are quite different from those of other bacteria in that they lack derivatives of diaminopimelic and muramic acids. Ordinary lipoprotein membranes burst or fall apart at high salt concentrations, but the halobacters' special lipids include derivatives of glyceroldiether that stabilize the membranes under high salt concentrations.

Analysis of the ribosomes and lipids of halobacters indicates that these bacteria are related more closely to the methanogens than to the other pseudomonads. Some classifiers place them with the methanogens in a separate kingdom, called Archaebacteria (see Phylum M-5).

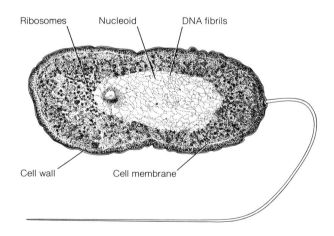

Ribosomes   Nucleoid   DNA fibrils

Cell wall   Cell membrane

A longitudinal (right) and a more transverse (left) view of *Pseudomonas multi-
vorans.* Pseudomonads are nearly ubiquitous in soils and water. TEM, bar = 1 μm.
[Photograph courtesy of S. Holt and T. G.
Lessie; drawing by I. Atema.]

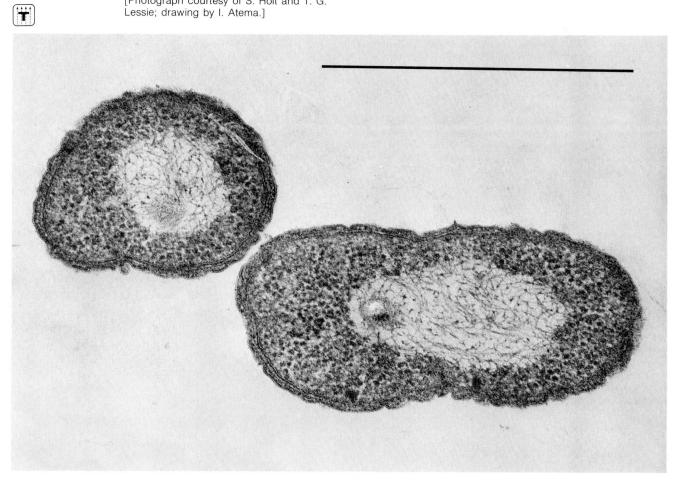

M-11 Aeroendospora

(Aerobic endospore-forming bacteria)

Bacillus
Sporolactobacillus
Sporosarcina

Greek *aer*, air; *endos*, within; *spora*, seed

Endospores formed within the cells of all members of this phylum are specialized reproductive structures resistant to heat and desiccation. A special compound, calcium dipicolinate, makes up from 5% to 15% of the dry weight of the spore. The core of the spore contains one copy of the parent bacterium's genetic material. It is enclosed in a cortex made of peptidoglycan (peptide units attached to nitrogenous sugars), surrounded by an outer layer, called the spore coat. Morphogenesis of the spore is quite complicated.

The majority of the bacteria belonging to the group are Gram positive and motile by means of peritrichous or laterally inserted flagella. Members of this phylum are either obligate or facultative aerobes; all are capable of oxygen respiration. At least three genera are recognized: *Bacillus,* a huge and important genus, *Sporosarcina,* and *Sporolactobacillus.*

The bacilli comprise some fifty species with hundreds of strains distributed in a multitude of habitats all over the world. Growing bacilli are rod shaped, as a rule, and their spores either elliptical or spherical. A bacillus "mother cell" produces only a single, airborne spore, which may land anywhere. In this dormant stage, bacilli can survive for years without water and nutrients. If nutrients and moisture are plentiful, the spore germinates and vegetative growth ensues until nutrient or water is depleted. As conditions become unfavorable for growth, sporulation sets in; the spore's position in the mother cell may be either central or terminal. Finally, the mother cell shrivels or disintegrates, releasing the spore.

Many strains of bacilli produce antibiotics during the active growth stage. Most species of bacilli produce acid and all can metabolize glucose. Several produce gas or acetoin as products of glucose catabolism. Many species can hydrolyze starch to glucose and some also produce pigments such as the red-brown or orange pulcherrinin and the brown or black melanin.

Many species break down tough plant substances such as pectin, polysaccharides, and even cellulose. The ability to degrade cellulose is unusual in the living world, which accounts in part for the persistence of forest litter, petroleum, peat, coal, and other organic-rich sediments. Cellulose-degrading bacilli have been isolated from the hindguts of wood-eating termites.

The requirements for growth vary widely among bacilli. Some have a high tolerance or even a requirement for salt; others do not. Vitamins and other complex growth factors are needed by certain strains, but not by others. Some can fix molecular nitrogen from the air, but most have more complex organic nitrogen requirements. Some may grow at a temperature as low as $-5°C$; some strains, called psychrophils, grow *optimally* at $-3°C$. Thermophils, found in hot springs, grow at temperatures above $45°C$.

Entire colonies of bacilli can be motile, rotating as a unit for unknown reasons and by unknown mechanisms. Some bacilli tend to grow in chains, forming colonies that are characterized by distinctive outgrowths.

The genus *Sporosarcina* has only a single species, *S. ureae.* Composed of a spherical cell from 1 μm to 2 μm in diameter, it forms tetrads (packets of four) and occasionally grows in distinctive cubical bundles that give the genus its name (Latin *sarcina* means packet). Urea, the metabolic product excreted by many animals, is converted by *S. ureae* to ammonium carbonate.

In appearance and metabolism, *Sporolactobacillus* is very similar to *Lactobacillus* (Phylum M-2, Fermenting bacteria), except that *Sporolactobacillus* produces endospores and can respire in the presence of oxygen.

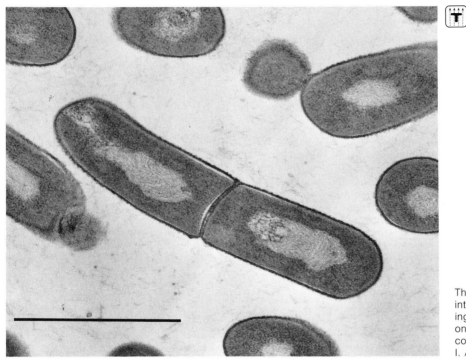

This bacillus has just completed division into two daughter cells. Such spore-forming bacilli are common both in water and on land. TEM, bar = 1 μm. [Photograph courtesy of E. Boatman; drawings by I. Atema.]

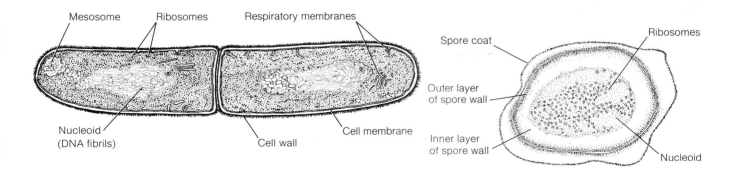

Mesosome   Ribosomes   Respiratory membranes

Nucleoid (DNA fibrils)   Cell wall   Cell membrane

Spore coat   Ribosomes

Outer layer of spore wall

Inner layer of spore wall   Nucleoid

Greek *mikros,* small; Latin *coccus,* berry

Aerococcus
Gaffkya
Micrococcus
Paracoccus
Planococcus
Sarcina
Staphylococcus

Micrococci are Gram-positive bacteria that require oxygen for growth. They are spherical cells, often found singly or grouped in pairs. Characteristically they divide in more than one plane to produce tetrads, irregular cubical packets of four cells. They do not form spores and many are not motile.

Micrococci are strictly or facultatively aerobic—some can ferment, but all respire, using oxygen as the terminal electron acceptor. They synthesize respiratory pigments called cytochromes and a class of quinones, also involved in respiration, called menaquinones. Most species metabolize sugars: glucose, for example, is oxidized either to acetate or completely to carbon dioxide and water. They metabolize glucose by the hexose monophosphate pathway, rather than by the Embden-Meyerhof pathway used by eukaryotes and many heterotrophic bacteria. Some species also oxidize smaller organic compounds, such as pyruvate, acetate, lactate, succinate, and glutamate, by the citric-acid or Krebs cycle, characteristic of mitochondria. Some micrococci can grow in hypersaline environments, such as 5% NaCl in water (sea water is about 3.4% total salts). They all break down hydrogen peroxide by using the enzyme catalase.

At least seven genera are recognized: *Planococcus, Aerococcus, Micrococcus, Sarcina, Gaffkya, Paracoccus,* and *Staphylococcus.*

Members of genus *Planococcus* have from one to four flagella. Their cells are small spheres 1 μm in diameter, often grouped in pairs, threes, or tetrads. They produce a water-insoluble lemon-yellow pigment and form smooth, glistening, slightly convex colonies when grown on solid-agar plates.

Aerococcus have a strong tendency to grow as tetrads. These small cocci, from 1 μm to 2 μm in diameter, are homofermentative—the product of their sugar fermentation is a single acid. They are microaerophils, growing best at oxygen concentrations lower than atmospheric.

The genus *Micrococcus* has several species, some motile. *M. luteus* is characterized by production of a yellow pigment and *M. roseus* by a reddish one. The cell walls of *M. luteus* are unique in that they contain L-lysine peptide units. Bacteria belonging to this genus have membrane-bound electron-transport multienzyme systems containing cytochromes *a, b,* and *c* as well as

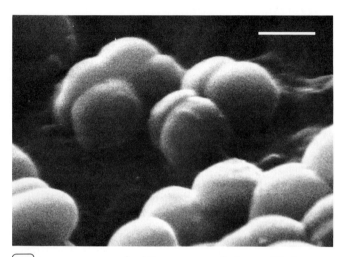

A *Micrococcus radiodurans.* Whole-mount SEM, ×18,000. [Courtesy of J. Troughton.]

yellowish and reddish carotenoid pigments. The optimum temperature for growth of most *Micrococcus* species is about 30°C.

Micrococcus roseus looks much like *M. radiodurans* pictured here. Some bacteriologists consider *M. radiodurans* simply a strain of *M. roseus* because they have similar proportions of guanine and cytosine in their DNA, similar pigments, and other biochemical features in common. However, *M. radiodurans* has a cell wall entirely different in composition from the walls of all other members of the genus; perhaps the difference is related to the extreme resistance of *M. radiodurans* to ultraviolet and gamma radiation.

Sarcina species form cubical packets of cells. They live under very acidic conditions and can ferment sugars.

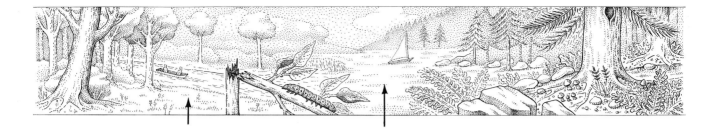

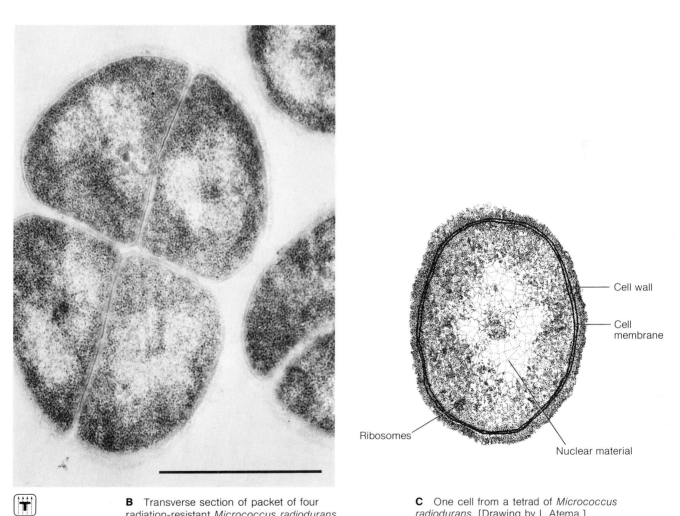

B Transverse section of packet of four radiation-resistant *Micrococcus radiodurans* cells. TEM, bar = 1 μm. [Courtesy of A. D. Burrell and D. M. Parry.]

C One cell from a tetrad of *Micrococcus radiodurans.* [Drawing by I. Atema.]

Cell wall

Cell membrane

Ribosomes

Nuclear material

M-13 Chemoautotrophic Bacteria

Ferrobacillus   *Nitrosomonas*
Macromonas   *Nitrosospira*
Methylococcus   *Nitrospina*
Methylomonas   *Sulfolobus*
Nitrobacter   *Thiobacillus*
Nitrococcus   *Thiobacterium*
Nitrocystis   *Thiomicrospira*
Nitrosococcus   *Thiospira*
Nitrosogloea   *Thiovelum*
Nitrosolobus

Chemoautotrophy, or chemolithotrophy, is metabolism that functions without sunlight and without preformed organic compounds—not a single vitamin, sugar, or amino acid. Thus, chemoautotrophic bacteria represent the pinnacle of metabolic achievement. They live on air, salts, water, and an inorganic source of energy. Provided with nitrogenous salts, oxygen, carbon dioxide (CO_2), and a proper energy source, such as the reduced gases ammonia (NH_3), methane (CH_4), or hydrogen sulfide (H_2S), they make all of their own nucleic acids and proteins, and derive their energy from the oxidation of the reduced gases. Some of these bacteria are capable of using organic compounds as food, but all of them can do without such compounds. Chemoautotrophic bacteria are crucial to the cycling of nitrogen, carbon, and sulfur through the world because they convert gases and salts unusable by animals and plants to usable organic compounds. The maintenance of the biosphere depends on such metabolic virtuosity, yet chemoautotrophy is strictly limited to bacteria.

Chemoautotrophic bacteria can be classified by the compounds that they oxidize to gain energy. We recognize at least three classes: oxidizers of nitrogen compounds, sulfur compounds, and CH_4. The class of chemoautotrophs that oxidize reduced nitrogen compounds includes seven genera belonging to two orders. One order, Nitrobacteriales, includes organisms that oxidize nitrite (NO_2^-) to nitrate (NO_3^-): *Nitrobacter, Nitrospina, Nitrocystis,* and *Nitrococcus,* which are distinguished from one another by their morphology.

Nitrobacters are short rods, many of them pear shaped or wedge shaped. Elaborate internal membranes extend along the periphery of one end of the cell. Old cultures of *Nitrobacter winogradskyi,* a widely distributed soil microbe, form a flocculent sediment made of gelatinous sheaths produced by the bacteria. Organic compounds and even ammonium salts inhibit the growth of this nitrobacter.

Nitrospinas are long slender rods that lack elaborate internal membranes. They are marine bacteria, strict aerobes, and strict chemoautotrophs—they cannot use preformed organic compounds at all.

Nitrococci are spherical cells containing distinctive internal membranes that form a branched or tubular network in the cytoplasm. They have the respiratory yellow proteins, cytochromes, but lack other pigmented compounds.

The other order of nitrogen-compound oxidizers contains the chemolithotrophs that oxidize NH_3 to NO_2^- for energy: *Nitrosomonas, Nitrosospira, Nitrosococcus,* and *Nitrosolobus.* These bacteria are strict chemolithotrophs and aerobes; they must live where both oxygen and ammonia exist, as at the edges of the anaerobic zone at the sedimentation interface in sea water, in soil, or deep in fresh water.

Nitrosomonas species are either ellipsoidal or rod shaped, occurring singly, in pairs, or as short chains. They are rich in cytochromes, which impart a yellowish or reddish hue to laboratory cultures. Internal membranes extend along the cell periphery. They grow at temperatures between 5°C and 30°C.

Nitrosospira is a genus of spiral-shaped freshwater microbes that lack internal membranes.

Nitrosococcus cells are spheres; they grow singly or in pairs, and often form an extracellular slime. Aggregates of cells attach to surfaces or become suspended in liquid.

Nitrosolobus cells are variously shaped, lobed cells that are motile by means of peritrichous flagella (scattered over the entire cell). They divide by constriction.

There are at least six genera of organisms currently recognized that grow by oxidizing inorganic sulfur compounds. Their cells contain sulfur granules and live in high concentrations of hydrogen sulfide or other oxidizable sulfur compounds; because they have not been grown in pure culture, however, any physiological conclusions about them must be tentative. The genera are of four distinct morphological types: nonmotile rods embedded in a gelatinous matrix (*Thiobacterium*), cylindrical cells having polar flagella (*Macromonas*), ovoid cells having peritrichous flagella (*Thiovelum*), and spiral cells having polar flagella (*Thiospira, Thiobacillus,* and *Sulfolobus*).

Thiobacillus, including eight species, is the best known genus of sulfur oxidizers—it has been grown in culture. Most of these rod-shaped Gram-negative cells are motile by means of a single polar flagellum. They derive energy from the oxidation of sulfur or its compounds, such as sulfide (S^{2-}), thiosulfate ($S_2O_3^{2-}$), polythionate, and sulfite (SO_3^{2-}). The final oxidation product is sulfate (SO_4^{2-}), but under certain conditions other sulfur compounds may accumulate. One species, *T. ferrooxidans,* also uses ferrous compounds as electron donors. Thiobacilli will grow on strictly inorganic media, using CO_2 to produce organic compounds.

Cells of the genus *Sulfolobus* were first isolated in culture in 1972. They have cell walls that lack peptidoglycan and are only facultatively autotrophic. They can use elemental sulfur as their energy source and fix CO_2, but they may also use glutamate, yeast extract, ribose, and other organic compounds. They are hot-springs bacteria, having a temperature optimum in the range between 70°C and 75°C and an ability to grow in waters as hot as 88°C. They die of freezing at temperatures below 55°C! They

grow at pH from 0.9 to 5.8, preferring acidic waters at pH from 2 to 3. Thus, the one well-described species is aptly named: *Sulfolobus acidocaldarius.*

Our third class in this phylum are the methylomonads, chemo-autotrophs that oxidize the reduced single-carbon compounds CH_4 or methanol (CH_3OH). Two genera may be distinguished, by morphology: *Methylomonas* (various Gram-negative rods) and *Methylococcus* (spherical cells usually appearing in pairs). Methylomonads cannot grow on complex organic compounds, but use CH_4 or CH_3OH as their sole source of both energy and carbon. In fact, many are inhibited from growing by the presence of other organic matter.

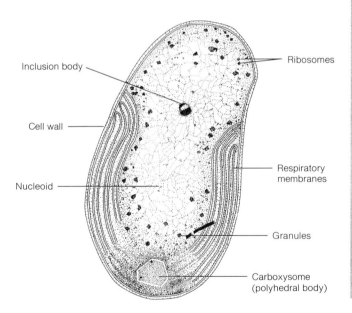

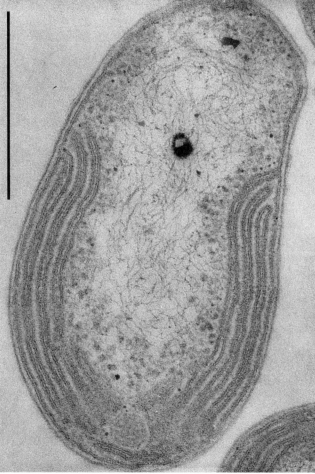

Nitrobacter winogradskyi. This specimen is young and thus lacks a prominent sheath. Carboxysomes are bodies in which are concentrated enzymes for fixing atmospheric CO_2. This species is named for the Russian Sergius Winogradsky, who pio- neered the field of microbial ecology. TEM, bar $= 0.5\ \mu$m. [Photograph courtesy of S. W. Watson, *International Journal of Systematic Bacteriology* 21:261; 1971. Drawing by I. Atema.]

55

M-14 Omnibacteria

(Enterobacteria and other facultatively aerobic Gram-negative heterotrophic bacteria)

Latin *omnis*, all; Greek *enteron*, gut

This is an enormous and extremely diverse group of bacteria, characterized by having alternative forms of metabolism. In the presence of oxygen, they respire aerobically. In the absence of oxygen they do not stop growing, as obligate aerobes do, but continue to respire, using compounds such as nitrate (NO_3^-) as the terminal electron acceptor. Respiration in the absence of oxygen leads to the concomitant reduction of NO_3^- to nitrite (NO_2^-), or of nitrogen (N_2) to nitrous oxide (N_2O). The cytochrome electron-transport pathway used with oxygen is the same as that used with NO_3^-. In many species, two respiration products can be excreted simultaneously—it depends on physiological and ecological conditions. All members of the phylum are chemoheterotrophic; that is, they require reduced organic compounds both for energy and for growth.

The omnibacteria may be divided broadly into two large groups: the class of enterobacteria, solitary, simple unicells, on the one hand, and several classes of complex morphological types (such as stalked, budding, and aggregated bacteria), on the other.

Enterobacteria have long been associated with human, plant, and animal diseases; many species have been isolated from intestinal tissue. A subclass, the enteric bacteria, comprises twelve famous genera of rod-shaped microbes, most of which have peritrichous flagella: *Escherichia, Edwardsiella, Citrobacter, Salmonella, Shigella, Klebsiella, Enterobacter, Serratia, Proteus, Yersinia (Pasteurella), Erwinia.*

Gram-negative rod-shaped unicellular bacteria grow rapidly and well. Although none produce spores, they seem to have remarkable persistence, waiting things out under conditions of adversity and vigorously taking advantage of new food sources. Any water sample in the world yields enterobacteria when incubated under the proper conditions of growth. Indeed, it is not too extreme to say that most life on Earth takes the form of facultatively anaerobic Gram-negative unicellular rod-shaped bacteria. They provide food for innumerable protists and are crucial to microbial food webs. More is known about *Escherichia coli* than about any other single organism on Earth: it is the chief object of study by molecular biologists.

A second subclass of the enterobacteria comprises mainly comma-shaped organisms most of which have single polar flagella. (Bacteria of this shape are called vibrios.) The subclass contains six genera: *Vibrio, Beneckea, Aeromonas, Pleisiomonas, Photobacterium*, and *Xenorhabditis.*

Members of the genus *Vibrio* are infamous because they cause cholera. They ferment carbohydrates to mixed products, including acids, but do not give off carbon dioxide and hydrogen. Members of the genus *Beneckea* are marine vibrio-shaped organisms that require salt. They are capable of growing fermentatively or by aerobic respiration on a broad range of carbon sources. Some species are bioluminescent. Only two other bioluminescent prokaryotic genera are known: *Photobacterium* and *Xenorhabditis. Photobacterium* species are famous for their associations with fish, which culture the bacteria in special pockets called light organs. The growth requirements of *Photobacterium* are generally far more restricted than those of the free-living marine genus *Beneckea. Xenorhabditis,* recently discovered, is known only from its associations with nematodes and nematode-eating insects.

Members of the genus *Aeromonas,* informally called aeromonads, are very common in ponds, lakes, and soils. They are coccoids or straight rods with rounded ends; most of them are motile by means of polar flagella. When growing anaerobically, they reduce nitrate to nitrite. Most of them contain cytochrome *c* and catalase, an enzyme that decomposes hydrogen peroxide to water and oxygen. As a group, they are very eclectic in their tastes and attack a wide variety of food sources, especially plant materials such as starch, casein, gelatin, dextrin, glucose, fructose, and maltose.

There is a large number of simple unicellular Gram-negative bacteria that we classify as enterobacteria but not in either of the two subclasses described above. These include *Zymomonas, Chromobacterium, Flavobacterium, Haemophilus, Actinobacillus, Cardiobacterium, Streptobacillus, Calymmatobacterium,* and several species of parasites or symbionts of the ciliate protists *Paramecium* (Phylum Pr-18, Ciliophora). The paramecium symbiotic bacteria, originally called kappa, lambda, sigma, and mu particles, have complex growth requirements. As a result, most are poorly known because they cannot be grown in culture. They were once thought to be cytoplasmic genes of the ciliate itself.

The enterobacteria are distinguished from each other by the carbohydrates they can attack (lactose, glutamic acid, arabinose, sugar, alcohols, citrate, tartrate, and other fairly small organic compounds). They are distinguished as well by their chemical abilities—for example, to hydrolyze urea, to produce gas from glucose, or to break down gelatin. They can also be distinguished by their sensitivity to specific bacteriophages, by their surface antigens, by their attraction to certain hosts, by

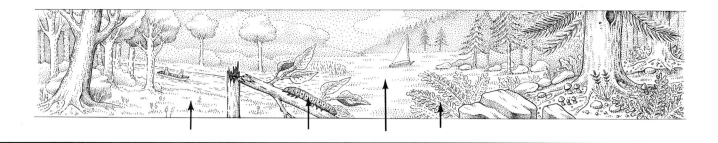

their pathogenicity, and by morphological traits, such as mode of motility.

Many enterobacteria produce colorful pigments—the violet, ethanol-soluble violacein of *Chromobacterium*, the red prodigiosin of *Serratia* and *Beneckea*, the red-yellow, orange, and brown pigments, some of which are carotenoids and some not, of *Flavobacterium* and *Aeromonas*. No one yet understands the functions of these pigments.

The second great group of omnibacteria is extremely heterogeneous and should probably be divided into several classes.

Appendages called prosthecae, or stalks, made of living material protrude from the cells of the prosthecate bacteria. *Caulobacter* and *Asticcacaulis* have single polar or subpolar prosthecae. Their life cycles superficially resemble those of invertebrate animals: a stalked sessile form divides to produce a motile form, which swims away; its offspring, in turn, is a sessile form.

The budding bacteria, or hyphomicrobia, reproduce by the outgrowth of buds that eventually swell to parent size. Hyphomicrobial colonies may form quite complex networks that resemble the mycelia of fungi. This sort of budding has also been found in the purple photosynthetic microbes (Phylum M-6, Anaerobic photosynthetic bacteria), which indicates that it probably evolved more than once among the bacteria.

The aggregated bacteria, which include *Sphaerotilus*, can be recognized by their distinctive clumps of cells. These microbes oxidize iron or manganese, which they then deposit around themselves as manganese and iron oxides. The bacteria are thought to gain some energy from these oxidations; however, there is still no proof that any of them are obligate lithotrophs. All prefer fixed nitrogen and carbon compounds; they grow faster if supplied with organic food.

Parasitic, nonflagellated bacteria of the genus *Neisseria* are infamous as the cause of gonorrhea and meningitis.

The acetic-acid bacteria, *Gluconobacter* and *Acetobacter*, oxidize ethanol to acetic acid—wine to vinegar. They form sheaths having a rectangular cross section.

The rickettsias and chlamydias are obligate intracellular parasites of vertebrates and arthropods. They have residual Gram-negative walls and their metabolism is limited, probably as a consequence of their parasitism. Chlamydias and rickettsias are responsible for all sorts of diseases: psittacosis of parrots and pigeons is caused by *Chlamydia psittaci*, and several species of *Rickettsia* can cause Rocky Mountain spotted fever.

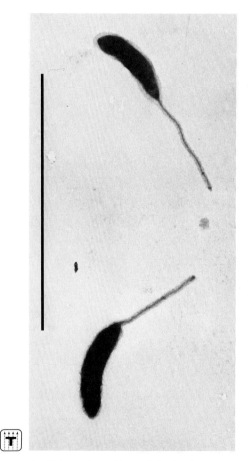

A Thin section of two stalked cells of *Caulobacter crescentua*, which, in nature, would be attached to rocks or other solid surfaces. These divide to form swarmer cells. TEM, negative stain, bar = 10 μm. [Courtesy of J. Staley.]

Acetobacter
Actinobacillus
Aerobacter
Aeromonas
Alcaligenes
  (Achromobacter)
Alginobacter
Asticcacaulis

Beneckea
Calymmatobacterium
Cardiobacterium
Caulobacter
Chlamydia
Chromobacterium
Citrobacter
Coxiella

Edwardsiella
Enterobacter
Erwinia
Escherichia
Flavobacterium
Gluconobacter
Klebsiella
Haemophilus

Neisseria
Neorickettsia
Photobacterium
Pleisiomonas
Proteus
Rickettsia
Rickettsiella
Salmonella

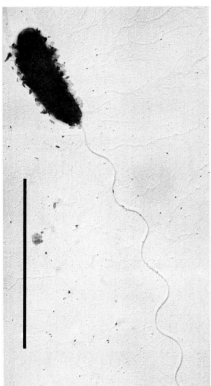

B   *Caulobacter* swarmer cell with long polar flagellum. The swarmer cells settle and metamorphose into stalked cells. TEM, negative stain, bar = 5 μm. [Courtesy of J. Des Rosier.]

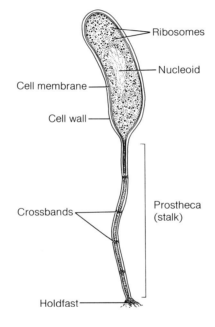

Ribosomes

Nucleoid

Cell membrane

Cell wall

Crossbands

Prostheca (stalk)

Holdfast

C   Stalked *Caulobacter.* [Drawing by I. Atema.]

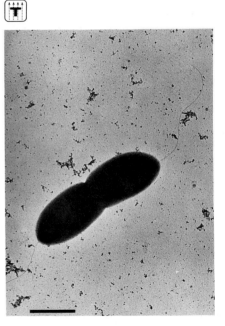

D   *Aeromonas punctata* (new strain) symbiotic in the gastrodermal cells of the green hydra, *Hydra viridis* (Phylum A-3, Cnidaria). The cell is dividing; a flagellum can be seen at each end. TEM (negative stain), bar = 1 μm. [Courtesy of J. F. Stolz.]

Serratia
Shigella
Sphaerotilus
Streptobacillus
Vibrio
Xenorhabditis
Yersinia (Pasteurella)
Zymomonas

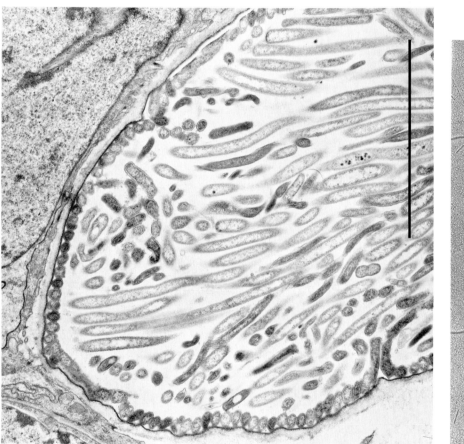

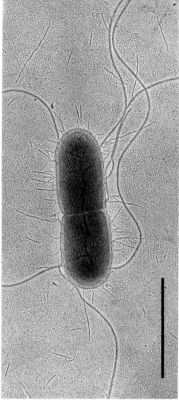

E An intact bacterial community from a pocket in the hindgut wall of the Sonoran desert termite *Pterotermes occidentis* (Phylum A-27). More than 10 billion bacteria per milliliter have been counted in these hindgut communities. Many are of unknown species. In our studies, 28 or 30 strains isolated were facultative aerobes, most of them motile and Gram negative—thus, by definition, omnibacteria. Notice that some of the bacteria line the wall of the gut, while others float free in the lumen. TEM, bar = 5 μm. [Courtesy of D. Chase.]

F *Escherichia coli* is peritrichously flagellated—the flagella emerge on all sides. A new cell wall has formed and the bacterium is about to divide. The smaller appendages, called pili, are known to make contact with other cells in bacterial conjugation. However, even many strains that do not conjugate have pili. TEM (shadowed with platinum), bar = 1 μm. [Courtesy of D. Chase.]

M-15   Actinobacteria

(Actinomycetes, Actinomycota)

Greek *aktis*, ray

Actinomyces	*Mycobacterium*
Arthrobacter	*Mycococcus*
Cellulomonas	*Nocardia*
Corynebacterium	*Propionibacterium*
Dermatophilus	*Streptomyces*
Micromonospora	*Thermoactinomyces*

This phylum includes the so-called coryneform bacteria and the actinobacteria *sensu stricto*. We shall consider these groups to be classes.

The coryneforms are unicellular Gram-positive organisms, straight or slightly curved rods with a tendency to form club-shaped swellings. Corynebacterial genera include long or spiny *Arthrobacter,* which have as many as twenty spines per cell, the cellulose-attacking *Cellulomonas,* and *Propionibacterium,* whose species produce propionic or acetic acids as products of sugar metabolism. The daughter cells of a fissioned coryneform or other actinobacterium typical remain attached in a Y- or V-shaped configuration.

The second class, the actinobacteria *sensu stricto,* includes microorganisms that were originally mistaken for fungi and called actinomycetes. They are distinguished by their funguslike morphology and the production of actinospores. This large and diverse group includes organisms that make extremely short multicellular filaments (such as *Mycobacterium,* the causative agent of tuberculosis) and many others that produce long and complex filaments; from these, resistant germinative structures, the actinospores, are released.

Actinospores are often, but inaccurately, called conidia. True conidia are the eukaryotic, haploid propagating spores of the ascomycote and basidiomycote fungi (Phyla F-2 and F-3). Actinospores differ also from bacterial endospores, which are formed within a parent cell and then released (see Phyla M-2, Fermenting bacteria, and M-11, Aeroendospora); in the development of an actinospore, the entire cell converts to a thick-walled, resistant form. Actinobacteria probably evolved the "fungal" habit of growing hyphae that form a visible mycelial mass long before true fungi evolved. Thus, actinospores, endospores, and fungal conidia are all convergent structures—they represent a response to similar environmental pressures.

All six families in the class of actinobacteria *sensu stricto* form true mycelia, but only two enclose their actinospores in external structures. In these two, the Frankiaceae and the Actinoplanaceae, actinospores are borne inside structures called sporangia, by analogy with fungi. The Frankiaceae are symbiotic in plant nodules, where, like *Rhizobium* (Phylum M-9), they fix atmospheric nitrogen.

In the Dermatophilaceae, represented by *Dermatophilus,* the mycelial filaments divide transversely and in at least two longitudinal planes to form coccoid, motile bacteria that swim away and form mycelial filaments. Some species form pathogenic lesions on human skin; others have been isolated from soil. The motile bacteria have been called zoospores, a name also applied to motile eukaryotic cells capable of continued asexual growth—for example, chytrids (Phylum Pr-26) and plasmodiophorans (Phylum Pr-24). The term ought to be restricted to eukaryotes. However, a multicellular mycelial habit alternating with zoospores, both stages strictly asexual, has evolved independently in prokaryotes and in several distinctly different groups of eukaryotic microorganisms. The tendency to form resistant structures in a "head" and then release them to be scattered by wind or by other organisms also has evolved many times in actinobacteria, in other prokaryotes, and in several protoctist and fungal groups.

The family Nocardiaceae includes the widely distributed genus *Nocardia.* Nocardias typically form mycelial filaments that fragment, yielding single nonmotile bacteria. They tend, especially in old cultures, to be Gram variable (some Gram-negative cells) rather than clearly Gram positive. The family contains some pathogenic forms and at least one species capable of nitrogen fixation. Most nocardias are very tenacious and survive, but do not grow, under many noxious conditions. If their entire developmental pattern is not observed, they can easily be mistaken for unicellular bacteria. Eventually, however, they betray their nocardial nature by forming filaments, mycelia, Y- and V-shaped cell groups, and actinospores.

Members of the family represented by *Streptomyces* form mycelia that tend to remain intact. From the mycelia grow quite remarkable and well-developed aerial spore-bearing structures and fuzz. Some are easy to confuse, at least superficially, with the smaller fungi. *Streptomyces* species form long chains of actinospores. This group is justly famous for its synthetic versatility in producing streptomycin and other antibiotics.

Bacteria of the family Micromonosporaceae, represented by *Micromonospora,* form spores singly, in pairs, or in short chains on either aerial or subsurface mycelia. The mycelia are branched and septate, and the spores are often brown.

A Colony of *Streptomyces rimosus* after a few days of growth on nutrient agar in petri plates. Bar = 10 mm. [Courtesy of L. H. Huang, Pfizer, Inc.]

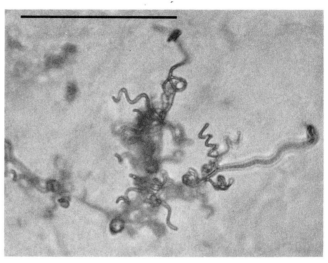

B Aerial trichomes (filaments) bearing actinospores of *Streptomyces*. LM, bar = 50 μm. [Courtesy of L. H. Huang, Pfizer, Inc.]

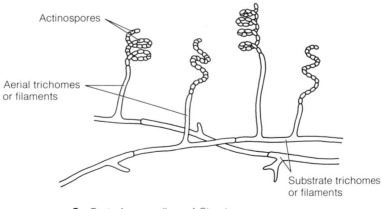

Actinospores

Aerial trichomes or filaments

Substrate trichomes or filaments

C Part of a mycelium of *Streptomyces*. [Drawing by R. Golder.]

M-16 Myxobacteria

Greek *myxa*, mucus

Alysiella
Angiococcus
Archangium
Beggiatoa
Chondrococcus
Chondromyces
Cytophaga
Flexibacter
Flexithrix
Herpetosiphon

Myxococcus
Podangium
Polyangium
Saprospira
Simonsiella
Sporocytophaga
Stelangium
Stigmatella
Synangium

The myxobacteria represent the acme of morphological complexity among the prokaryotes. Most bacteriologists recognize two groups—traditionally called orders, but classes in our scheme—of these gliding or waving bacteria: the Myxobacteriales, which form upright multicellular fruiting bodies, and the Cytophagales, which glide but lack such fruiting bodies.

Individual myxobacteria are unicellular Gram-negative rods, usually less than 1.5 μm in diameter, although they may be as long as 5 μm; they often aggregate into complex colonies that show distinctive behavior and form. The cells are typically embedded in slime, polysaccharides of the cells' own making. Individual cells divide by ordinary bacterial binary fission. Although they unmistakably move by gliding along surfaces, the part of the bacteria in contact with surfaces shows no organelles of motility, such as flagella. Numerous tiny intracellular fibrils seen in electron micrographs of some species have been correlated with motility, but the means of motility in these organisms is not really understood.

When nutrients or water are depleted, myxobacteriales aggregate and form upright structures composed of extracellular excretion products and many cells. Bacterial cells within these "fruiting bodies" enter a resting stage—they are called myxospores. In some species, these cells become encapsulated, thick walled, and shiny; in others, they seem to be quite like active, vegetative bacteria. Some, like *Polyangium violaceum*, form brightly colored fruiting structures. Others form branched stalks; these tiny "trees" may be barely visible to the unaided eye. Some form thick-walled, darkly colored, spore-filled cysts, called sporangioles, that open when wetted to release huge numbers of individual gliding bacteria; the gliding cells move together to form migrating colonies. The entire life cycle of such myxobacteria is uncannily analogous to that of the slime molds (Phyla Pr-22 and Pr-23). Vegetative cells may also enter the myxospore stage outside a fruiting body.

All myxobacteriales are obligate aerobes; none grows by fermentation. They typically produce enzymes capable of hydrolyzing macromolecules such as proteins, nucleic acids, fatty-acid esters, and complex polysaccharides; some may even hydrolyze cellulose. Some may also actively lyse (cause to break open) protoctists or bacteria by secreting digestive and other degradative enzymes into their surroundings. By osmotrophy, the myxobacteria then absorb the nutrients released by disintegration of their prey.

Myxobacteriales commonly inhabit soils; no marine forms have been isolated. Many are difficult to culture through their entire life cycles. It is likely that many species, especially those in the tropics, are not yet known to science.

The cytophagales are unified only by their ability to glide at some stage in their life cycle and by the fact that they can grow heterotrophically. They are a mixed bag and will probably be reclassified. At present, six families are recognized. One, the Cytophagaceae, constitutes a natural group of straight rods or rigid helical filaments, gliding forms that produce orange pigments, carotenoids. The family includes *Cytophaga,* which breaks down agar, cellulose, or chitin, *Flexibacter,* which is able to metabolize less tough carbohydrates, such as starch and glycogen, and *Herpetosiphon,* which can break down cellulose, but not agar or chitin, and has a sheath around its cells. Three filamentous genera are *Flexithrix* and helical *Saprospira,* neither of which forms cysts, and *Sporocytophaga,* which forms small resistant cells called microcysts.

Two other families, the Beggiatoaceae, represented by the free-living *Beggiatoa,* and the mouth-dwelling Simonsiellaceae, are negatively defined—they have no resting stages and not enough carotenoid pigment to be visibly colored. The Beggiatoaceae, obligate aerobes, include some strikingly beautiful, widely distributed filamentous bacteria found in sulfur-rich zones, collectively called the sulfuretum. Some (mostly filamentous sheathed bacteria) can grow chemolithotrophically by oxidizing sulfides to sulfates. Often they are arranged in rosettes, as in our illustration. The sheaths of beggiatoas contain conspicuous sulfur granules, and their cells often contain poly-β-hydroxybutyrate-rich or phosphorus-rich granules called volutin.

The Simonsiellaceae comprise two genera of gliding bacteria, *Simonsiella* and *Alysiella,* both of which are found in the mouths of vertebrates. These have been cultured in the laboratory, but only on complex media such as blood or serum. The cells of *Simonsiella* are associated in short flat filaments reminiscent of the hormogonia of filamentous cyanobacteria (Phylum M-7). These filaments fragment into shorter units, which can glide away to resume growth elsewhere; the terminal cells of the filaments have rounded ends. The cells of *Alysiella* are arranged in pairs within flat filaments; the terminal cells are not rounded. How they live or what they are doing in the mouths of animals is not known.

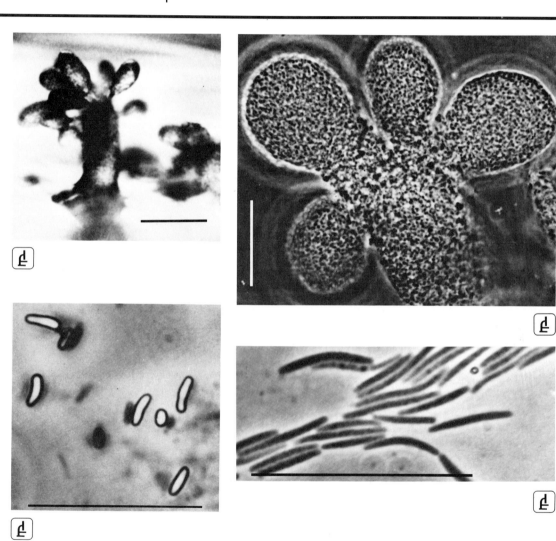

A The fruiting bacterium *Stigmatella aurantiaca,* which grows on remains of vegetation in soil. Top left: The fruiting body. LM, bar = 100 μm. Top right: The fruiting body as seen in a squash preparation in a permanent mount. Phase-contrast LM, bar = 50 μm. Bottom left: Myxospores. LM, bar = 50 μm. Bottom right: Growing vegetative cells, which glide in contact with solid sufaces. LM, bar = 50 μm. [Photographs courtesy of H. Reichenbach and M. Dworkin. In *The Prokaryotes,* M. Starr (ed.), New York: Springer-Verlag, in press.]

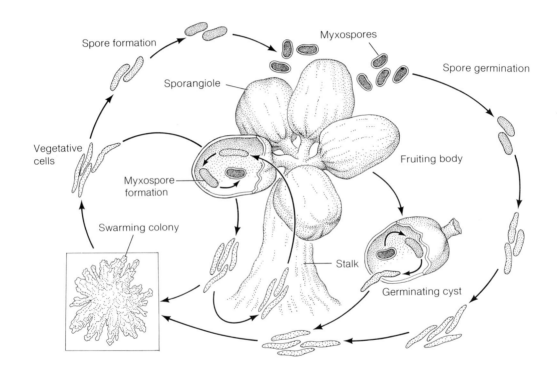

B Life cycle of *Stigmatella aurantiaca.*
[Drawing by L. Meszoly; labeled by
M. Dworkin.]

General

Buchanan, R. E., and N. E. Gibbons, eds., *Bergey's manual of determinative bacteriology,* 8th ed. Williams & Wilkins; Baltimore, Maryland; 1974.

Fenner, F., B. R. McAuslan, C. A. Mims, J. Sambrook, and D. O. White, *The biology of animal viruses,* 2nd ed. Academic Press; New York; 1974.

Gillies, R. R., and T. C. Dodds, *Bacteriology illustrated,* 2nd ed. Williams & Wilkins; Baltimore, Maryland; 1968.

Stanier, R. Y., E. A. Adelberg, and J. L. Ingraham, *The microbial world,* 4th ed. Prentice-Hall; Englewood Cliffs, New Jersey; 1976.

Starr, M. P., H. Stolp, H. G. Trüper, A. Balows, and H. G. Schlegel, eds., *The prokaryotes: A handbook on habitats, isolation and identification of bacteria,* 3 volumes. Springer-Verlag; Heidelberg and New York; in press.

Wolfe, S., *The biology of the cell.* Wadsworth; Belmont, California; 1972.

M-1   Aphragmabacteria

Bove, J. M., and J. F. Duplan, eds., *Mycoplasms of man, animals, plants and insects.* INSERM; Bordeaux, France; 1974.

Madoff, S., ed., *Mycoplasma and the L forms of bacteria.* Gordon and Breach; New York; 1971.

Whitcomb, R. F., "The genus *Spiroplasma.*" *Annual Review of Microbiology* 34:677–709; 1980.

M-2   Fermenting bacteria

Brock, T. D., *Biology of microorganisms,* 3rd ed. Prentice-Hall; Englewood Cliffs, New Jersey; 1979.

Stanier, R. Y., E. A. Adelberg, and J. L. Ingraham, *The microbial world,* 4th ed. Prentice-Hall; Englewood Cliffs, New Jersey; 1976.

Prévot, A., and V. Fredette, *Manual for the classification and determination of the anaerobic bacteria.* Lea & Febiger; Philadelphia; 1965.

M-3   Spirochaetae

Breznak, J. A., "Biology of nonpathogenic host-associated spirochetes." *Critical Reviews in Microbiology* 2:457–489; 1973.

Brock, T. D., *Biology of microorganisms,* 3rd ed. Prentice-Hall, Englewood Cliffs, New Jersey; 1979.

Canale-Parola, E., "Motility and chemotaxis of spirochetes." *Annual Review of Microbiology* 32:69–99; 1978.

Poindexter, J. S., *Microbiology, an introduction to protists,* Macmillan; New York; 1971.*

M-4   Thiopneutes

Brock, T. D., *Biology of microorganisms,* 3rd ed. Prentice-Hall; Englewood Cliffs, New Jersey; 1979.

Frobisher, M., R. D. Hinsdill, K. T. Crabtree, and C. R. Goodheart, *Fundamentals of microbiology,* 9th ed. W. B. Saunders; Philadelphia; 1974.

Stanier, R. Y., E. A. Adelberg, and J. L. Ingraham, *The microbial world,* 4th ed. Prentice-Hall; Englewood Cliffs, New Jersey; 1976.

M-5   Methaneocreatrices

Doetsch, R. N., and T. M. Cook, *Introduction to bacteria and their ecobiology.* University Park Press; Baltimore, Maryland; 1973.

Mitchell, R., *Introduction to environmental microbiology.* Prentice-Hall; Englewood Cliffs, New Jersey; 1974.

Mah, R. A., D. M. Ward, L. Baresi, and L. Glass, "Biogenesis of methane." *Annual Review of Microbiology* 31:309–341; 1977.

M-6   Anaerobic photosynthetic bacteria

Brock, T. D., *Biology of microorganisms,* 3rd ed. Prentice-Hall; Englewood Cliffs, New Jersey; 1979.

Clayton R. B., and W. R. Sistrom, eds., *The photosynthetic bacteria.* Plenum Press; New York; 1978.

Stanier, R. Y., E. A. Adelberg, and J. L. Ingraham, *The microbial world,* 4th ed. Prentice-Hall; Englewood Cliffs, New Jersey; 1976.

---

*The word *protists* in this title refers to both prokaryotic and eukaryotic microorganisms; this practice has been discouraged in recent years.

M-7 Cyanobacteria

Brock, T. D., *Biology of microorganisms,* 3rd ed. Prentice-Hall; Englewood Cliffs, New Jersey; 1979.
Fogg, G. E., W. D. P. Stewart, P. Fay, and A. E. Walsby, *The blue-green algae.* Academic Press; London and New York; 1973.
Stanier, R. Y., E. A. Adelberg, and J. L. Ingraham, *The microbial world,* 4th ed. Prentice-Hall; Englewood Cliffs, New Jersey; 1976.

M-8 Chloroxybacteria

Lewin, R. A., "Prochlorophyta as a proposed new division of algae." *Nature* 261:697–698; 1976.
Lewin, R. A., "*Prochloron,* type genus of the prochlorophyta." *Phycologia* 16:217; 1977.
Whatley, J. M., P. John, and F. R. Whatley, "From extracellular to intracellular: The establishment of mitochondria and chloroplasts." *Proceedings of the Royal Society of London* (series B) 204:165–187; 1979.

M-9 Nitrogen-fixing aerobic bacteria

Alexander, M., *Microbial ecology.* John Wiley and Sons; New York; 1971.
Brock, T. D., *Biology of microorganisms,* 3rd ed. Prentice-Hall; Englewood Cliffs, New Jersey; 1979.
Frobisher, M., R. D. Hinsdill, K. T. Crabtree, and C. R. Goodheart, *Fundamentals of microbiology,* 9th ed. W. B. Saunders; Philadelphia; 1974.

M-10 Pseudomonads

Brock, T. D., *Biology of microorganisms,* 3rd ed. Prentice-Hall; Englewood Cliffs, New Jersey; 1979.
Frobisher, M., R. D. Hinsdill, K. T. Crabtree, and C. R. Goodheart, *Fundamentals of microbiology,* 9th ed. W. B. Saunders; Philadelphia; 1974.
Stanier, R. Y., N. J. Palleroni, and M. Douderoff, "The aerobic pseudomonads: a taxonomic study." *Journal of General Microbiology* 43:159–271; 1966.

M-11 Aeroendospora

Brock, T. D., *Biology of microorganisms,* 3rd ed. Prentice-Hall; Englewood Cliffs, New Jersey; 1979.
Frobisher, M., R. D. Hinsdill, K. T. Crabtree, and C. R. Goodheart, *Fundamentals of microbiology,* 9th ed. W. B. Saunders; Philadelphia; 1974.
Poindexter, J. S., *Microbiology: An introduction to protists.* Macmillan; New York, 1971.*

---

*The word *protists* in this title refers to both prokaryotic and eukaryotic microorganisms; this practice has been discouraged in recent years.

M-12 Micrococci

Brock, T. D., *Biology of microorganisms,* 3rd ed. Prentice-Hall; Englewood Cliffs; New Jersey, 1979.
Poindexter, J. S., *Microbiology: An introduction to protists.* Macmillan; New York; 1971.*
Stanier, R. Y., E. A. Adelberg, and J. L. Ingraham, *The microbial world,* 4th ed. Prentice-Hall; Englewood Cliffs, New Jersey; 1976.

M-13 Chemoautotrophic bacteria

Doetsch, R. N., and T. M. Cook, *Introduction to bacteria and their ecobiology.* University Park Press; Baltimore, Maryland; 1973.
Sieburth, J. M., *Sea microbes.* Oxford University Press; New York; 1979.
Stanier, R. Y., E. A. Adelberg, and J. L. Ingraham, *The microbial world,* 4th ed. Prentice-Hall; Englewood Cliffs, New Jersey; 1976.

M-14 Omnibacteria

Baumann, P., and L. Baumann, "Biology of the marine enterobacteria: genera *Beneckea* and *Photobacterium.*" *Annual Review of Microbiology* 31:39–61; 1977.
Brock T. D., *Biology of microorganisms,* 3rd ed. Prentice-Hall; Englewood Cliffs, New Jersey; 1979.
Poindexter, J. S., *Microbiology: An introduction to protists.* Macmillan; New York; 1971.*
Sieburth, J. M., *Sea microbes.* Oxford University Press; New York; 1979.

M-15 Actinobacteria

Brock, T. D., *Biology of microorganisms,* 3rd ed. Prentice-Hall; Englewood Cliffs, New Jersey; 1979.
Hawker, L. E., and A. H. Linton, eds., *Micro-organisms: Function, form, and environment.* American Elsevier; New York; 1971.
Slack, J. M., and M. A. Gerencser, *Actinomyces, filamentous bacteria: Biology and pathogenicity.* Burgess; Minneapolis, Minnesota; 1975.
Veldkamp, H., "Saprophytic coryneform bacteria." *Annual Review of Microbiology* 24:209–240; 1970.

M-16 Myxobacteria

Doetsch, R. N., and T. M. Cook, *Introduction to bacteria and their ecobiology.* University Park Press; Baltimore, Maryland; 1973.
Kaiser, D., C. Manoil, and M. Dworkin, "Myxobacteria: cell interactions, genetics and development." *Annual Review of Microbiology* 33:595–639; 1979.
Lechevalier, H. A., and D. Pramer, *The microbes.* J. B. Lippincott; Philadelphia; 1971.

PROTOCTISTA

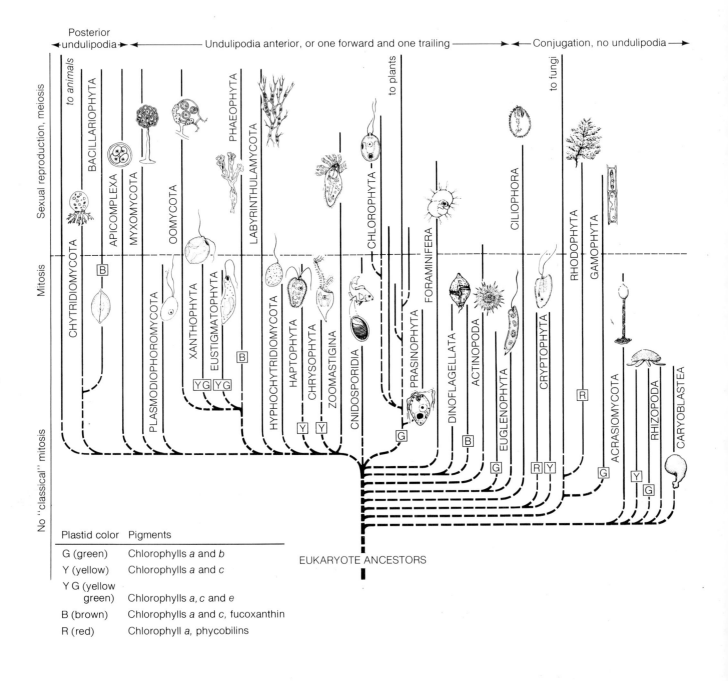

Posterior
◄─undulipodia─► ◄──────── Undulipodia anterior, or one forward and one trailing ────────► ◄─ Conjugation, no undulipodia ─►

to animals

to plants

to fungi

Sexual reproduction, meiosis

BACILLARIOPHYTA

PHAEOPHYTA

CHLOROPHYTA

CILIOPHORA

RHODOPHYTA

GAMOPHYTA

APICOMPLEXA

MYXOMYCOTA

OOMYCOTA

LABYRINTHULAMYCOTA

FORAMINIFERA

CHYTRIDIOMYCOTA

B

Mitosis

PLASMODIOPHOROMYCOTA

XANTHOPHYTA

EUSTIGMATOPHYTA

B

HYPHOCHYTRIDIOMYCOTA

HAPTOPHYTA

CHRYSOPHYTA

ZOOMASTIGINA

CNIDOSPORIDIA

PRASINOPHYTA

DINOFLAGELLATA

ACTINOPODA

EUGLENOPHYTA

CRYPTOPHYTA

R

ACRASIOMYCOTA

RHIZOPODA

CARYOBLASTEA

Y G   Y G

Y   Y

No "classical" mitosis

G

B

G

R Y

G

Y

G

EUKARYOTE ANCESTORS

Plastid color	Pigments
G (green)	Chlorophylls *a* and *b*
Y (yellow)	Chlorophylls *a* and *c*
Y G (yellow green)	Chlorophylls *a*, *c* and *e*
B (brown)	Chlorophylls *a* and *c*, fucoxanthin
R (red)	Chlorophyll *a*, phycobilins

PROTOCTISTA

Greek *protos*, very first; *ktistos*, to establish

Kingdom Protoctista is defined by exclusion: its members are neither animals (which develop from a blastula), plants (which develop from an embryo), fungi (which lack undulipodia and develop from spores), nor prokaryotes. They comprise the eukaryotic microorganisms and their immediate descendants: all nucleated algae (including the seaweeds), undulipodiated (flagellated) water molds, the slime molds and slime nets, and the protozoa. Protoctist cells have all the characteristically eukaryotic properties, such as aerobiosis and respiration in mitochondria, and most have 9 + 2 undulipodia at some stage of the life cycle.

Why *protoctist* rather than *protist?* Since the nineteenth century, the word *protist,* whether used informally or formally, has come to connote a single-celled organism. In the last two decades, however, the basis for classifying single-celled organisms separately from multicellular ones has weakened. It has become evident that multicellularity evolved many times from unicellular forms—many multicellular organisms are far more closely related to certain unicells than they are to any other multicellular organisms. For example, the ciliates (Phylum Pr-18, Ciliophora), which are unicellular microbes, include at least one species that forms a sorocarp, a multicellular spore-bearing structure; euglenoids (Phylum Pr-6), chrysophytes (Phylum Pr-4), and diatoms (Phylum Pr-11) also have multicellular derivatives.

We have adopted the concept of protoctist propounded in modern times by H. F. Copeland in 1956 (see the bibliography at the end of this chapter). The word had been introduced by John Hogg in 1861 to designate "all the lower creatures, or the primary organic beings;—both *Protophyta,* . . . having more the nature of plants; and *Protozoa* . . . having rather the nature of animals." Copeland appreciated, as several scholars in the nineteenth century had, the absurdity of calling giant kelp by a word, *protist,* that had come to imply unicellularity and, thus, smallness. He proposed an amply defined Kingdom Protoctista to accommodate certain multicellular organisms as well as the unicells that may resemble their ancestors—for example, kelp as well as the tiny brownish cryptophyte alga *Nephroselmis.* The kingdom thus defined also solves the problem of blurred boundaries that arises if the unicellular organisms are assigned to the various multicellular kingdoms.

We propose 27 protoctist phyla. This number is a matter of taste rather than tradition, because there are no established rules for defining protoctist phyla. Our groupings are debatable; for example, some argue that the Caryoblastea (Phylum Pr-1), comprising a single species of nonmitotic amoeba, ought to be placed in Phylum Rhizopoda (Pr-3) with the other amoebas, or that the cellular and plasmodial slime molds (Phyla Pr-22 and Pr-23) should be placed together. Some believe that the chytrids, hypho-chytrids, and oomycotes belong to the fungi and that the chlorophytes belong to the plants. Some insist that chaetophorales and prasinophytes, which here are considered chlorophytes (Phylum Pr-15), ought to be raised to phylum status. There are arguments for and against these views.[*] Our system has the advantage of defining the three multicellular kingdoms precisely, but the disadvantage of grouping together as protoctists amoebae, kelps, water molds, and other eukaryotes that have little in common with each other.

Protoctists are aquatic: some primarily marine, some primarily freshwater, and some in watery tissues of other organisms. Nearly every animal, fungus, and plant—perhaps every species—has protoctist associates. Some protoctist phyla, such as Apicomplexa (Pr-19) and Cnidosporidia (Pr-20), include hundreds of species all of which are parasitic on other organisms.

No one knows really how many species of protoctists there are; thousands have been described in the biological literature.[†] Water molds and plant parasites have traditionally been dealt with by the mycological literature, parasitic protozoa by the medical literature, algae by the botanical

---

[*] For details of alternative kingdom systems, see R. H. Whittaker and L. Margulis, "Protist classification and the kingdoms of organisms." *BioSystems* 10:3–18; 1978.

[†] Georges Merinfeld estimates that there are more than 65,000 species (personal communication).

literature, free-living protozoa by the zoological literature, and so forth. Inconsistent practices of describing, naming, and defining species have led to a confusion that this book attempts to dispel. Another reason for ignorance is that the group of eukaryotic microbes is large, with much diversity in tropical regions, whereas protozoologists and phycologists are scarce and concentrated in the north temperate zones. Furthermore, distinguishing species of free-living protoctists often requires time-consuming genetic and ultrastructural studies. Funding for such studies is limited because most protoctists are not sources of food and cause no diseases, thus they are of no direct economic importance.

The protoctists show remarkable variation in cell organization, patterns of cell division, and life cycle. Some are oxygen-eliminating photoautotrophs; others are ingesting or absorbing heterotrophs. In many species, the type of nutrition depends on conditions: when light is plentiful, they photosynthesize; in the dark, they feed. However, although protoctists are far more diverse in life style and nutrition than animals, fungi, or plants are, they are far less diverse metabolically than the bacteria.

Increasing knowledge about the ultrastructure, genetics, life cycle, developmental patterns, chromosomal organization, physiology, metabolism, and protein amino-acid sequences of the protoctists has revealed many differences between them and the animals, fungi, and plants. It has even been suggested that the major protoctist groups, here called phyla, are so distinct from each other as to deserve kingdom status, and that nearly 20 kingdoms ought to be created to accommodate them.* With due respect for their differences, a recognition of their common eukaryotic heritage, and a sense of humility toward both their complexity and our ignorance, we present our 27 protoctist phyla.

---

*G. F. Leedale, "How many are the kingdoms of organisms?" *Taxon* 23:261–270; 1974.

Greek *karyon*, nucleus or kernel; *blastos*, bud or sprout

Caryoblasteas are giant cells visible to the naked eye; they are probably the most primitive of all living eukaryotes. Enigmatic organisms, they are classified as eukaryotes for minimal reasons—they have membrane-bounded nuclei. However, they lack nearly every other cell inclusion characteristic of eukaryotes; they have no endoplasmic reticulum, Golgi bodies, mitochondria, chromosomes, or centrioles. The phylum comprises only one, well-studied species, *Pelomyxa palustris*. It appears to be a large amoeba with many nuclei; upon close inspection, one can see that it lacks mitochondria and, what is more shocking, that its nuclei do not divide by anything resembling the mitotic process. No chromosomes, mitotic spindle, or centrioles are present. The nuclei seem to divide directly, as bacteria do: new membranes form and two nuclei appear where there had been only one.

Until recently it was thought that microtubules and undulipodia were also absent, but in the late 1970s some subpellicular microtubules were seen in thin section as well as an unemerged (intracellular) nonfunctional 9+2 undulipodium. These observations suggest that *Pelomyxa*, like other eukaryotes, evolved from undulipodiated ancestors, but the microtubules and undulipodium apparently have nothing to do with nuclear or cell division. Needless to say, no gametes are formed, and sexuality in any fashion is absent.

Pelomyxa palustris has been discovered in only one habitat, mud on the bottom of freshwater ponds. In fact, nearly all studies have been made on organisms taken from Elephant Pond at Oxford University. Elephant Pond, which is in the yard behind the University Museum, is so named because in the nineteenth century the discarded carcasses of elephants were thrown there by taxidermists preparing museum exhibits. At the muddy bottom of the small pond, *Pelomyxa* feeds on algae and bacteria; there it grows, divides, and survives the English winters, yet no scientist has been able to grow it in the laboratory. Those who study *P. palustris* must be content to collect it from time to time from ponds.

Pelomyxa seems to be microaerophilic—it requires lower concentrations of oxygen than most eukaryotes do. This giant amoeba has two types of bacterial endosymbionts, which may be functional equivalents of mitochondria of other eukaryotes. The endosymbionts of one type lie in a regular ring around each nucleus and are therefore called perinuclear bacteria. The other symbionts lie scattered in the cytoplasm; they are a different species of bacterium having a characteristic wall structure. *Pelomyxa* will die if it is treated with antibiotics to which its endosymbiotic bacteria are sensitive. Before the amoeba dies, lactic acid and

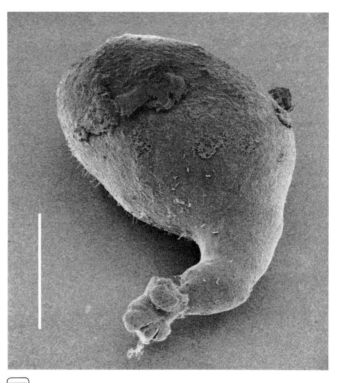

other metabolites accumulate in the cytoplasm; it is thought that the healthy symbiotic bacteria remove the lactic acid from the cytoplasm and metabolize it. In any case, the ever-present bacterial partners seem to be required. *Pelomyxa* has vacuoles and stores glycogen, a polysaccharide that is stored also by many types of animal cells.

Whether *Pelomyxa*-like eukaryotic microbes were ancestral to the other protoctists is difficult to determine. In any case, the absence of any vestige of mitosis sets the species distinctly apart from the other eukaryotes.

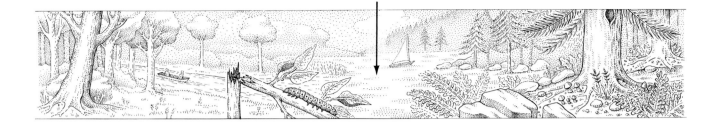

Pelomyxa palustris. SEM, bar = 100 μm.
[Photograph courtesy of E. W. Daniels,
The biology of amoebae, K. W. Jeon (ed.);
Academic Press; 1973. Drawings by
R. Golder.]

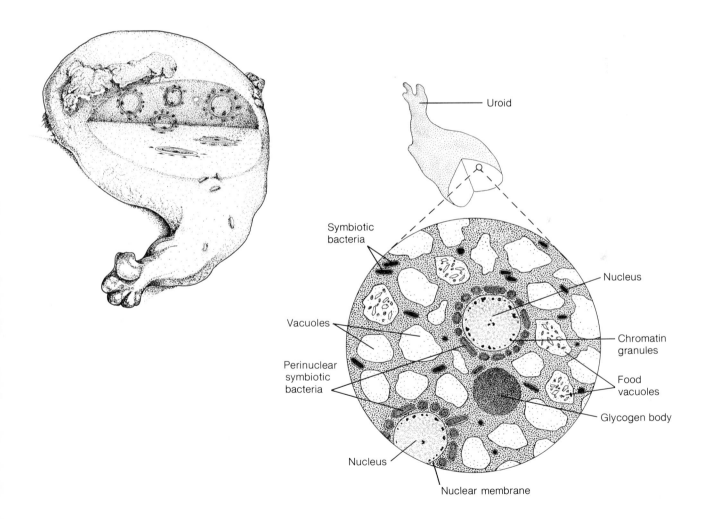

Pr-2   Dinoflagellata

Greek *dinos*, whirling, rotation, or eddy; Latin *flagellum*, whip

Amphidinium   *Polykrikos*
Ceratium   *Prorocentrum*
Cystodinium   *Protopsis*
Dinothrix   *Warnowia*
Erythropsidium
Gonyaulax
Gymnodinium
Nematodinium
Noctiluca
Peridinium

Of the several thousand known species of dinoflagellates, nearly all are marine planktonic forms. They are especially abundant in warm seas. Most are single celled, but some form colonies. An occasional species is parasitic or epiphytic on marine animals or other organisms.

Some dinoflagellates produce powerful toxins that are accumulated by fish and marine invertebrates. The sometimes toxic red tides are colorful blooms of marine microbes, many of which are dinoflagellates such as *Gonyaulax tamarensis*

Many dinoflagellates are bioluminescent. Some of these cause the twinkling of lights in the waves of the open ocean at night. *Noctiluca miliaris,* "a thousand night lights," is a large carnivorous cell bearing an immense feeding tentacle, which it uses to sweep in various microorganisms.

Dinoflagellates often enter into symbiotic associations with marine coelenterates, such as corals and sea anemones, and with clams. In fact, the most common intracellular photosynthetic symbiont in the reef communities of the world is the dinoflagellate *Gymnodinium microadriaticum.*

Dinoflagellates have membrane-bounded nuclei, two undulipodia (one of them rests in a characteristic groove, or girdle, encircling the cell). The undulipodia lie at a right angle to each other. Often, when they both move, the cell whirls. Most dinoflagellates have a rigid wall, called a test, made of plates embedded in the plasma membrane of the cell. In most, the plates are composed of cellulose and are encrusted with silica.

Some dinoflagellates are photoautotrophs, others are heterotrophs. If photosynthetic, they contain brownish plastids. The pigments of these plastids generally include chlorophyll a and c_2 (sometimes c_1 as well), β and γ carotene, several xanthins (diadinoxanthin, diatoxanthin, dinoxanthin, fucoxanthin, neofucoxanthin, and neoperidinoxanthin), and a carotenoid peculiar to dinoflagellates, called peridinin. Photosynthetic dinoflagellates store starch.

Although dinoflagellates are undoubtedly eukaryotic, their nuclear organization is so idiosyncratic that they have been called *meso*karyotic (between prokaryotic and eukaryotic). The DNA of nearly all other eukaryotes is organized into fibrils, 10 nm wide, that are complexed with histone proteins. The DNA fibrils of dinoflagellates are only 2.5 nm wide and are complexed with very tiny quantities of a peculiar basic protein rather than with the four or five common histones; in this sense, dinoflagellate chromatin is organized like that of bacteria.

The stages of mitosis (interphase, prophase, metaphase, anaphase, and telophase) are absent in dinoflagellates, and their

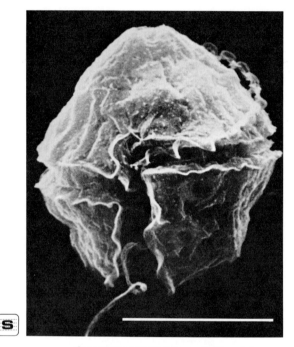

Gonyaulax tamarensis, a dinoflagellate from the Pacific Ocean. SEM, bar = 50 μm. [Photograph courtesy of F. J. R. Taylor; drawings by R. Golder (external view) and E. Hoffman (cutaway view); information from L. Leoblich (external view) and F. J. R. Taylor.]

chromatin is always condensed into chromosomes, which stand out and stain brightly throughout the division cycle. This is particularly strange because the condensed-chromosome stages in animals and plants are just those in which the genome is turned off: RNAs and proteins are not synthesized while chromatin is condensed. However, the genes in the condensed chromatin of dinoflagellates are not turned off—the condensed chromosomes continue to direct the synthesis of macromolecules. In some spe-

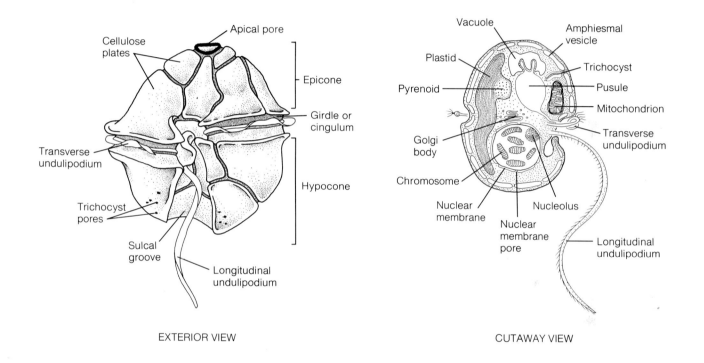

EXTERIOR VIEW

CUTAWAY VIEW

cies, microtubules penetrate the nucleus during division; the kinetochores, which in plants and animals are attached directly to chromosomes, in dinoflagellates are embedded in the nuclear membrane, and the chromatin is segregated to daughter cells by its attachment to the membrane. The pattern of cell division differs greatly from one dinoflagellate species to another as if, within the phylum, mitosis has evolved in its own peculiar fashion. The majority of dinoflagellates show no evidence of having discovered sex, although prior to cyst formation, some do engage in fusion, which seems to entail mating and gene exchange of some kind.

Dinoflagellates have speciated in some remarkable ways. *Protopsis, Warnowia,* and *Nematodinium* have an eyespot consisting of a layer of light-sensitive bodies containing carotenoid pigments overlain by a clear zone. The sedentary species

Erythropsidinium pavillardii has a quite complex ocellus that it apparently uses to detect the approach of prey. The ocellus includes a lens and a fluid-filled chamber underlain by a light-sensitive pigment cup. The lens can change shape, the pigment cup can move freely about the lens, and the whole ocellus can be protruded from the cell to point in different directions. It is apparent during the development of *E. pavillardii* that the pigment cup is derived from a chloroplast.

The tests of many dinoflagellates are very complex; each has its characteristic apical pore, epicone, and hypocone (see drawing). Many dinoflagellates also form hard, resistant cysts. Hystrichospheres, objects well known to micropaleontologists, are actually the tests of fossil dinoflagellates. The fossil record of these organisms goes at least to the base of the Cambrian; there is some evidence for them even earlier, in the late Proterozoic.

Pr-3 Rhizopoda

Greek *rhiza*, root; *pous*, foot

Acanthamoeba
Amoeba
Arcella
Centropyxis
Chlorarachnion
Chrysarachnion
Difflugia
Entamoeba
Hartmannella

Hyalodiscus
Mayorella
Paramoeba
Thecamoeba
Xenophyophora

The organisms in this group are all single-celled amoebas, either naked or having shells, called tests. Amoebas are distributed world wide in both fresh and marine waters, and they are especially common in soil. Many are parasitic on animals, and pass from host to host, or from the soil or fodder to host. They are among the most simple of the protoctists: they reproduce by direct division into two offspring cells of equal volume. All are microscopic, yet some are very large for single cells, as much as hundreds of micrometers long. As defined here, the amoebas all lack undulipodia at all stages in their life cycle. They also lack meiosis and any sort of sexuality. They differ from Caryoblastea (Phylum Pr-1) in that they have mitotic spindle microtubules and chromatin granules, which in some species, form chromosomes. In these species, the anaphase and telophase stages of cell division have been observed. The nuclear membrane persists well into the mitotic division stages; in some amoebas, the nuclear membrane does not disperse at all during division.

Even though amoebas lack centrioles, kinetosomes, and undulipodia, they show cell motility of various kinds. They are well known for their pseudopods (false feet). These are flowing cytoplasmic extensions that are used for forward locomotion and to surround and engulf food particles. Where they have been studied, actinomyosin proteins (the ''nonmuscle actins'') have been shown to be associated with pseudopodial movements. Such movements are sensitive to variations in the concentration of calcium ion (Ca^{2+}).

This phylum contains five classes: Tubulina, Thecina, Flabellina, Conopodina, and Acanthopodina.

The Tubulinas are uninucleate, cylindrical, naked amoebas. They are grouped into three families. The Amoebidae, which include the famous *Amoeba proteus,* tend to be polypodial—they have many feeding, changing, flowing pseudopods at one time. The Hartmannellidae, on the other hand, are monopodial—they form one pseudopod at a time. Some form desiccation-tolerant resting cysts; the cell inside each cyst is binucleate. The Entamoebidae are also monopodial. Their nucleolus, which contains the ribosomal precursors, is organized into a body or several bodies called endosomes. Amoebas are probably the most ancient phylum to have endosomes, which are also found in the nuclei of other protoctists (for example, euglenoids, Phylum Pr-6). That the Entamoebidae form cysts is of great importance,

because they are nearly all parasites. Some, such as *Entamoeba histolytica,* are responsible for amoebic dysenteries. The cysts enable the amoebas to resist digestion by their animal hosts. The amoebal nuclei can divide inside the cysts without accompanying cytoplasmic division; this leads to four, eight, or even more nuclei per encysted cell. The cysts can germinate in the digestive tracts of animals, or they can pass out in the feces.

Thecina amoebas seem to roll their wrinkled surface as they move. They form a rather obscure group of free-living forms having various mitotic patterns.

The Flabellina form spatula-shaped pseudopods in which flowing endoplasm seems to erupt. Some, such as *Hyalodiscus,* are fan shaped.

The Conopodina are shaped like fingers. When they move, they are longer than they are wide; some float on water, where they extend slender radiating pseudopods. *Mayorella* and *Paramoeba,* both marine genera, belong to the family Paramoebidae, the only family in the class Conopodina. *Paramoeba eilhardi* contains two distinctive bodies called *nebenkörper,* which are packages of benign, bacterium-like symbionts. *P. eilhardi* can also be attacked and killed by certain marine bacteria that are only able to grow and divide in its nucleus.

The Acanthopodina have finely tipped subpseudopodia; that is, each pseudopod extends smaller pseudopods of its own. The cell as a whole may be disc shaped. *Acanthamoeba* itself forms a polyhedral or thickly biconvex cyst having a wall that contains cellulose. The Echinoamoebas are members of this group; more or less flattened when they move, they have tiny finely-pointed pseudopods that look like spines.

Although naked amoebas constantly change shape, the range of shapes that each takes on is genetically limited and species specific. Many naked amoebas correspond to shelled forms that are thought to have derived from them. To construct their tests, amoebas glue together sand grains, bits of carbonate particles, and other inorganic detritus, depending on what is available. Some of these tests, such as those of *Arcella,* are distinctive enough to be recognized in the fossil record. Such tests give the Rhizopoda a fossil record that extends back well into the Paleozoic Era. Also, some of the pre-Phanerozoic microfossils called acritarchs have been interpreted as tests of shelled amoebas.

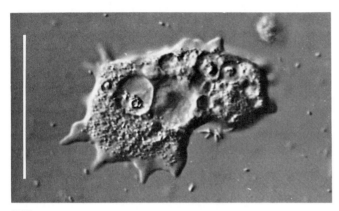

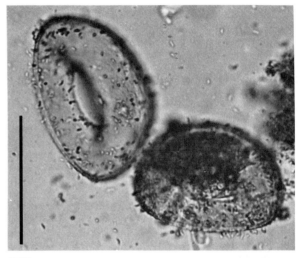

A *Mayorella penardi,* a living naked amoeba from the Atlantic Ocean. LM (differential interference contrast microscopy), bar = 50 μm. [Courtesy of F. C. Page, *An illustrated key to freshwater and soil amoebae,* Scient. Publ. Freshwat. Biol. Ass. no. 34; 1976.]

B Two empty shells of the freshwater amoeba *Arcella polypora.* LM, bar = 100 μm. [Courtesy of F. C. Page.]

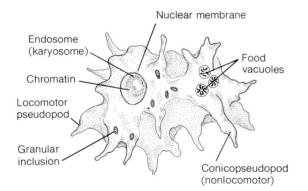

Nuclear membrane

Endosome (karyosome)

Chromatin

Locomotor pseudopod

Granular inclusion

Food vacuoles

Conicopseudopod (nonlocomotor)

C Structure of *Mayorella penardi* seen from above. [Drawing by E. Hoffman.]

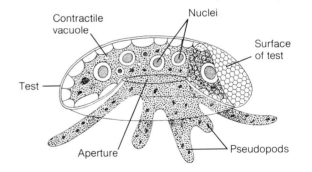

Contractile vacuole

Nuclei

Surface of test

Test

Aperture

Pseudopods

D Structure of *Arcella polypora,* cutaway view. [Drawing by R. Golder.]

Pr-4 Chrysophyta

Greek *chrysos*, golden; *phyton*, plant

Chromulina
Chrysobotrys
Chrysocapsa
Dinobryon
Echinochrysis
Epipyxis
Gonyostomum
Hydrurus
Mallomonas
Monas

Nematochrysis
Ochromonas
Phaeothamnion
Raphidomonas
Reckertia
Rhizochrysis
Sarcinochrysis
Synura
Thallochrysis
Thaumatomastix

Trentonia
Vacuolaria

The chrysophytes form a large and complex group of algae whose plastids contain golden-yellow pigments. They are heterokonts—at some stage of their life cycle, their cells have two anteriorly attached undulipodia of unequal sizes. (See Phylum Pr-9, Xanthophyta, for discussion of other heterokont groups.) Chrysophytes are ubiquitous in fresh, temperate waters, such as lakes and ponds. Except for one widespread group of marine plankton, the silicoflagellates, most are not marine. Of the primarily unicellular protoctist phyla (Pr-1 to Pr-11), chrysophytes have the strongest tendency to multicellularity. They are morphologically even more diverse than the famous green algae (Phylum Pr-15, Chlorophyta). Some of them form large, complicated, branching colonies; nevertheless, however complex they become, the cells retain the characteristic ultrastructure and golden-yellow plastids that permit their unequivocal identification as chrysophytes.

Chrysophytes lack germ cells; neither a sexual stage nor meiosis has ever been documented reliably in them. They do form swarmer cells, however, heterokont zoospores that swim away to develop into new colonies. In an alternative form of reproduction, an entire colony will divide into two or more containing hundreds or thousands of cells. The offspring colonies simply float away from each other to establish themselves at new sites. Waves and other mechanical disturbances in the water apparently enhance this mode of growth.

Chrysophytes interact strongly with minerals, especially with silica and iron. In particular, the marine planktonic silicoflagellates form tests, or shells, from silica scavenged from sea water. These tests show complex and often strikingly beautiful patterns that can be used to distinguish genera. The scavenging activity of chrysophytes and of diatoms (Phylum Pr-11), actinopods (Phylum Pr-16), and glass sponges (Phylum A-2) keeps soluble silica in the surface waters of the ocean at a concentration so low that it cannot be detected by chemical means. How silicoflagellates make dissolved silica into elaborate tests is poorly known because of the difficulty of capturing the organisms alive and rearing them in the laboratory. Some tests have been preserved as fossils. Thanks to these fossils, chrysophytes are known from as long ago as the early Paleozoic, about 500 million years ago.

The freshwater chrysophytes fall into three groups: the Chrysomonadales, Chrysosphaerales, and Chrysotrichales. In traditional literature these are orders, but we consider them classes here.

The Class Chrysomonadales contains two orders: Chrysomonadineae, mainly unicellular forms and their direct descendants, and Chrysocapsineae, which are all multicellular, forming

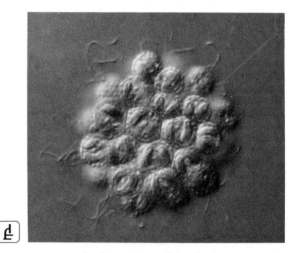

A *Synura* sp., a living freshwater colonial chrysophyte from Massachusetts. LM; each cell is about 18μm in diameter. [Courtesy of S. Golubic and S. Honjo.]

B A siliceous surface scale from a member of the colony shown in (A). SEM; the scale is about 1 μm long. [Courtesy of S. Golubic and S. Honjo.]

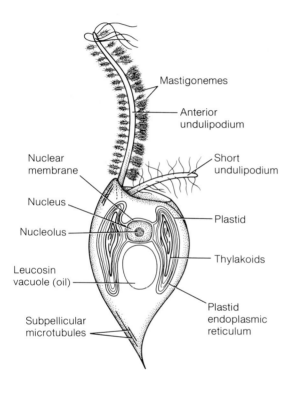

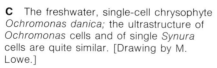

C The freshwater, single-cell chrysophyte *Ochromonas danica;* the ultrastructure of *Ochromonas* cells and of single *Synura* cells are quite similar. [Drawing by M. Lowe.]

Labels on figure:
Mastigonemes
Anterior undulipodium
Nuclear membrane
Short undulipodium
Nucleus
Plastid
Nucleolus
Leucosin vacuole (oil)
Thylakoids
Subpellicular microtubules
Plastid endoplasmic reticulum

lular genera and clearly related colonial genera. Apparently, the evolutionary transition from unicellularity to multicellularity occurred independently many times in the Chrysomonadineae.

The class Chrysosphaerales contains many species; typically, they form complex, differentiated, spherical colonies that release heterokont zoospores.

The Chrysotrichales include *Nematochrysis, Thallochrysis,* and *Phaeothamnion,* among many others. The members of this class show every sort of variation on the theme of the filamentous habit. Some genera are composed of long intertwined threads of algae; close inspection reveals each compartment of a thread to be a typical chrysophyte cell. Others form flaccid, highly branched treelike structures reminiscent of branching cyanobacteria (Phylum M-7), green algae (Phylum Pr-15), or yellow-green algae (Phylum Pr-9). This sort of branching organization evidently evolved independently many times in different groups of microbes.

Freshwater chrysophytes that must overwinter or survive desiccation typically make structures called statocysts—the cellulosic membrane of the cell becomes endogenously silicified within an outer coat. Sometimes the statocyst coat becomes heavily encrusted with iron minerals. Within the statocyst are often seen two conspicuous granules of a storage oil, called leucosin or chrysolaminarin. At the top of the statocyst is a pore, often surrounded with a tapering, conical collar and plugged with material that dissolves when the germinating amoeboid chrysophyte emerges. The cyst walls may be uneven in thickness. Some fossil chrysophyte cysts have been preserved well enough to be identified by micropaleontologists; others may be among the pre-Phanerozoic acritarchs, abundant spheroidal microfossils that are difficult to identify.

An obscure group of large unicellular algae that, like chrysophytes, contain plastids with chlorophyll *a* and *c* are misnamed the Chloromonads. These mud, lake, and marine dwellers are not green algae at all. (They may be raised to phylum status.) A better name for the group is Raphidomonads, named after the best known genus *Raphidomonas.* They bear two undulipodia, apically inserted, one forward and one trailing. *Vacuolaria virescens* is a naked ovoid or pear-shaped cell that forms spherical cysts with thick envelopes of mucilage.

Like chrysophytes raphidomonads have large nuclei at the anterior end of their cells and store food as fat. *Thaumatomastix* and *Reckertia* are colorless genera, lacking plastids, but have the raphidomonad cell organization. The group is arbitrarily placed as chrysophyte class until more information about them becomes available.

either compact spherical colonies or loosely branched ones. The well-known genera *Synura, Chromulina, Mallomonas, Ochromonas, Chrysobotrys, Monas, Rhizochrysis,* and *Dinobryon* are members of the Chrysomonadineae. Each of these genera is characteristic of a large group, usually considered a family (for example, the Synuraceae). In each family, there are both unicel-

Pr-5 Haptophyta

(Coccolithophorids)

Greek *haptein*, fasten; *phyton*, plant

Calcidiscus
Calciosolenia
Calyptrosphaera
Chrysochromulina
Coccolithus
Discosphaera
Emiliania
Gephyrocapsa
Hymenomonas

Pontosphaera
Prymnesium
Rhabdosphaera
Syracosphaera

These tiny golden motile algae have been seen, like the prover-bial elephant, as two very different kinds of organisms: as chrysophyte plankton (Phylum Pr-4) and as coccolithophorids. Coccoliths are microscopic disclike calcium carbonate struc-tures of reknown to paleontologists; coccolithophorids are the planktonic microorganisms that produce and bear the coccoliths as overlapping surface plates. In the last decade, ultrastructural and developmental studies have united the two views: the golden chrysophytelike algae (photograph A) and the coccolithophorid (B) are different stages in the life cycles of the same organisms.

Haptophytes are primarily marine organisms, although some freshwater genera are known. Although they have golden-yellow plastids, their cell structure differs enough from that of chryso-phytes to justify placing them in a different phylum. Unlike the chrysophytes, haptophytes have only the very weakest tendency to become multicellular. No sexual stages are known.

Haptophytes are distinguished by their haptonemes, scales, and coccoliths. The haptoneme is a thread, often coiled, that may be used as a holdfast to anchor the free-swimming protist to a stable object. Each cell has one haptoneme, generally at its anterior end. Like an undulipodium, it is derived from a special kinetosomelike structure and is microtubular in transverse sec-tion. The microtubules are 24 nm in diameter; six doublets of microtubules are arranged in a circle whose center is occupied by a single microtubule or none at all. Thus, the haptoneme is evidently a specialized modification of the ubiquitous 9 + 2 undulipodium (see Figure 2 in the Introduction). With the hap-toneme at the anterior end of the cell are two undulipodia and, generally, a Golgi body.

In the transformation of the free-swimming stage into the rest-ing, coccolithophorid stage, coccoliths and scales develop in-side the cell: calcium carbonate crystals precipitate on conspicu-ous scales, which are made of organic polymers inside the Golgi apparatus. The scales and the coccoliths that form on them bear intricate patterns that are species or genus specific. The gradu-ally assembling coccoliths are transported to the edge of the cell by microtubule-mediated processes. They are deposited, in some cases with exquisite regularity, on the surface of the cell. These coccolithophorid stages are often very resistant and per-mit tolerance of conditions that would be prohibitive to the grow-ing swimming forms of haptophytes.

Coccolithophorids have produced great quantities of particu-late calcium carbonate continuously for 100 million years or so since the Cretaceous; they have contributed significantly to the chalk deposits of the world. Because they are distinctive, they serve as stratigraphic markers; several hundred morphotypes or

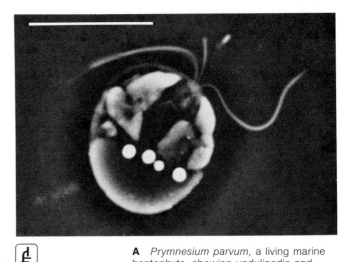

A *Prymnesium parvum,* a living marine haptophyte, showing undulipodia and haptoneme. LM, bar = 10 μm. [Courtesy of I. Manton and G. F. Leedale, *Archiv für Microbiologie* 45:285–303; 1963.]

fossil species have been studied by geologists. Thus, more is known about the morphology of fossils than of extant forms. The correlation between the haptonemid and the coccolithophorid stage has still not been made for many species. There is much need for work on the entire life cycle so that the development of calcium carbonate skeletal patterns can be seen in the context of the biology of the live haptophytes.

Haptophytes typically have two golden-yellow plastids (chrysoplasts) surrounded by a plastid endoplasmic reticulum that is continuous with the nuclear membrane. The plastids con-tain chlorophylls a, c_1, and c_2, but lack chlorophylls b and e. In addition to β carotene, they have α and γ carotenes. They also have fucoxanthin, an oxidized isoprenoid derivative that is also found in brown algae (Phylum Pr-12) and diatoms (Phylum Pr-11); this compound is probably the most important determinant of the brownish-yellow color. They do not store starch; rather, like eu-glenoids (Phylum Pr-6), they form a glucose polymer with the β-1-3 linkage of the monosaccharides. This white storage material called paramylon, is stored within pyrenoids between the thyla-koids of the plastids.

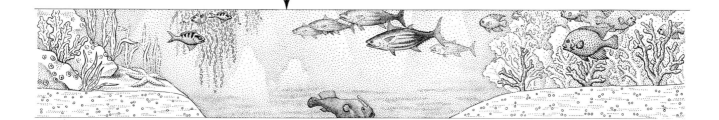

B *Emiliania huxleyi*, a coccolithophorid from the Atlantic. That coccolithophorids are resting stages of haptophytes has been realized only in the last decade. SEM, bar = 1 μm. [Courtesy of S. Honjo.]

C *Prymnesium parvum*, the free-swimming haptonemid stage of a haptophyte. The surface scales shown here are not the ones on which coccoliths form. [Drawing by R. Golder.]

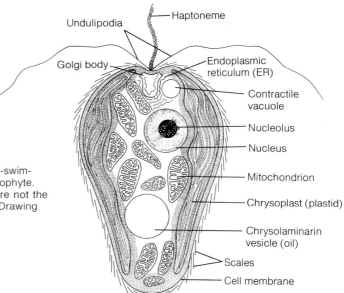

Undulipodia — Haptoneme

Golgi body — Endoplasmic reticulum (ER)

Contractile vacuole

Nucleolus

Nucleus

Mitochondrion

Chrysoplast (plastid)

Chrysolaminarin vesicle (oil)

Scales

Cell membrane

Pr-6 Euglenophyta

(Euglenoid flagellates)

Greek *eu*, true; *glene*, eyeball, socket of joint; *phyton*, plant

Astasia
Colacium
Dinema
Distigma
Euglena
Heteronema
Peranema
Phacus
Trachelomonas

Although most euglenoid flagellates are unicellular and live in fresh and stagnant water, some are marine, some are colonial, and some are parasitic. About 800 species have been described. Most euglenoids are photosynthetic, but many lack chloroplasts and hence are limited to heterotrophy. Even the photosynthetic euglenoids under certain conditions eat dissolved or particulate food.

The euglenoid flagellates in some classifications are placed with the green algae (Phylum Pr-15, Chlorophyta) because most of them contain grass-green chloroplasts. However, they differ in many ways from the chlorophytes, even in their chloroplast pigmentation. Like those of chlorophytes and plants, the plastids of euglenoids contain chlorophylls *a* and *b* only. They also contain β carotene and these carotenoid derivatives: alloxanthin, antheraxanthin, astaxanthin, canthoxanthin, diadinaxanthin, diatoxanthin, echinenone, neoxanthin and zeazanthin. Some of these, such as diadinaxanthin and diatoxanthin, are not present in chlorophytes and plants. Unlike chlorophytes and plants, euglenoids lack rigid cellulosic walls. Instead, they have pellicles, finely sculptured outer structures made of protein. These are usually very flexible, so that euglenoids typically can change shape easily. Unlike chlorophytes and plants, they do not store starch, but paramylon, a glucose polymer with a β-1-3 linkage of the monosaccharides.

Euglenas often have been used as an example of the plant-animal dilemma. They can feed and move by virtue of one mastigonemate, anteriorly directed undulipodium. They also contain a second one, called the "preemergent flagellum"; it probably takes part, by transmission of signals from the nearby eyespot, in the reaction of the organisms to light. In two-kingdom schemes, these traits make euglenoids animals. However, most of them live photosynthetically on water, sunlight, oxygen, carbon dioxide, and vitamin B_{12}; this ability has classically characterized them as plants. In our classification, they are neither plants nor animals, but protoctists, which are distinct in several ways.

Euglenoid reproduction is never sexual; all attempts to find meiosis and gametogenesis in euglenoids has failed. The nuclei of different individuals of the same species may have different amounts of DNA. The nuclei contain large karyosomes, or endosomes, structures homologous to the nucleoli of other cells: like nucleoli, the endosomes are composed of RNA and protein com-

bined in bodies that are precursors to the ribosomes of the cytoplasm. Euglenoids have a nearly invisible mitotic spindle, composed of a few intranuclear microtubules. Many lack distinct, countable chromosomes, in cell division, their chromatin granules do not move in a single mass as in standard anaphases. The chromatin granules do not split at metaphase; rather, no metaphase plate is formed and each granule autonomously proceeds to one of the nuclear poles, where the newly replicated undulipodia are located. Prior to nuclear division, the two undulipodial apparatuses move toward the same end of the cell, causing the cell to distort in a way characteristic of dividing euglenoids. The cell then divides lengthwise.

The species *Euglena gracilis* has proved to be a fine tool for analyzing cell organelles. Euglenas can be found both with and without chloroplasts, which may be permanently lost or temporarily "turned off." Thus, the effect of light, chemical inhibitors, temperature, and many other agents on the development of chloroplasts and other organelles can be beautifully observed. For example, if the cells are placed in the dark, the chloroplasts regress; after a little growth, the euglenoids turn white and become entirely dependent on external food supplies for growth. These "animal" cells can be reconverted into "plants." If dark-grown white euglenas are reilluminated, they turn light green within hours. Their chloroplasts go through a series of developmental changes induced by the light, and after three days or so they recover their bright green color.

Euglena gracilis is the only species known that can be genetically "cured" of chloroplasts without killing the organism. If *E. gracilis* cells are treated with ultraviolet light or with a number of other treatments to which the chloroplast genetic system is more sensitive than the nucleocytoplasmic, the genetic entities responsible for chloroplast development can be lost permanently. The euglenas then lose all their plantlike characteristics and become irreversibly dependent on food. By this treatment, the metabolism of the nucleocytoplasm can be studied in detail separately from that of the plastid.

The growth of *Euglena gracilis* absolutely requires the presence of vitamin B_{12}, although unbelievably tiny quantities are sufficient. In fact, euglenas can detect some 10^{-13} g/ml of this vitamin; they are far more sensitive than chemical tests. Thus, the growth of euglenas has been used as a commercial assay for the vitamin.

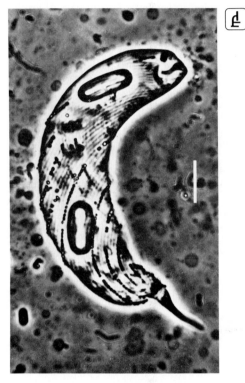

Euglena spirogyra, a freshwater euglenoid from England. LM (phase contrast), bar = 10 μm. [Photograph courtesy of G. F. Leedale (1967). Drawing by R. Golder, information from G. F. Leedale.]

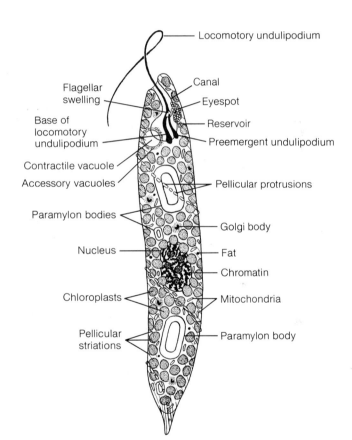

Locomotory undulipodium

Flagellar swelling

Canal

Eyespot

Base of locomotory undulipodium

Reservoir

Preemergent undulipodium

Contractile vacuole

Accessory vacuoles

Pellicular protrusions

Paramylon bodies

Golgi body

Nucleus

Fat

Chromatin

Chloroplasts

Mitochondria

Pellicular striations

Paramylon body

Pr-7 Cryptophyta

(Cryptomonads)

Greek *kryptos*, hidden; Latin *monas*, unit; Greek *phyton*, plant

Chilomonas
Chroomonas
Cryptomonas
Cyanomonas
Cyathomonas
Hemiselmis
Hillea
Nephroselmis

Cryptomonads, or cryptophytes, are found all over the world in moist places. Some commonly form algal blooms on beaches, whereas others thrive as intestinal parasites in domesticated animals. These widely differing habitats have led very different types of scientists (for example, marine botanists and parasitologists) to study them. This has led to confusions in terminology, lack of communication among experts, and a general ignorance of their existence by most biologists.

Like euglenoids (Phylum Pr-6), cryptophytes may be pigmentless animallike protozoa or brightly pigmented and photosynthetic plantlike algae. They are common in fresh water and are found primarily as free-living single cells. They are unlike euglenoids, however, in details of cell structure and division, as well as in the nature of their photosynthetic pigmentation (if present).

Cryptophytes bear two anterior undulipodia inserted in a characteristic way along the oral groove, or crypt. In the carnivorous members of the group, which eat bacteria or other protists, the crypt is typically lined with trichocysts and microbes. Trichocysts are special structures that expel poisons, which subdue and kill the microbial prey. Even some members of the photosynthetic genera have trichocysts and ingest particulate food through the oral groove.

Pigmented cryptophytes as a rule contain chlorophyll c_2 in addition to chlorophyll *a*. Unlike most algae, they lack β carotene and zeaxanthin, but contain α carotene, cryptoxanthin, and alloxanthin. Many cryptophytes having this pigmentation are green or yellowish green. *Cyathomonas* and many others also contain characteristic ratios of phycocyanin and phycoerythrin, and so tend to be redder or bluer. In general, phycocyanin pigments are strictly limited in nature: they are found in all cyanobacteria (Phylum M-7) and red algae (Phylum Pr-13, Rhodophyta), and also in miscellaneous anomalous algae such as *Cyanophora paradoxa*, organisms that are thought to harbor blue-green symbionts instead of containing plastids. It seems likely, because the peculiar cryptophyte cell structure is combined with plastids of various colors, that cryptophytes became photosynthetic secondarily. More than once in the course of their evolution they acquired photosynthetic prokaryote symbionts that became chloroplasts or rhodoplasts, depending on the prokaryote. By this reckoning, some common cryptomonads, such as *Chilomonas*, have retained their heterotrophic life and voracious habits and never acquired photosynthesis at all.

Meiotic sexuality and gametogenesis is unknown in cryptophytes. Many have been grown and observed in the laboratory: apparently, they simply divide into two offspring cells. Just prior to cell division, new kinetosomes and undulipodia appear with a

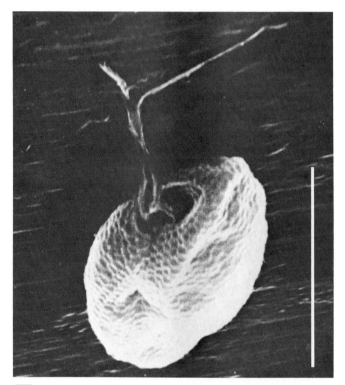

A *Cyathomonas truncata,* a freshwater cryptomonad, SEM, bar = 5 μm.

new crypt in proximity to the old one. The new oral structure then turns upside down and migrates to the opposite end of the cell. In the meantime, chromatin inside the closed nuclear membrane forms small knobby chromosomes, which segregate into two bundles at opposite sides of the nucleus. The nucleus divides, cytokinesis ensues, and two offspring cells with a plane of mirror symmetry between them separate. This type of asexual reproduction distinguishes cryptophytes, regardless of their nutritional mode, from other protists. It was first documented, beautifully, by Karl Bêlar in 1926.

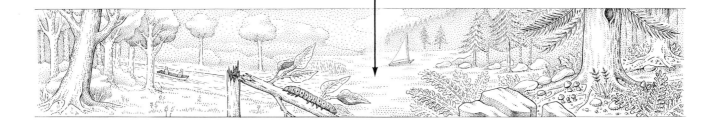

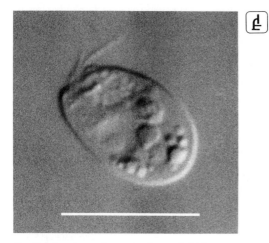

B *Cyathomonas truncata*, live cell. LM, bar = 5 μm. [Courtesy of F. L. Schuster.]

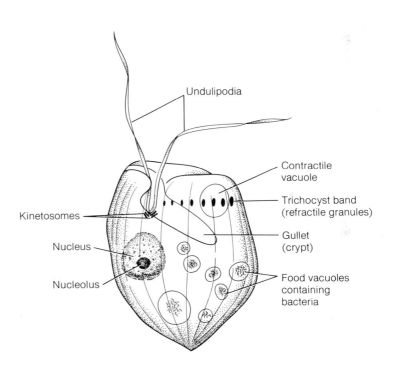

Undulipodia

Contractile vacuole

Trichocyst band (refractile granules)

Gullet (crypt)

Kinetosomes

Nucleus

Nucleolus

Food vacuoles containing bacteria

C *Cyathomonas truncata*. [Drawing by M. Lowe.]

Pr-8 Zoomastigina
(Animal flagellates, zoomastigotes)

Greek *zoion*, animal; *mastix*, whip

All zoomastigotes are unicellular. Each bears at least one undulipodium; some bear thousands. They may be either free living or parasitic, sexual or asexual, but all are heterotrophic and lack plastids. Some, had they plastids, would be classified, on the basis of their cell structure, among the algae.

In two-kingdom classifications, zoomastigotes have been placed in the order Protomonadina in the class Flagellata in the phylum Protozoa in the kingdom Animalia. In our classification, Phylum Zoomastigina consists of eight classes: amoeboflagellates, opalinids, choanoflagellates, bicoecids, kinetoplastids, diplomonads, pyrsonymphids, and parabasalids. This classification is tentative—the first five classes may not be related to each other at all, but may belong to other, different phyla. However, the diplomonads, pyrsonymphids, and parabasalids can be related to each other directly by their ultrastructure as revealed by the electron microscope.

The amoeboflagellates, or Class Schizopyrenida, are freshwater or parasitic microbes distinguished by their ability to change from an undulipodiated to an amoeboid stage and back again. This transformation is best studied in *Naegleria,* because it can be cultured. It is induced by the depletion of nutrients. When *Naegleria* amoebas are suspended in distilled water, they develop kinetosomes, grow undulipodia from them, and soon elongate into a swimming form. They quickly take off in search of food (bacteria). After they find it, they lose their undulipodia and return to a completely amoeboid life style.

The opalinids are characterized by the falyx, a unique structure composed of several rows of hundreds of close-set kinetosomes. The falyx lies along the anterior part of the long axis of the cell, and undulipodia emerge from it. As in euglenoids (Phylum Pr-6), a patterned proteinaceous skin, or pellicle, lies just beneath the plasma membrane. Opalinids lack a cytostome ("cell mouth"), and so absorb dissolved nutrients directly through their pellicle. They have sexual stages: dissimilar haploid flagellated gametes fuse to form diploid cells. Nuclear division can occur without cytoplasmic division, giving rise to bi- or multinucleate cells. Opalinids are parasites in the large intestine, or rectum, of fish, reptiles and amphibians, mostly of tailless amphibians such as frogs and toads. They have flattened bodies and swim with spiral movements. Four genera are recognized. The *Protoopalina* are considered the most primitive; each cell has two large nuclei. *Zelleriella* cells also have two large nuclei, which lie on opposite sides of the cell's long axis. *Cepedea* species are very large multinucleate cells, as long as 2.8 mm. They have small nuclei and a lengthy falyx. *Opalina* species, not as large, are also multinucleate and have an extensive falyx; their pellicle is pleated.

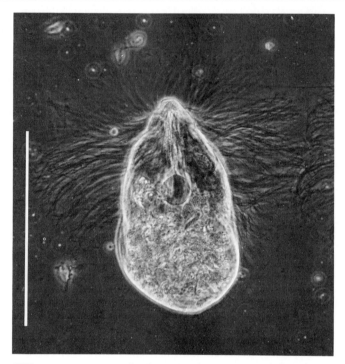

A *Staurojoenina* sp., a wood-digesting hypermastigote from the hindgut of the dry-wood termite *Incisitermes (Kalotermes) minor,* LM (stained preparation), bar = 50 μm.

Some choanoflagellates, such as *Monosiga,* are colorless. Others, such as *Desmarella,* contain green plastids. They are distinguished by having a lorica, a hard structure from which the single undulipodium emerges. The lorica has ribs, called choastes. The undulipodium is tapered and smooth; it can be retracted into the cell. Many choanoflagellates stand on a peduncle, a cell stalk that contains longitudinal fibers. The organization of their cells is remarkably similar to that of the body cells of sponges (Phylum A-2, Porifera). In fact, it is generally thought that choanoflagellates, some of which are organized into colonies, are direct ancestors of the sponges, but not of any other metazoan phyla.

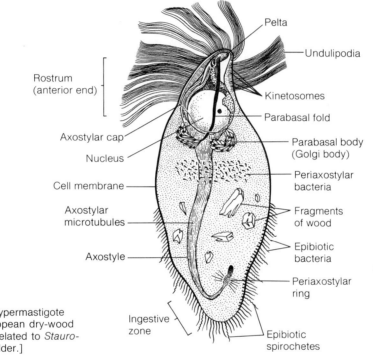

B *Joenia annectens,* a hypermastigote from the hindgut of a European dry-wood termite. *Joenia* is closely related to *Staurojoenia.* [Drawing by R. Golder.]

Labels on figure:
Pelta
Undulipodia
Kinetosomes
Parabasal fold
Parabasal body (Golgi body)
Periaxostylar bacteria
Fragments of wood
Epibiotic bacteria
Periaxostylar ring
Epibiotic spirochetes
Rostrum (anterior end)
Axostylar cap
Nucleus
Cell membrane
Axostylar microtubules
Axostyle
Ingestive zone

Bicoecids, our fourth class, have a shell and two undulipodia. One undulipodium extends forward from the cell and is mastigonemate—it bears numerous tiny appendages called tinsel or flimmers. The other lies along the surface of the cell and is smooth. The undulipodial insertion of these colorless flagellates is so like that of some of the chrysophytes (see the drawing of *Ochromonas danica,* Phylum Pr-4) that at least some chrysophytes may have evolved from bicoecids by acquiring plastids. Alternatively, bicoecids may have evolved from chrysophytes by losing plastids. They tend to form colonies. Sexuality has not been reported in this group.

Kinetoplastids include the bodos, mainly free-living biundulipodiated cells common in stagnant water, and the infamous parasitic trypanosomes. All contain a special large mitochondrion called the kinetoplast. Sexuality, although widely sought, has never been found in any member of the class. Many genera, such as *Trypanosoma* and *Crithidia,* contain important pathogens of and parasites on man and domestic animals. Elephantiasis, sleeping sickness, and Chagas' disease are caused by kinetoplastids. Marked changes—for example, the elongation or near disappearance of the kinetoplast—take place when the cells move from vertebrate blood to insect salivary gland, and when they move back. A few multicellular derivatives are known. For example, *Cephalothamnion cyclopum*, a bodo having a characteristically backward-directed undulipodium, forms colonies of several dozen cells attached to a common stalk.

Pyrsonymphids live in the intestines of wood-eating cockroaches and termites. They lack mitochondria and have ribbon-shaped organelles composed of hundreds of microtubules, which are sometimes connected to each other by bridges and arranged in elaborate patterns. The genera include *Notila*, *Oxymonas*, *Pyrsonympha*, and *Saccinobaculus*.

Parabasalids also are parasitic or symbiotic in the intestines of insects. Apparently, these microbes digest cellulose, from which they derive sugars both for themselves and for their hosts. Particles of wood are taken up through a sensitive posterior zone. In various stages of digestion, they are often seen in the cytoplasm. Parabasalids lack mitochondria; they bear at least four undulipodia, an axostyle, and conspicuous parabasal bodies. The membranous and granular ultrastructure of the parabasal body, whose presence distinguishes parabasalids from pyrsonymphids, suggests a similarity to the Golgi apparatus of animal cells and the dictyosomes of plant cells, and hence an involvement in the synthesis of RNA and protein. Sexuality is known, but, because the organisms live in the intestines of insects, no detailed study of it has been possible. In some species, the entire haploid adult is seen to transform into a sexually receptive gamete. Two gametes fuse into a gametocyst in which, it is thought, meiosis takes place—meiotic products, haploid adults, emerge from the cyst.

The class comprises three orders: Trichomonadida, Polymonadida, and Hypermastigida. The trichomonads, polymonads, and pyrsonymphids are called informally the polymastigotes ("many whips"). In particular, trichomonads typically bear from four to about sixteen undulipodia. The undulipodia are often associated with supernumerary nuclei, two undulipodia per nucleus. On the basis of distinctive cell structure, in particular, the manner of insertion of the undulipodia and fibers in the cortical cell layer (just beneath the plasma membrane), four families of trichomonads have been described: Monocercomonadidae (*Hexamastix, Monocercomonas, Histomonas*), Devescovinidae (*Devescovina, Metadevescovina*), Trichomonadidae (*Trichomonas, Trichomitus*), and Calonymphidae (*Calonympha, Snyderella*). The Order Hypermastigida, informally called hypermastigotes, have hundreds and even hundreds of thousands of undulipodia attached to special bands. The mitotic spindle, which grows out from these bands, is external to the nuclear membrane. Hypermastigotes generally have one or a few nuclei, unlike some of the trichomonads (Family Calonymphidae), which have very many, sometimes more than a thousand.

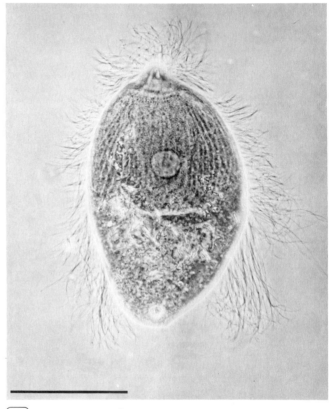

C The hypermastigote *Trichonympha ampla* from the Sonoran desert dry-wood termite *Pterotermes occidentis*. LM, bar = 100μm. [Courtesy of D. Chase.]

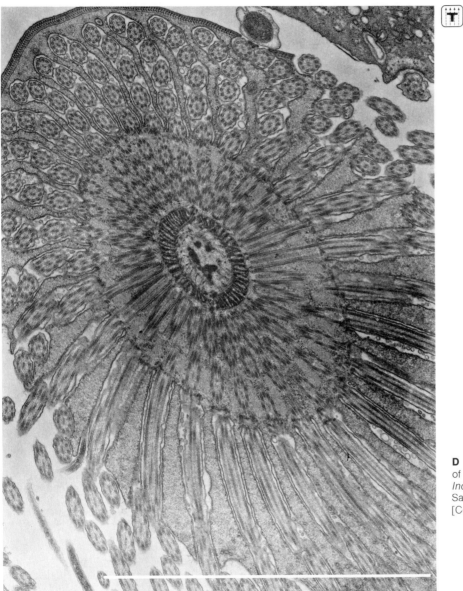

D Transverse section through the rostrum of a *Trichonympha* sp. from the termite *Incisitermes* (*Kalotermes*) *minor* from near San Diego, California. TEM, bar = 5 μm. [Courtesy of D. Chase.]

Pr-9 Xanthophyta

Greek *xanthos*, yellow; *phyton*, plant

Botrydiopsis *Ophiocytium*
Botrydium *Tribonema*
Botryococcus *Vaucheria*
Characiopsis
Chloridella
Gonyostomum
Heterodendron
Mischococcus

Xanthophytes, like eustigmatophytes (Phylum Pr-10), are yellow green in color. However, the unique organization of their cells and their tendency to form all sorts of strange colonial forms suggest that they are related to eustigs only by pigmentation. Their photosynthetic organelles, xanthoplasts, probably share ancestry with those of the eustigs, but the nonplastid portion of the cell is far more like that of the chrysophytes (Phylum Pr-4). Some phycologists prefer to group our three algal phyla Xanthophyta, Chrysophyta, and Phaeophyta (Phylum Pr-12), but not haptophytes (Phylum Pr-5), in a single phylum called Heterokonta on the basis of the morphology of the nonplastid part of the cells. For example, in all these groups, one of the two anteriorly inserted undulipodia is directed forward and mastigoneme, that is, hairy. The other undulipodium, which trails behind, is smooth. The nonplastid parts of the cells of these three groups probably did share common ancestry. In fact, by their production of heterokont zoospores, certain undulipodiated "molds," such as oomycotes (Phylum Pr-27) and plasmodiophorans (Phylum Pr-24), also may be related to xanthophytes. However, the differences among these groups seem to us marked enough to justify the raising of each to phylum status. Within each phylum, there is a good deal of uniformity.

The plastid pigmentation of the xanthophytes, like that of the eustigs, consists of chlorophylls a, c_1, c_2, and e. Several xanthins are found in the best studied members: cryptoxanthin, eoxanthin, diadinoxanthin, and diatoxanthin; heteroxanthin and β carotene have been detected by spectroscopic methods. Oils are the storage products of photosynthesis—starch is absent. At least some of the glucose monomers in these oils are linked by β-1-3 linkages.

Xanthophytes typically have pectin-rich cellulosic walls made of overlapping discontinuous parts. Many cells are covered with scales characteristic of the species. In winter and under other adverse conditions, many species form cysts that are encrusted with iron or in which silica is embedded.

Xanthophytes are highly successful in fresh waters. They are found in a variety of highly structured multicellular and syncytial forms; many produce amoebas or undulipodiated zoospores. Although complex differentiation patterns are known, meiotic sexual cycles have never been reported. There are four major subgroups, each having many genera, that have been traditionally reported as orders. We raise them here to the status of class. They are the Heterochloridales, Heterococcales, Heterotrichales, and Heterosiphonales.

The Heterochloridales contain the morphologically least complex xanthophytes. They comprise two major groups, here called

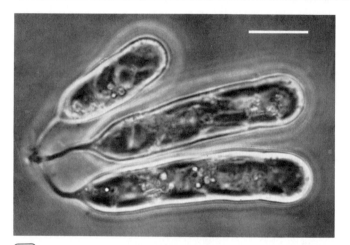

A Vegetative cells of *Ophiocytium arbuscula,* a freshwater xanthophyte from alkaline pools in England. LM (phase contrast), bar = 10 μm. [Courtesy of D. J. Hibberd.]

orders: Heterochlorineae (motile unicells) and Heterocapsineae (palm-shaped flattened colonial forms). *Botryococcus,* a very common pond scum organism, is perhaps the best-known member of the second group.

The Heterococcales, which include *Ophiocytium,* are coccoid cells. Their genera take various colonial forms—filamentous, branched, or bunched—reminiscent of the colonies formed by coccoid cyanobacteria (Phylum M-7), chrysophytes (Phylum Pr-4), and chlorophytes (Phylum Pr-15). However, the single coccoid units have an internal organization characteristic of the xanthophytes. The well-known genus *Botrydiopsis* looks like a bunch of grapes.

The Heterotrichales include many complex multicellular organisms, most of them being variations on the theme of filaments. They include the highly branched, flaccid, tree-shaped alga called *Heterodendron.*

The Heterosiphonales contain the most morphologically complex xanthophytes. They can be quite formidable in appearance. For example, *Botrydium* develops a collapsible balloonlike multi-

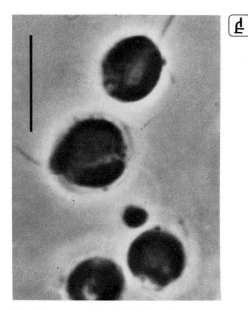

B Living zoospores of *Ophiocytium majus*. LM, bar = 10 μm. [Courtesy of D. J. Hibberd and G. F. Leedale (1971).]

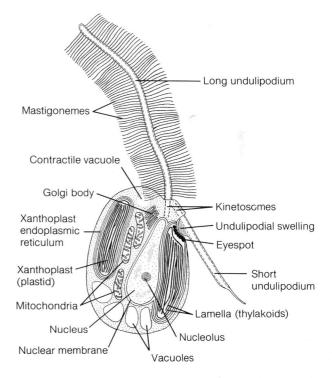

C Zoospore of *Ophiocytium arbuscula*. [Drawing by R. Golder.]

cellular thallus in drying muds; it looks superficially like a chytrid (Phylum Pr-26), may become encrusted with calcium carbonate, and may grow to nearly a meter in length. It has an extensive system of branched rhizoids in which resistant, hard-shelled cysts develop. The cysts, when wetted, can germinate into unicellular, heterokont zoospores typical of xanthophytes. The zoospores disperse, germinate, and grow into thalli, completing the life cycle. There is no evidence for the production of males or females: it seems that all zoospores are competent by themselves for further development.

Although there are fewer than one hundred well-documented species, this phylum of algae, best known as unsightly messes in muddy water, probably has many other members that have not yet come under scrutiny.

Pr-10 Eustigmatophyta

Greek *eu*, true, original, primitive; *stigma*, brand put on slave (refers to eyespot) mark, brand, spot; *phyton*, plant

Because they are yellowish green in color, form immotile coccoid vegetative cells, and propagate by motile elongated asexual zoospores, until recently eustigmatophytes were lumped together with the xanthophytes. Electron microscopic studies, however, have revealed a distinctive eyespot and organization of the eustigmatophyte cell. Their morphology justifies recognizing them as a unique set of photosynthetic motile protoctists. Very few genera are known to be in Eustigmatophyta for sure, but this is due far more to lack of study at the ultrastructural level than to a paucity of these organisms. At present, *Pleurochloris, Polyedriella, Ellipsoidion,* and *Vischeria* are the major genera of "eustigs," as they are fondly called. Although there are some multicellular eustigs, the majority are independent single cells. They live primarily in fresh waters.

In the pigmentation of their plastids, eustigs are indeed very much like the true yellow-green xanthophytes. Their plastids, called xanthoplasts, contain chlorophyll *a* as all oxygen-eliminating organisms do; in addition, they contain chlorophylls c_1, c_2, and *e*. They lack chlorophyll *b*. They contain β-carotene and several oxygenated carotenoids—which ones depend on the genus. Violaxanthin is commonly present; epoxanthin, diadinoxanthin, and diatoxanthins may also be present. Eustigs store glucose not as starch but as a solid material (not yet chemically identified) that lies outside the plastid. In some vegetative cells, a conspicuous polyhedral crystalline body comprises the pyrenoid of the plastid; it is typically attached to the thylakoids by a thin stalk.

Although the pigments of eustigs are like those of xanthophytes, their cell organization is not at all. Most eustigs have only a single hairy anterior undulipodium, at the base of which is a conspicuous undulipodial swelling, "T" shaped in transverse section. An adjacent swelling filled with drops of carotenoids forms the eyespot; it probably communicates somehow with the undulipodium to direct the cell to optimally lighted environments. The eyespot is not associated with the plastid, nor is it membrane bounded. Some eustigs (for example, *Ellipsoidion*) have a second, smooth undulipodium. The anterior undulipodium of xanthophytes lacks a swelling. There is a swelling on the posteriorly directed second undulipodium, and it is apposed to a specialized portion of the plastid.

The yellow-green plastid is single and long; it lies in the center or at the posterior end of the cell and fills some two-thirds of its volume. The thylakoids are stacked inside rather like the grana of plant plastids. Nearly all the endoplasmic reticulum (ER) of

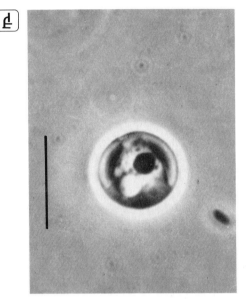

A Vegetative cell of *Vischeria* (*Polyedriella*) sp. LM, bar = 10 μm. [Courtesy of D. J. Hibberd (Hibberd and Leedale, 1972).]

eustigs is associated with the plastid. The xanthoplast ER is not associated with the nuclear membrane as it is in many other algae, and there is little developed free ER in the cytoplasm. This morphological arrangement suggests an integrated metabolic relationship between the products of photosynthesis and the nucleocytoplasm-directed biosyntheses. The cell wall is entire—that is, it goes all the way around the cell—and in some cases it contains silica deposits. Cell division is direct into two offspring cells, and sexual processes are not known to occur within the group.

These planktonic algae are eaten by other protists and by zooplankters; they are at the base of aquatic food chains. They have hardly been studied, however, so that very little is known about their natural history.

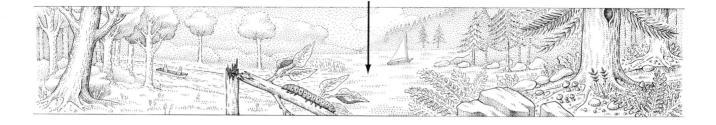

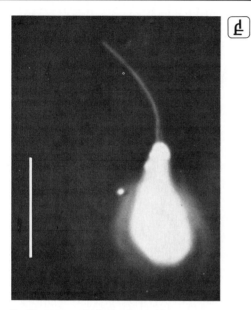

B Zoospore of *Vischeria* sp. LM, bar = 10 μm. [Courtesy of D. J. Hibberd (Hibberd and Leedale, 1972).]

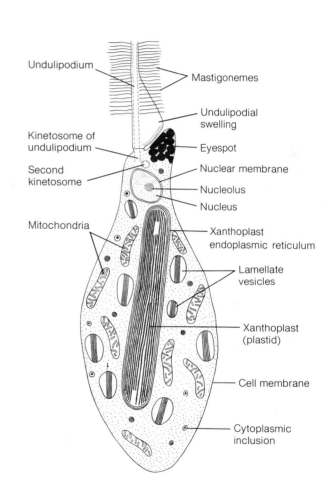

Undulipodium

Mastigonemes

Kinetosome of undulipodium

Undulipodial swelling

Second kinetosome

Eyespot

Nuclear membrane

Nucleolus

Nucleus

Mitochondria

Xanthoplast endoplasmic reticulum

Lamellate vesicles

Xanthoplast (plastid)

Cell membrane

Cytoplasmic inclusion

C Zoospore of *Vischeria* sp. [Drawing by R. Golder.]

Pr-11 Bacillariophyta
(Diatoms)

Latin *bacillus*, little stick; Greek *phyton*, plant

Achnanthes	*Fragilaria*	*Tabellaria*
Amphipleura	*Hantschia*	*Thalassiosira*
Arachnoidiscus	*Melosira*	
Asterionella	*Meridion*	
Biddulphia	*Navicula*	
Coscinodiscus	*Nitschia*	
Cyclotella	*Pinnularia*	
Cymbella	*Planktoniella*	
Diatoma	*Rhopalodia*	
Eunotia	*Surirella*	

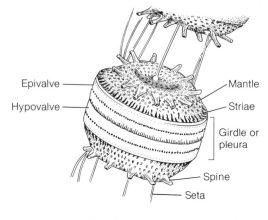

A *Thalassiosira nordenskjøldii.*
[Drawing by E. Hoffman.]

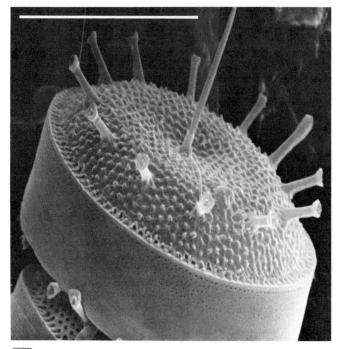

  B *Thalassiosira nordenskjøldii,* a marine diatom from the Atlantic Ocean. SEM, bar = 10 μm. [Courtesy of S. Golubic.]

There may be as many as 10,000 living species of these beautiful aquatic protists; all are single cells or form simple filaments or colonies. Hundreds also are known from the fossil record as far back as the Cretaceous.

Diatoms have shells or tests made of two parts, called valves. Each valve is composed of pectic organic materials impregnated with silica, hydrated SiO_2, in an opaline state. The valves may be extremely elaborate and beautiful; their elegantly symmetrical patterns are used to test lenses for optical aberrations. Diatoms require dissolved silica for growth; they are so good at removing silica from natural waters that they can reduce the concentration to less than a part per million, less than can be detected by chemical techniques.

Diatoms are very widely distributed in the photic zones of the world; they are an important group at the base of marine and freshwater food chains. Some species are found in hypersaline ponds and lagoons; others in clear fresh water. All species under study are obligate photosynthesizers, although some also require organic substances, such as vitamins, for growth.

There are two great classes of diatoms, the Centrales and the Pennales. The Centrales, or centric diatoms, have radial symmetry, like both *Thalassiosira* and *Melosira* in our illustrations. The Pennales, or pennate diatoms, have bilateral symmetry; many of them are boat shaped or needle shaped. Pennate diatoms have a slit between the valves, called the raphe. Cytoplasmically pro-

duced slime is exuded from the raphe and the diatom moves in slow gliding fashion. The centric diatoms always lack a raphe and they are not motile. In spite of the correlation of the raphe and movement, the detailed mechanism of gliding in the pennate diatoms is not known. The centric diatoms usually have numerous small plastids, whereas in the pennate diatoms the plastids are fewer, as a rule, and larger.

Diatoms are generally brownish in color; for years, they were classified with the golden-yellow algae, the chrysophytes (Phylum Pr-4). The chrysoplasts (plastids) of diatoms contain the pigments chlorophyll *a*, chlorophyll *c*, β carotene, and xanthophylls, including fucoxanthin, lutein, and diatoxanthin. The photosynthetic food reserve of both diatoms and chrysophytes is the oil

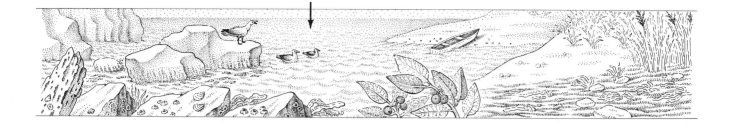

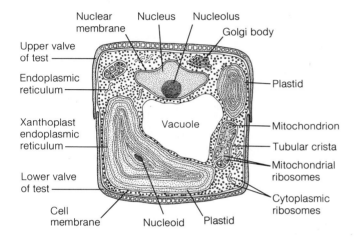

Nuclear membrane
Nucleus
Nucleolus
Golgi body
Upper valve of test
Endoplasmic reticulum
Xanthoplast endoplasmic reticulum
Vacuole
Lower valve of test
Cell membrane
Nucleoid
Plastid
Plastid
Mitochondrion
Tubular crista
Mitochondrial ribosomes
Cytoplasmic ribosomes

C *Melosira* sp., a centric diatom. [Drawing by L. Meszoly.]

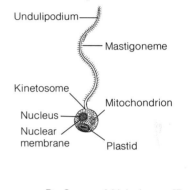

Undulipodium
Mastigoneme
Kinetosome
Mitochondrion
Nucleus
Nuclear membrane
Plastid

D Sperm of *Melosira* sp. [Drawing by L. Meszoly.]

chrysolaminarin. Nevertheless, in life cycle, cell structure, and division, the diatoms differ enormously from the other golden-yellow algae.

Diatoms are sexually advanced. Like animals, they spend most of their life cycle in the diploid state, and meiosis occurs just before the formation of haploid gametes. After fertilization, the diploid zygote develops into the familiar diatom. The diatoms make up such an easily distinguished natural group that, in the light of modern information, it seems mandatory to make them a phylum separate from the other organisms that have golden plastids.

The centric diatoms comprise four major groups, here recognized as orders. The discoids include *Thalassiosira;* this colonial diatom, extrudes threads, or setae, of chitin, which hold individual diatoms together in chains. The solenoids are elongate cylindrical or subcylindrical in shape. The biddulphioids are box shaped; most have horns or other decorations on their tests. The rutilarioids have naviculoid valves; that is, they depart from strict radial symmetry by having a boatlike shape with pointed ends that have radial or irregular markings.

The pennate diatoms are classified into four groups (here, orders) according to the presence and development of the raphe. The Araphideae lack a true raphe; the Raphidioideae show the beginnings of a raphe; the Monoraphideae have a fully developed raphe, but only on one of the valves; the Biraphideae have a fully differentiated raphe on each valve.

Although diatoms can reproduce sexually, more often they reproduce vegetatively by mitotic division into two daughter cells. Each daughter cell retains one valve of the parent shell; each produces one new valve, which fits into the parental valve. Hence, each offspring is slightly smaller than its parent. This tendency for diatoms to decrease in size is counteracted by auxospore formation. An auxospore is not really a spore in the sense that it can resist adverse conditions, but an enlarged, shell-less diatom formed when a protoplast is released from its rigid test. The size of the auxospore is a characteristic of each species. Freed from the inexorable sequence of shrinking, the auxospore secretes two new, large valves.

In the pennate diatoms, auxospore formation usually follows fertilization and zygote formation. Haploid male gametes, bearing single anterior undulipodia, fertilize immotile haploid female protoplasts; the valves of the tests sometimes open for the purpose. The zygote becomes the large auxospore; asexual reproduction by mitotic division follows. In the centric diatoms, auxospores may be formed without the intervention of meiosis and fertilization.

Pr-12 Phaeophyta

(Brown algae)

Greek *phaios*, dusky, brown; *phyton*, plant

Alaria
Ascophyllum
Chordaria
Cutleria
Dictyota
Ectocarpus
Fucus
Laminaria
Macrocystis
Nereocystis

Pelagophycus
Postelsia
Sargassum
Scytosiphon
Sphacelaria
Tilopteris
Zonaria

The phaeophytes are the brown seaweeds. They nearly all are marine. They are the largest protoctists; the giant kelps, for example, are as much as 100 meters long. About 1500 species have been described; all are photosynthetic. Most phaeophytes live along rocky coastal sea shores, especially in temperate regions. They dominate the intertidal zone, where they form great beds of seaweed. They are the primary producers for many communities of invertebrate animals and microbes.

Reproduction in this group is mostly sexual. Although details of the life cycle vary with genus, most phaeophytes follow a common pattern. They form eggs and biundulipodiated sperm. The sperm have one forward mastigonemate (hairy) undulipodium and one trailing smooth (whiplash) one. Thus, the morphology of phaeophyte sperm resembles that of the heterokont phyla, the chrysophytes (Phylum Pr-4) and the xanthophytes (Phylum Pr-9). (Indeed, the brown algae are thought to have evolved from single-celled chrysophytes.) The fertile egg of phaeophytes typically germinates in response to light; rhizoids grow out away from the source of light. The diploid organism that arises from a fertilized egg is called the sporophyte.

The sporophyte differentiates into a blade ("leaf"), a stipe ("stem") and holdfast ("root"); it can grow to enormous size. However, these phaeophyte thalli lack the complex tissue organization characteristic of land plants. Some do develop sieve tubes for conducting water and photosynthate (in many cases, mannitol); however, the tubes are analogous, not homologous, to the sieve tubes of plants. Sporangia form on the sporophyte blades. These sporangia release zoospores, not gametes; meiosis may precede the release of zoospores. Although they resemble the phaeophyte sperm, zoospores do not have to be fertilized in order to grow into a multicellular organism.

Some zoospores develop into male "plants" that will later produce sperm; others develop into female "plants" that will produce eggs. If meiosis has preceded zoospore formation, the haploid plants that grow from the zoospores are called gametophytes—male and female gametophytes produce sperm and eggs directly from their haploid tissue, in structures called gametangia. The female gametangia are called oogonia; the male gametangia are called antheridia. These structures are analogous but not homologous to similar structures in rhodophytes (Phylum Pr-13), chlorophytes (Phylum Pr-15), and plants, illustrating the convergent trends so common in the reproductive morphology of protoctists.

The alternation of haploid and diploid generations in phaeophytes is analogous but not homologous to the alternation of generations in chlorophytes and plants. In some phaeophytes,

A Thallus of *Fucus vesiculosus* taken from rocks on the Atlantic seashore. Bar = 10 cm. [Courtesy of W. Ormerod.]

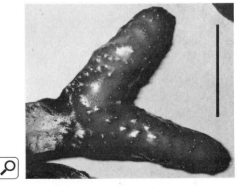

B Receptacles on sexually mature apices of *Fucus vesiculosus* thallus. Bar = 1 cm. [Courtesy of W. Ormerod.]

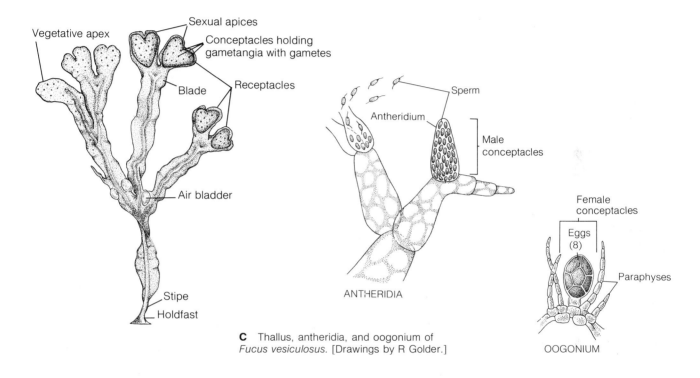

C Thallus, antheridia, and oogonium of *Fucus vesiculosus.* [Drawings by R Golder.]

the gametophytes are indistinguishable from the sporophytes except for their chromosome number. In others, the generations are so dissimilar that at first they were thought to be entirely different organisms.

Brown algae contain a distinct set of photosynthetic pigments: their plastids, called phaeoplasts, contain chlorophylls *a* and *c* but never *b.* Fucoxanthin is generally the prominent carotenoid derivative. This xanthophyll is responsible for the brown or olive-drab color of the thallus. The carbohydrate stored as food is called laminarin (named after *Laminaria,* a well-known kelp genus). Phaeophytes also store lipids. Many are capable of synthesizing elaborate organic compounds called secondary metabolites. The usefulness of the rapidly growing seaweeds as sources of drugs, food, and fuel is yet to be determined.

Although most frequently found along seashores, some brown algae, such as the famous *Sargassum* found in the Sargasso Sea, form immense floating masses far offshore. These provide the substratum for unique communities of marine animals and microbes, and lead to rates of primary productivity in the open oceans far higher than would be possible if the phaeophytes were absent. Attempts are being made to increase the ocean's productivity by planting *Macrocystis* on rope rafts off the coast of California.

Our example, *Fucus vesiculosus,* often called rockweed, is very common on temperate rocky sea coasts. As in other members of the order Fucales, the thallus is flattened—all branching is in one plane. It is used as a source of iodine. The sexes may be on the same or on separate plants. The gametangia are contained in dark dotlike bodies, called conceptacles, scattered on the surface of heart-shaped swellings called receptacles. Within a female conceptacle, each oogonium produces one, two, four, or eight eggs. Oogonia containing ripe eggs are released into the water, where they are fertilized by sperm produced by antheridia in male conceptacles.

Pr-13 Rhodophyta

(Red algae)

Greek *rhodos,* red; *phyton,* plant

Agardhiella	*Gelidium*
Bangia	*Goniotrichum*
Batrachospermum	*Hildebrandia*
Callophyllis	*Lemanea*
Chantransia	*Lithothamnion*
Chondrus	*Nemalion*
Corallina	*Polysiphonia*
Dasya	*Porphyra*
Erythrocladia	*Porphyridium*
Erythrotrichia	*Rhodymenia*

Rhodophytes and phaeophytes (Phylum Pr-12) are the largest and most complex of the protoctists, although the largest rhodophytes are somewhat smaller and less complicated than the largest phaeophytes. Rhodophytes commonly inhabit the edges of the sea, and are cosmopolitan in distribution. In the tropics, particularly, they abound on beaches and rocky shores. About 4000 species are known, the vast majority of which are marine. There are two classes, the Florideae and the Bangiales. In the Florideae, larger in size and in number of species, tissues are made of cells connected by pit connections, protoplasmic strands between cell walls. In the Bangiales, pit connections are rare.

The red algae form a natural group—all species display several traits that characterize the phylum. None have undulipodia at any stage in the life cycle, yet all reproduce sexually. Reproduction is öogamous: a large egg cell is formed in a special female organ. This organ, the oogonium, bears a long neck, which is receptive to the male gamete. The antheridium, or male organ, produces a single male "sperm." Because it lacks an undulipodium, it is incapable of locomotion. After male gametes are released near the female, at least one attaches to the neck of the female structure and moves down it to fertilize the egg. The detailed physiology of this process is poorly known.

After fertilization, meiosis may take place immediately to form haploid spores. Alternatively, no meiosis takes place and diploid spores, called carpospores, are formed. They are often formed in bunches of threads that grow out of the oogonium. In some of the more elaborate life cycles, the carpospores are formed in a special organ (cystocarp) that establishes a connection with the oogonium. The carpospores of some species develop into complex little "plants" that bear organs called tetrasporangia. Meiosis takes place in the tetrasporangia; the four meiotic products, the tetraspores, are released into the sea. They germinate into haploid thalli, which eventually produce oogonia or antheridia. Predominantly haploid or alternating haploid and diploid life cycles are common in the red algae; the details of most of the cycles have yet to be worked out.

All rhodophytes have reddish plastids, called rhodoplasts, which contain chlorophyll *a* and phycobiliproteins—the bluish phycocyanin and allophycocyanin and the reddish phycoerythrin. These phycobiliproteins contain open porphyrin rings very similar in molecular structure to the bile pigments of animals; some seem to be identical to the phycobiliproteins in some cyanobacteria (Phylum M-7). All rhodoplasts lack chlorophylls *b* and *c.* The product of photosynthesis in the rhodophytes is a starchlike polymer, a solid polysaccharide called floridean starch.

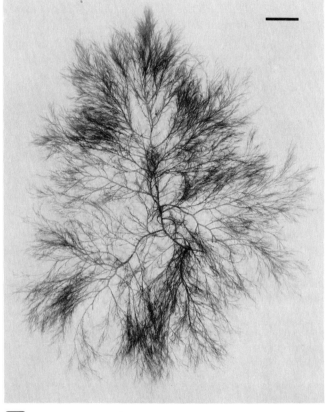

A *Polysiphonia harveyi* from rocky shore, Atlantic Ocean. Bar = 1 cm. [Courtesy of G. Hansen.]

Many red algae become encrusted with calcium carbonate: *Lithothamnion* looks like reddish circular crust on rocks, and *Corallina* looks like an encrusted tree. A single genus may have both calcifying and noncalcifying species. The propensity for calcification has produced a good fossil record for the phylum; mineralized forms of coralline algae first appear in the lower Paleozoic.

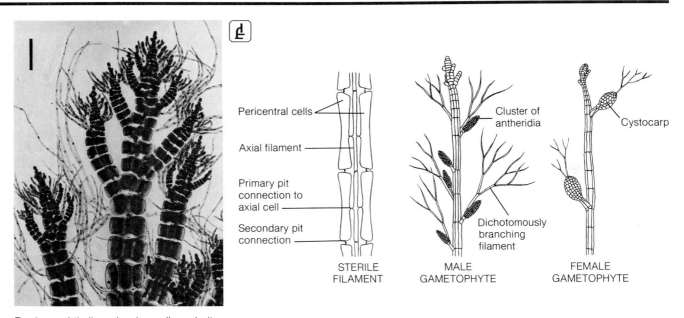

Pericentral cells

Axial filament

Primary pit
connection to
axial cell

Secondary pit
connection

STERILE
FILAMENT

Cluster of
antheridia

Dichotomously
branching
filament

MALE
GAMETOPHYTE

Cystocarp

FEMALE
GAMETOPHYTE

B Apex of thallus, showing cells and pit connections. LM, bar = 0.1 mm. [Courtesy of G. Hansen.]

C Sterile and sexually mature apices of thalli. [Drawings by R. Golder.]

Rhodophytes show a marked parallelism of forms with other groups of algae, the chrysophytes (Phylum Pr-4), the chlorophytes (Phylum Pr-15), and the phaeophytes (Phylum Pr-12). There are heterotrichous filaments (*Chantransia*), prostrate discs (*Erythrocladia*), cushions (*Hildebrandia*), elaborate erect structures (*Batrachospermum*), compact tissuelike types (*Lemanea*), and delicate, many-branched forms (*Polysiphonia* and *Porphyra*).

Agar, the substance used to harden the broth upon which colonies of microorganisms are grown so that they can be isolated and studied, is extracted from red algae. Polysaccharides from these seaweeds are also used in the manufacture of ice cream and other food products. The leafy dulce of New England's rocky shore is dried and eaten whole. Most rhodophytes have not yet been put to use by people.

D Apex of male thallus. [Drawing by R. Golder.]

Pr-14 Gamophyta

(Conjugating green algae)

Greek *gamos*, marriage; *phyton*, plant

Bambusina	*Micrasterias*
Closterium	*Mougeotia*
Cosmarium	*Netrium*
Cylindrocystis	*Penium*
Desmidium	*Spirogyra*
Gonatozygon	*Staurastrum*
Genicularia	*Temnogyra*
Hyalotheca	*Zygnema*
Mesotaenium	*Zygogonium*

Gamophytes are green algae. They have symmetrical cells, complex chloroplasts, which are usually aligned down the long axis of the cell, and no undulipodia in any stage of their life cycle. Most gamophytes have one large and conspicuous nucleus in each cell. They are found in ponds, lakes, and streams; no truly marine forms have been reported. To reproduce, the haploid vegetative cells either divide directly (asexually) or produce equal amoeboid gametes that fuse to form a zygote. This usually develops into a resistant and conspicuous structure called a zygospore. Zygotic meiosis occurs within the zygospore, and haploid algal cells eventually emerge.

In their pigmentation, gamophytes are similar to all the other green algae: they have chlorophylls *a* and *b*, and most are grass green in color. For this reason, they are often classified with the other chlorophytes (Phylum Pr-15).

There are two classes in the phylum Gamophyta as presented here: Euconjugatae (true conjugating algae) and Desmidioideae (the desmids).

The Euconjugatae, which are generally filamentous forms, consist of two orders: Mesotaenioideae (for example, *Cylindrocystis, Mesotaenium,* and *Netrium*) and Zygnemoideae. The latter, the zygnemids, are the best known.

Zygnema, Spirogyra, and *Mougeotia* form pond scums, stringy masses of long unbranched filaments. In *Zygnema* and *Spirogyra,* the chloroplast is helically wound along the length of the long cylindrical cell; in *Mougeotia,* a large barrel-shaped chloroplast extends the full length of the cell. These organisms grow rapidly by mitosis, the filaments breaking off fragments that start new filaments, thus forming a bloom or scum in a few days.

During sexual union, two filaments come to lie side by side. Protuberances grow and join to form conjugation tubes that link cells in opposite filaments. Through the conjugation tubes the cells of the "male" filament flow, chloroplasts and all, to fuse with the cells of the "female" filament. Each fusion results eventually in a dark, spiny zygote in a chamber of the female filament.

After a period of dormancy, the zygotes are released into the water; they undergo meiosis and germinate to produce new haploid filaments.

There are three families of desmids in our classification; each is named for its best-known genus: Penieae (*Penium*), Closterieae (*Closterium*), and Cosmarieae (*Cosmarium*). Several thousand species are known. Most are single cells—more precisely, they are pairs of cells whose cytoplasms are joined at an isthmus (Greek *desmos* means bond). The isthmus is the location of the single shared nucleus. Some desmids are colonial. In many desmids, the chloroplasts are lobed or have plates or processes that extend from the center toward the periphery of the cell. The outer layers of the cell wall are typically decorated with spines, knobs, granules, or other protrusions arranged in lovely designs. These outer layers are composed of cellulose and pectic substances, and in many cells are impregnated with iron or silica; the inner layer, on the other hand, is composed of cellulose only and is structureless at the light-microscopic level of magnification.

The outermost layer of the desmid cell is a mucilaginous sheath, sometimes thin and sometimes very thick and well developed. It is secreted through pores in the cell wall. The slow gliding movement of desmids is thought to be due to the secretion of this mucilage.

In the typical desmid, the two "half cells" are mirror images of each other, and each has its own chloroplast and nucleus. In asexual reproduction, the two partners simply separate (after the nucleus divides) and each one grows a new partner. During sexual conjugation, the two partners also separate; both leave their shells and fuse outside, either with each other or with a liberated protoplast from another desmid, to form a dark, spiny zygote reminiscent of the zygotes of the Euconjugatae and zygomycotes (Phylum F-1). In *Desmidium cylindrium,* only one partner, the "male," leaves its shell; conjugation takes place inside the shell of the "female," and that is where the zygote is lodged.

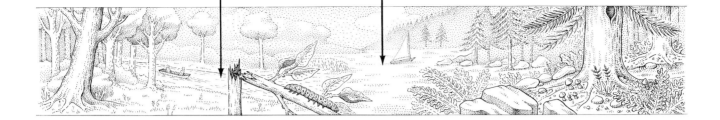

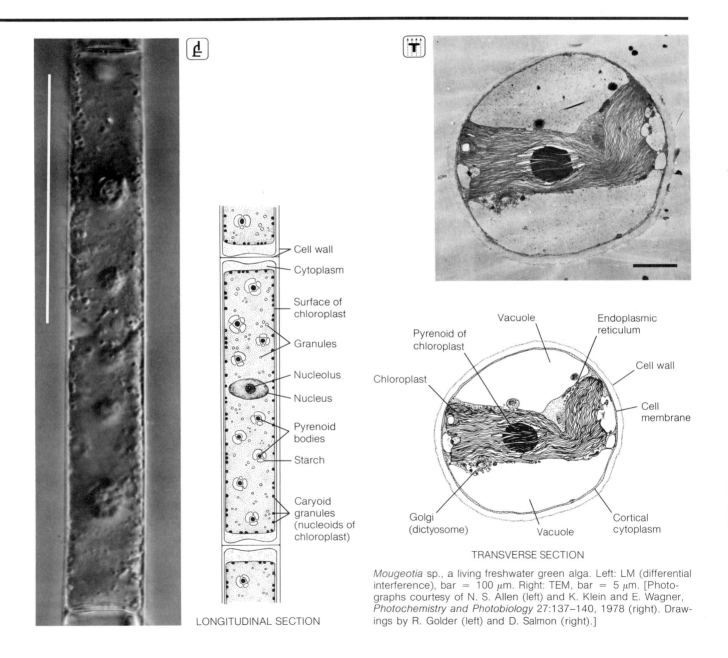

Cell wall

Cytoplasm

Surface of chloroplast

Granules

Nucleolus

Nucleus

Pyrenoid bodies

Starch

Caryoid granules (nucleoids of chloroplast)

LONGITUDINAL SECTION

Vacuole

Endoplasmic reticulum

Pyrenoid of chloroplast

Cell wall

Chloroplast

Cell membrane

Golgi (dictyosome)

Vacuole

Cortical cytoplasm

TRANSVERSE SECTION

Mougeotia sp., a living freshwater green alga. Left: LM (differential interference), bar = 100 μm. Right: TEM, bar = 5 μm. [Photographs courtesy of N. S. Allen (left) and K. Klein and E. Wagner, *Photochemistry and Photobiology* 27:137–140, 1978 (right). Drawings by R. Golder (left) and D. Salmon (right).]

Pr-15   Chlorophyta

Greek *chloros*, yellow green; *phyton*, plant

Acetabularia
Bryopsis
Caulerpa
Chaetomorpha
Chara
Chlamydomonas
Chlorella
Chlorococcum
Chlorodesmis
Cladophora
Codium

Cylindrocapsa
Derbesia
Dunaliella
Enteromorpha
Fritschiella
Gonium
Halimeda
Lamprothamnium
Microspora
Nitella
Nitellopsis

Oedogonium
Oocystis
Pandorina
Penicillus
Platymonas
Prasinocladus
Protococcus
Pseudobryopsis
Pseudotrebouxia
Pyramimonas
Spongomorpha

Stigeoclonium
Tetraspora
Tolypella
Trebouxia
Udotea
Ulothrix
Ulva
Urospora
Volvox

Chlorophytes are green algae that form zoospores (or gametes) having cup-shaped grass-green chloroplasts and at least two anterior undulipodia of equal length. About 7000 species have been described. Within the phylum, several evolutionary lines have led from unicellular forms to multicellular organisms. Most botanists agree that somewhere in this extremely diverse group lie the ancestors of the higher plants.

In this book, Phylum Chlorophyta excludes the gamophytes (Phylum Pr-14), which lack undulipodia, but unites the Siphonales, Charales, and prasinophytes with the chlorophytes *sensu stricto,* because all of these green algae have undulipodia at some stage. Although this is a somewhat arbitrary plan, it emphasizes the tendency of the unicellular, biundulipodiated algae to give rise to impressive and cohesive classes of reproductively and morphologically complex "water plants." Some of them are at least periodically resistant to desiccation; that one or several such algae were the progenitors of the land plants seems incontrovertible.

Chlorophytes are a major component of the phytoplankton; it has been estimated that they fix more than a billion tons of carbon in the oceans and freshwater ponds every year. Their chloroplasts are grass green; they contain chlorophylls *a* and *b* as well as these carotenoid derivatives: astaxanthin, canthaxanthin, flavoxanthin, loraxanthin, neoxanthin, violaxanthin (which tends to convert to zeaxanthin in dark-grown algae placed in the light), and the xanthophyll, echinenone. Starch, the α-1-4-linked glucose polymer, is the food reserve stored by most green algae.

The cell walls of green algae, like those of land plants, are composed of cellulose and pectins, or of polymers of xylose (*Bryopsis* and *Caulerpa*) or mannose (*Acetabularia*) linked to protein. In many genera, the walls are encrusted with calcium carbonate, silica, and, less frequently, other minerals such as iron oxides.

Sexuality is rampant in this group: there is a trend from isogamy, in which two motile gametes conjugate and fuse, toward oogamy, in which a large immotile egg is fertilized by a small motile sperm. The sperm is very much like the individual adults (*Chlamydomonas* and *Dunaliella*), zoospores, or isogametes of many species in the phylum. In *Acetabularia,* a diploid zygote is the product of fertilization. It immediately undergoes meiosis to regenerate the haploid stage in the life cycle.

Within the Chlorophyta as presented here, there are nine classes: Chlorococcales, Volvocales, Ulotrichales, Oedogoniales, Chaetophorales, Cladophorales, Siphonales, Charales, and Prasinophycales. Within each of these classes, except the last,

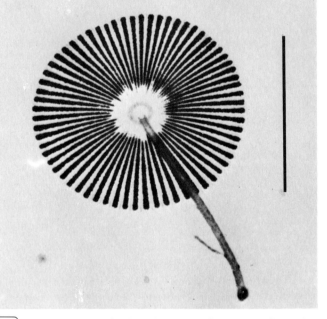

A *Acetabularia mediterranea,* a living alga from the Mediterranean Sea. Bar = 1 cm. [Courtesy of S. Puiseux-Dao.]

can be seen trends from unicellular forms to various types of complex colonies.

The Chlorococcales are a very diverse and probably polyphyletic class. Eight families have been recognized, the most famous of which are Chlorococcaceae, which include the ubiquitous tree-scum alga *Chlorococcum,* and Chlorellaceae, which include both symbiotic and free-living *Chlorella* species. Chlorellas grow like weeds in the laboratory; they are frequently used to study the physiology of photosynthesis. The "water nets" called Hydrodictyaceae are another family in this class.

The Volvocales include *Chlamydomonas.* Probably more is known about the genetic control of mating, undulipodia, photosynthesis, and mitochondrial metabolism in *Chlamydomonas* than in any other protoctist.

The Ulotrichales are primarily filamentous (*Microspora* and

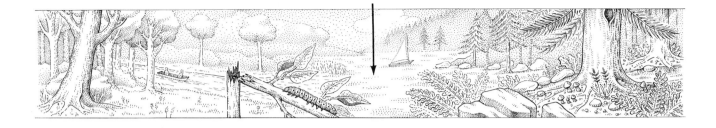

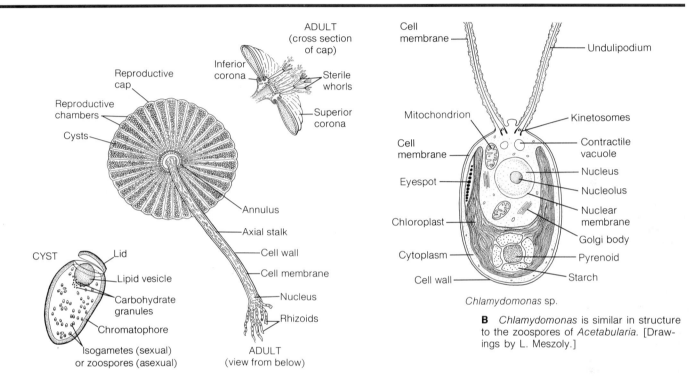

B *Chlamydomonas* is similar in structure to the zoospores of *Acetabularia*. [Drawings by L. Meszoly.]

Chlamydomonas sp.

Cylindrocapsa) or thalloid (*Ulva,* or sea lettuce, and the extremely common estuarine form *Enteromorpha*).

The Oedogoniales produce zoospores having an unusual ring of undulipodia; they show a unique method of cell division and an elaborate style of sexual reproduction. The relationship of *Oedogonium* and other Oedogoniales to the other chlorophytes is not well understood. They may have evolved from the Ulotrichales.

The Chaetophorales (for example, *Stigeoclonium* and *Fritschiella*) are mainly branched multicellular algae; they are differentiated into a prostrate and an upright thallus.

The Cladophorales may be unbranched (*Urospora* and *Chaetomorpha*) or branched (*Cladophora* and *Spongomorpha*); typically they are composed of multinucleate elongate cells.

The Siphonales comprise many species of common green seaweeds, such as *Codium* and *Acetabularia*. Many of them are quite large, but all are syncytial: no cell membranes form, so that millions of nuclei and chloroplasts share the same cytoplasm.

The Charales are multicellular and live in fresh or brackish water. Their cells are long and either uninucleate or multinucleate. *Nitella* and *Chara,* delicate pondweeds, are favorite experimental organisms.

The prasinophytes, which are unicellular, differ a good deal from all the other classes. They lack the typical chlorophyte gametes and sexual life cycle, and their cell structure is not like that of the standard chlorophyte. For example, the prasinophytes have an anterior pit or groove, from which emerge four undulipodia, and they bear scales on their cell surface. Only a few genera have been described, including *Platymonas,* a regular tissue symbiont of the green photosynthetic flatworm *Convoluta roscoffensis*. It is likely that further work will raise these green algal protists to phylum status.

Pr-16 Actinopoda

Greek *actinos*, ray; *pous*, foot

Acantharia
Acanthocystis
Acanthometra
Actinophrys
Actinosphaerium
  (Echinosphaerium)
Challengeron
Ciliophrys
Clathrulina
Collozoum

Heterophrys
Pipetta
Sticholonche
Thalassicola
Zygacanthidium

The actinopods, heterotrophic protists, are distinguished by their long, slender, projecting cytoplasmic axopods, also called axopodia. Each of these fine projections is stiffened by a bundle of microtubules; this bundle is called an axoneme. Each axoneme has an often quite elaborate arrangement of microtubules characteristic of each actinopod group, and the microtubules are often cross linked. There are four groups, which we recognize as classes. These are the marine planktonic Polycystina and Phaeodaria, the marine, usually planktonic Acantharia, and a heterogeneous group, the Heliozoa, which are primarily freshwater microbes. The two or three first classes are often united as the class or order Radiolaria, but electron-microscopic studies have shown that the classes Acantharia, Polycystina, and Phaeodaria are products of evolutionary convergence and are only remotely related to each other.

Acantharians, generally spherical organisms, have a unique radially symmetrical skeleton composed of rods of crystalline strontium sulfate ($SrSO_4$). The skeleton usually has ten diametrical (twenty radial) spines, spicules, inserted according to a precise rule, Müller's law, discovered by Johannes Müller in the nineteenth century. The acantharian cell may be conceived as a globe from whose center the spicules radiate and pierce the surface at fixed latitudes and longitudes. If there are twenty spicules, then there are five quartets, one "equatorial," two "polar," and two "tropical," that pierce the globe at the latitudes 0°, 30°N, 30°S, 60°N, and 60°S. For the equatorial and both polar quartets, the longitudes of the piercing points are 0°, 90°W, 90°E, 180°; for the tropical quartets, 45°W, 45°E, 135°W, and 135°E. Even in acantharians that do not have the general shape of a globe, these orientations are strictly observed, although some spicules are thicker and longer than the others. Some species have much more than 20 spicules, as many as several hundred, but they are always grouped by some elaboration of Müller's law.

The acantharian organism is made of distinct layers. The innermost layer is the central mass of the cell. It appears to be coarsely granulated and contains many small cell nuclei. Immediately surrounding the central mass is a perforated, flimsy network of microfilaments called the central capsule membrane. Through the central capsule membrane, the central mass extends several kinds of cytoplasmic outgrowths: cytoplasmic sheaths surrounding the skeletal spines; reticulopods, cross-connected netlike pseudopods lacking axonemes; filopods, very thin pseudopods stiffened by one or very few microtubules; and a number of axopods (usually 54, but in some acantharians there may be several hundred) arising from axoplasts between the spines.

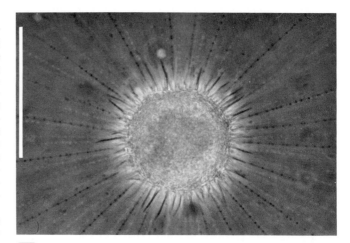

A *Acanthocystis aculeata*, a freshwater centrohelidian heliozoan, showing axopods and spines. LM, bar = 50 μm. [Courtesy of C. Bardele.]

At the periphery of the organism is the cortex, a thin, flexible layer of microfilaments, which may be arranged in intricate designs. The cortex is underlain by a network of reticulopods. Where the strontium sulfate skeletal spines pass through, the cortex is pushed out, like a tent stretched out over tent poles. At these points are filaments called myonemes that apparently control the tension of the cortex and bind it to the skeletal rods.

The delicate axopods increase the amount of cell surface exposed to the sea. They retard sinking and perhaps allow a more efficient scavenging of scarce nutrients from the water. Prey, generally protists and small animals, stick to the axopods. Cytoplasm from the axopods then engulfs the prey and cytoplasmic flow transports it down the axopods toward inner parts of the cell, where the particulate food is digested.

Acantharians produce many small swarmer cells, each containing a drop of oil reserve and a crystal and bearing two 9 + 2 undulipodia, probably one trailing and the other directed forward. The undulipodia originate from kinetosomes in the anterior

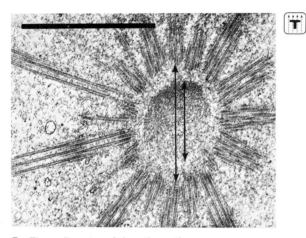

B The cell center of *Acanthocystis penardi,* showing the microtubules of the axonemes and the central organelle (short arrow) of the centroplast (longer arrow). TEM, bar = 1 μm. [Courtesy of C. Bardele.]

C A generalized polycystine actinopod in cross section. Thick black lines = skeleton; stippled areas = cytoplasm; N = nucleus. [Drawing by L. Meszoly, information from Georges Merinfeld.]

Mitochondrion

Nucleus

Oil droplet

D Generalized swarmer cell as can be found in some acantharian actinopods. [Drawing by L. Meszoly.]

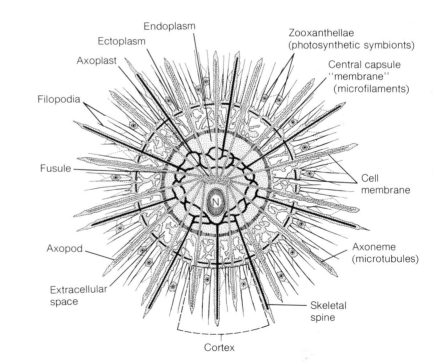

Endoplasm

Ectoplasm

Axoplast

Filopodia

Fusule

Axopod

Extracellular space

Zooxanthellae (photosynthetic symbionts)

Central capsule "membrane" (microfilaments)

Cell membrane

Axoneme (microtubules)

Skeletal spine

Cortex

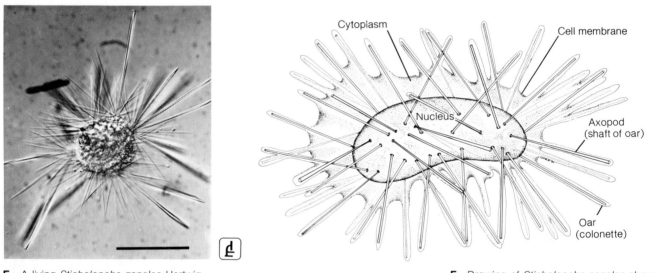

Cytoplasm

Cell membrane

Nucleus

Axopod (shaft of oar)

Oar (colonette)

E A living *Sticholonche zanclea* Hertwig taken from the Mediterranean off Ville Franche sur Mer Marine Station. LM, bar = 100μm. [Courtesy of M. Cachon.]

F Drawing of *Sticholonche zanclea* showing the placement of the microtubular oars on the nuclear membrane. [Based on electron micrographs of A. Hollande, M. Cachon, and J. Valentin, *Protistologica* 3:155–170, 1967. Drawing by L. M. Reeves.]

part of the swarmer cell. Some acantharians round up to form a special cyst, in which they undergo mitotic divisions; the swarmers develop and are later released from the cyst. Very little about the development process is known because swarmers have been devilishly difficult to culture in the laboratory. Meiotic sex has never been seen.

Most acantharians harbor large numbers of symbiotic haptophytes (Phylum Pr-5), which live and grow in the cell. They are grass green in color, and photosynthetic; they permit the acantharians to live partly like phytoplankton, obtaining their energy and food by photosynthesis in the nutrient-poor open ocean. The hosts provide nitrogen- and phosphorus-rich wastes for their haptophyte symbionts.

The two classes Polycystina and Phaeodaria have often been grouped together as "Radiolaria" because of their superficial resemblance. Polycystines and phaeodarians often have strik-ingly beautiful opaline skeletons made of hydrated amorphous silica; they are extremely common in tropical waters. Along with diatoms, silicoflagellates, and sponges, they are responsible for the depletion of dissolved silica in surface waters.

Polycystines and phaeodarians differ in many ways. The polycystine skeleton is made of opal (hydrated amorphous silica); the phaeodarian skeleton is made of silica plus often a large quantity of organic substances of unknown nature. The polycystine skeletal elements look solid under the light microscope; however, electron microscopy often reveals an ultrastructure of tiny canals and pores in their skeletons. The skeletal elements of phaeodarians are evidently hollow even under the light microscope: their spines are tubular and the continuous shells of many species have a bubbly "styrofoam" ultrastructure barely visible by light microscopy and very conspicuous under the electron microscope. Crystals (but not skeletal components) of $SrSO_4$ are secreted by some adult polycystines in their endoplasm and perhaps by all of them in their undulipodiated swarmers, whereas $SrSO_4$ is unknown in phaeodarians.

In both polycystines and phaeodarians, the capsule enclosing the central mass of cytoplasm is not a flimsy microfibrillar net with open meshes as in the acantharians, but is made of massive organic material. In the polycystines, the capsule (probably composed of mucoproteins or mucopolysaccharides) is made of numerous juxtaposed plates, like the pieces of a jigsaw puzzle separated from each other by narrow slits. The phaeodarian capsule, however, is a single continuous structure. The polycystine capsule can grow in diameter during the life of the organism (whether by lateral growth of the single plates at their rims, by intercalation of new plates, or by both processes is not known). The phaeodarian capsule cannot increase in diameter once it has formed; it can only thicken its wall.

The axonemes of the polycystine axopods studied so far are all made of parallel microtubules aligned in geometrical arrays, and there are bridges between microtubules. With the exception of a few species that lack axopods entirely, there are always many such axopods per cell. Polycystines usually have one axoplast from which all axonemes originate, but some groups have other arrangements—for example, individual axoplasts (one per axoneme) all located next to the nucleus. In phaeodarians, only two axonemes penetrate the capsule; they originate from separate axoplasts just inside the capsule. The microtubules in the basal part of these axonemes are not linked by bridges. Light microscopy reveals a cortex of many thin peripheral pseudopods, which are perhaps branches of the two axopods. No polycystine axoneme is known to branch.

The orifices of the polycystine capsule, called fusules, are complex mufflike structures filled with a dense plug that permits the passage of the axonemal microtubules (if these originate inside the endoplasm), but that hampers the circulation of cytoplasm between the endoplasm and the extracapsular pseudopodial network. The phaeodarian capsule normally has only three orifices, of two kinds: a wide, complex astropyle, an opening that ensures easy contact and exchange between the endoplasm and whatever cell parts lie outside the capsule (it does not seem to serve for the passage of any axoneme); and two (rarely more) parapyles, simpler than the polycystine fusules, which allow the passage of the two thick axonemes of the cell. Inside the capsule at each parapyle there is a cup-shaped axoplast from which an

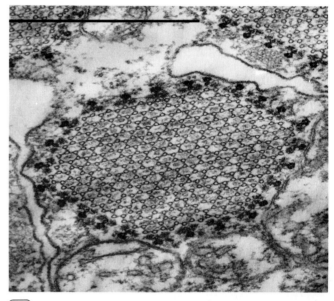

G Transverse section through an oar of *Sticholonche zanclea*. SEM, bar = 1 μm. [Courtesy of J. Cachon and M. Cachon.]

axoneme originates. Outside the capsule of many phaeodarians, in front of the astropyle, is a mass of apparently predigested food, the phaeodium. The polycystines have no phaeodium.

In the phaeodarian endoplasm are numerous strange tubes, called rodlets, about 200 nm wide, having a complex repeating ultrastructure. Their role is unknown (perhaps they take part in the secretion of the capsule). No such rodlets are known in the polycystines.

Many polycystines have zooxanthellae or zoochlorellae, symbiotic yellow or green symbiotic algae. The phaeodarians have no algal symbionts.

Most polycystines and all phaeodarians have only one nucleus, large and polyploid. Only the phaeodarian nucleus undergoes an extraordinary equational division, which superficially resembles classical mitosis and during which two monstrous "equatorial plates" are formed, each having more than 1000 chromosomes.

Class Polycystina is divided into the orders Spumellaria and Nassellaria. Spumellarians have fusules scattered all over their central capsule membrane; thus, their axopods radiate in all directions. The organism is usually spherical, ellipsoidal, or flattened, and so, naturally, is the skeleton. Some spumellarians form large colonies in which hundreds of individual organisms are embedded in a common mass of jelly. The fusules of nassellarians, which never form colonies, are clustered at one pole of the capsule membrane; their axopods are grouped in a conical bunch that leaves the cell at that pole. The cell is typically egg-shaped, and the skeleton usually resembles a basket, open at the pole from which the axopods radiate outward.

Heliozoans are primarily planktonic and freshwater forms, although estuarine, marine, and benthic species are also known. Many use their axopods to catch prey. The axopods radiate out into the water, surrounded along their length by plasma membrane. In some heliozoans, the axonemes grow out directly from the endoplasm; in others, each axoneme grows out from its own axoplast located next to the nucleus. In a group called the centrohelidians, all the axonemes arise from a single axoplast, called a centroplast, whose center often contains a clearly defined organelle.

The rowing actinopod illustrated here, *Sticholonche zanclea* Hertwig, has been an enigma for taxonomists. Its peculiar skeleton, the placement of its axopods on the nuclear membrane, and the hexagonal pattern of the axopods in cross section have justified its placement as the only species in an isolated order (Sticholonchidea). That order was originally thought to be radiolarian (as suggested by A. Hollande, M. Cachon, and J. Valentin in 1967), but it is more likely a marine heliozoan (as suggested by M. Cachon recently). *Sticholonche* is found rowing in the Mediterranean with the splendor of a Roman galley: it has microtubular oars and sets of moveable microfibrillar "oar locks." Unfortunately, it cannot grow and reproduce in the laboratory.

Many heliozoans have siliceous or organic surface scales or spines. In a few species, a spherical organic or siliceous cage has been seen, enclosing the entire cell. The cage has bars arranged in a repeating hexagonal pattern through which the axopods penetrate.

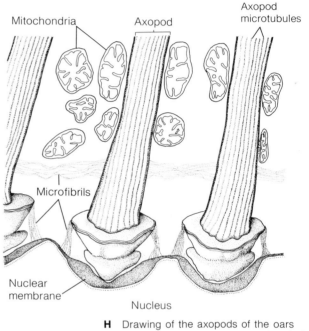

H Drawing of the axopods of the oars (colonettes) of *Sticholonche* showing their relationship to the nucleus (central capsule) and the mitochondria, [Drawing by L. M. Reeves.]

Except in the order Desmothoraca (for example, *Clathrulina elegans*), reproduction by zoospores or swarmer cells is unknown in heliozoans. Reproduction is generally asexual, producing two or several daughter cells by binary or multiple fission or budding. In several multinucleate species, the nuclear and cytoplasmic divisions are not synchronized. In uninucleate forms, the axopods withdraw so that the organism does not move or feed during cell division. Retraction is caused by disassembly of the microtubules in the axonemes.

A kind of autogamy (self-fertilization) has been reported in some heliozoans. A mature cell forms one or more cysts inside the cell. Meiosis apparently occurs in the cysts, and certain nuclei degenerate. Two of the final meiotic products in each cyst then fuse—their haploid nuclei form a new single diploid nucleus. The only surviving product of the two meiotic divisions and the fusion emerges from the cyst as a mature heliozoan. Whether or

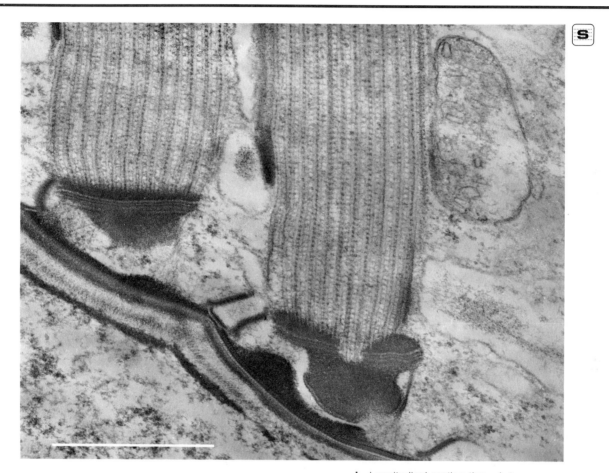

S

I Longitudinal section through two oar-locks and the attached oars of *Sticholonche zanclea*. SEM, bar = 1 μm. [Courtesy of J. Cachon and M. Cachon.]

not this inbred sort of sex is common in heliozoans is simply not known because of the paucity of study. It is possible that, in nature, heterogamy (fusion of nuclei from *different* individuals) occurs more frequently. In *Actinophrys,* two cells (but not their nuclei) may fuse just before they undergo autogamy. Gametes originating from one of the two cells have been seen to fuse with gametes originating from the other. Cell fusion is common in heliozoans, but whether it often entails sexual processes is not known.

Pr-17 Foraminifera

Latin *foramen*, little hole, perforation; *ferre*, to bear

Allogromia
Discorbis
Elphidium
Fusulina
Glabratella
Globigerina
Iridia
Miliola

Nodosaria
Rotaliella
Textularia

A Adult agamont stage of *Globigerina* sp., an Atlantic foraminiferan. SEM, bar = 10 μm. [Courtesy of G. Small.]

Foraminifera, affectionately known as forams, are exclusively marine organisms. The smallest are about 20 μm in diameter, whereas the largest ones may grow to several centimeters in diameter. The vast majority are tiny and live in the sand or attached to algae, stones, or other organisms. However, two families of free-swimming planktonic forams (Globigerinidae and Globorotalidae) are very important in the economy of the sea. Present in large numbers, they are food for many invertebrate animals.

Forams have pore-studded shells (called tests) composed of organic materials, usually reinforced with minerals. Most are made of neatly cemented granules of calcium carbonate deposited from sea water; others are made of sand grains. Some forams, by mechanisms that are still unknown, choose echinoderm plates or sponge spicules to construct their tests. The test and the organism itself may be brilliantly colored—salmon, red, or yellow brown.

The simplest forams have tests of only a single chamber; the tests of most, however, are multichambered. A typical test looks like a clump of blobs, of partial spheres. Openings in the test permit thin cytoplasmic projections, the microtubule-reinforced podia, to emerge. If the podia anastomose (link up to form nets), they are called reticulipodia; straight or branched podia are called filopodia. The podia are used for swimming, gathering materials for tests, and feeding. Forams are omnivorous: they eat algae, ciliates (Phylum Pr-18), actinopods (Phylum Pr-16), and even nematodes (Phylum A-14) and crustacean larvae (Phylum A-27). Many forams, probably most of the ones that live in shallow water, harbor photosynthetic symbionts—dinoflagellates (Phylum Pr-2), chrysophytes (Phylum Pr-4), and diatoms (Phylum Pr-11).

Although some foram genera (for example, *Textularia*) have only been caught reproducing asexually, by budding into multiple offspring, the others that have been well studied, some dozen species, show a remarkable complexity of life cycle. The known cycles are variations on the theme of *Rotaliella*. Meiosis takes place in the agamont, a fully adult diploid organism that produces and releases smaller haploid forms called agametes. These disperse and grow by mitotic cell divisions into a second kind of adult, called gamonts. The gamonts reproduce sexually, their offspring being agamonts.

The alternation of the diploid agamont and haploid gamont generations is obligatory in the forams that have been studied, just as alternation of generations is obligatory in many plants, such as bryophytes (Phylum Pl-1) and ferns (Phylum Pl-4). In fact, forams are the only heterotrophic protoctists that have an alternation of morphologically distinct free-living adult generations. What complicates matters is that, unlike all other organisms except the ciliates (Phylum Pr-18), forams show a striking nuclear dimorphism. The agamonts of *Rotaliella roscoffensis,* for example, contain four diploid nuclei. Three of these nuclei, called the generative nuclei, reside in a chamber separate from that in which a larger nucleus remains. This nucleus, called the somatic nucleus, never undergoes meiosis; it eventually becomes pycnotic (it stains heavily) and disintegrates. The three generative nuclei, on the other hand, give rise to twelve haploid products by meiosis. These products become the nuclei of the small haploid agametes. Later, in the gamonts, pairs of nuclei, apparently of opposite sex, fuse to form diploid zygotes—in effect, each gamont fertilizes itself, but neither egg nor sperm is formed.

Foram shells have contributed greatly to the sediment on the bottom of marine basins, especially since the Triassic Period, 230 million years ago. There are fossil giant forams of great fame. Some, such as *Lepidocyclina elephantina,* had shells as much as 1.5 cm thick. *Camerina laevigata* (also known as

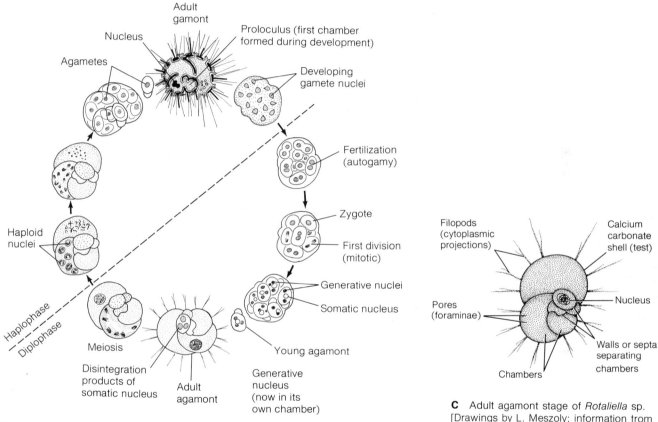

B Life cycle of *Rotaliella roscoffensis*.

C Adult agamont stage of *Rotaliella* sp. [Drawings by L. Meszoly; information from K. Grell.]

Nummulites, the "coin stone") is a large (10 cm wide) foram that lived in warm shallow waters from the Jurassic Period to the Miocene Epoch (some 18 million years ago). Rocks bearing Miocene forams, many of them easily visible to the naked eye, abound on the shores of the Mediterranean. It is from such "nummulitic" limestone that the pyramids of Egypt were constructed.

The abundance of foram tests and their detailed architecture make them excellent stratigraphic markers; geologists use them to identify geographically separate sediment layers of the same age. Because the tests are often found in strata that cover oil deposits, knowledge of their distribution is helpful in petroleum exploration. Unfortunately, although foram life-cycle stages often correspond to extreme changes in shell morphology, paleontologists assume that each morphotype represents a different species. Protozoologists and geologists have different aims and terminologies; thus, the taxonomy of the foraminiferans is in rather a mess. The bewildering complexity of these organisms and their life cycles assures both groups of scientists much taxonomic work for a long time.

Pr-18 Ciliophora

(Ciliates)

Latin *cilium,* eyelash or the lower eyelid;
Greek *phorein,* to bear

Balantidium
Blepharisma
Carchesium
Didinium
Dileptus
Ephelota
Euplotes
Gastrostyla
Halteria
Nyctotherus

Paramecium
Prorodon
Sorogena
Spirostomum
Stentor
Tetrahymena
Tokophyra
Uroleptus
Vorticella

Ciliates are heterotrophic microbes; with few exceptions, they are all unicellular. A characteristic ciliate is covered with cilia, short, moveable whiplike extensions whose attached ends are embedded in a tough, fibrillar outer cortex of the cell. Nearly all ciliates possess two very different types of nuclei, small micronuclei and larger macronuclei. Often there are several of each kind. About eight thousand freshwater and marine species have been described in the literature of biology. There are probably many more than that in nature. Of the known species, nearly all eat bacteria or water rich in dissolved nutrients.

The fine, threadlike cilia are undulipodia, having the typical nine-fold symmetrical array of microtubules. Traditionally, cilia have been distinguished from eukaryotic flagella by being shorter and more numerous. Cilia are often modified to perform specialized locomotory and feeding functions. The most usual modification is the grouping of cilia into bundles (cirri) or sheets (membranelles). In various species, cirri or membranelles function as mouths, paddles, teeth, or feet. In ciliates the undulipodia are embedded in an outer proteinaceous layer about 1 μm thick containing complex fibrous connections between the kinetosomes. This layer of ciliar substructure (infraciliature) taken together comprises the cortex.

Of the two types of nuclei in each ciliate cell, only the micronuclei, which contain apparently normal chromosomes, divide by mitosis. The huge macronuclei, which develop from precursor micronuclei by a series of complex steps, do not contain typical chromosomes. Instead, the DNA is broken into a great number of little chromatin bodies; each body contains hundreds or even thousands of copies of only one or two genes. The macronuclei are absolutely required for growth and reproduction. They divide by elongating and constricting—no coordinated longitudinal divisions of chromosomes occur. They take part in routine cellular functions, such as the production of messenger RNA to direct protein synthesis. The micronuclei are essential for those sexual processes unique to ciliates, but are not required for growth or reproduction.

Ciliates usually reproduce by transverse binary fission, dividing across the short axis of the cell to form two equal offspring. However, certain stalked and sessile species, such as some suctorians, asexually bud off "larval" offspring. These are "born": small rounded offspring, covered with cilia, emerge through "birth pores" of their entirely different-looking, stalked "mother."

Most ciliates can undergo in some form a sexual process called conjugation. The conjugants, two cells of compatible mating types, or "sexes," remain attached to each other for as long as many hours. Each conjugant retains some micronuclei and donates others to its partner. There follows a series of nuclear fusions, divisions, and disintegrations that results in the two conjugants becoming "identical twins," as far as their micronuclei are concerned. The conjugants eventually separate and undergo a complex sequence of maturation steps. Although the micronuclei of the two exconjugants are now genetically identical (each conjugant having contributed equally), each new cell retains the cytoplasm and cortex of only one of the original conjugants. Thus, because cytoplasmic and cortical inheritance in ciliates can be definitively distinguished from nuclear, these organisms are widely used in the analysis of eukaryotic cell genetics.

Three classes of ciliates have been recognized: kinetofragminophorans, oligohymenophorans, and polyhymenophorans. The kinetofragminophorans are covered with few or many generally unmodified cilia. They include the tentacled suctorians and the entodiniomorphs found in the rumen of animals such as cows and sheep.

The oligohymenophorans possess a few specialized ciliary structures, such as cirri or membranelles, in addition to their body cilia. This class includes the most studied genera, *Paramecium* and *Tetrahymena.*

The polyhymenophorans, distinguished by their many complex ciliary structures, include *Gastrostyla* (pictured here) and several other species, such as *Stentor* and *Euplotes,* that are important research organisms. They also include a large order called the tintinnids. These marine ciliates form shell-like structures from sand and organic cements. Their ancestors have left evidence in the fossil record since the Cretaceous Period, some 100,000,000 years ago. Very few other ciliates fossilize.

Because of their various ciliary modifications, their rapid and controllable growth rates, and the ease with which they can be handled in the laboratory, ciliates are valuable for anatomical, genetic, and neurophysiological studies of single cells. Except for the parasite *Balantidium,* which occasionally grows in the human gut, ciliates cause no disease and are of no obvious economic importance.

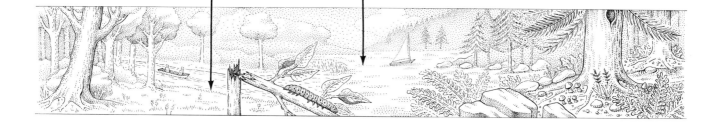

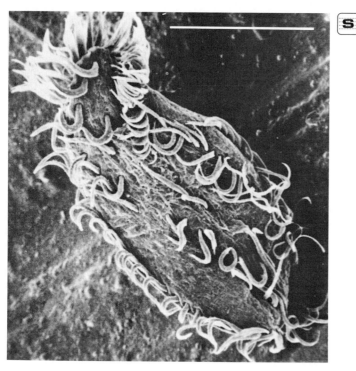

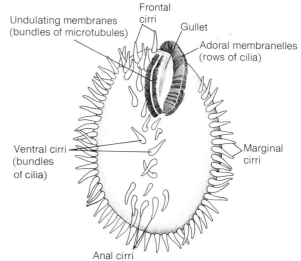

Gastrostyla steinii, a polyhymenophoran ciliate. The adoral membranelles, consisting of rows of cilia, are used in feeding. They sweep particulate food (bacteria and small ciliates) into the gullet. SEM, bar = 100 μm. [Photograph courtesy of J. Grim, *Journal of Protozoology* 19:113–126; 1972. Drawing by R. Golder.]

Undulating membranes (bundles of microtubules)

Frontal cirri

Gullet

Adoral membranelles (rows of cilia)

Ventral cirri (bundles of cilia)

Marginal cirri

Anal cirri

Pr-19   Apicomplexa

(Sporozoa or Telosporidea)

Latin *apex,* summit; *complexus,* an embrace, enfolding

Babesia
Coccidia
Coelospora
Eimeria
Gregarina
Haemoproteus
Haplosporidium
Isospora
Plasmodium

Schizocystis
Selenidium
Toxoplasma

The Apicomplexa are heterotrophic microbes. Like the cnidosporidians (Phylum Pr-20), all are spore-forming parasites of animals. The spores, in this case, are not heat- and desiccation-resistant cells like bacterial spores, but small infective bodies that permit dissemination and transmission of the species from host to host.

The phylum is named for the "apical complex," a distinctive arrangement of fibrils, microtubules, vacuoles, and other cell organelles at one end of each cell. Thus, the apicomplexans form a natural group that is probably monophyletic. There are two classes:

Class Sporozoasida
  Subclass Gregarinasina: gregarines (*Gregarina*)
  Subclass Coccidiasina: coccidians (*Eimeria, Isospora*)
    and hemosporidians (*Plasmodium, Haemoproteus*)

Class Piroplasmasida (*Babesia*)

Apicomplexans reproduce sexually, with alternation of haploid and diploid generations. Both diploids and haploids can also undergo schizogony, a series of rapid mitoses not alternating with cell growth. Schizogony produces small infective spores.

In fertilization, a small, undulipodiated male gamete (the microgamete) having a highly structured apical complex fertilizes a large female gamete (the macrogamete). Zygote formation is followed by the formation of a thick-walled oocyst. The oocyst, rather than the infective spores, is the desiccation-, heat-, and radiation-resistant stage. The oocysts serve to transmit the microbes to new hosts. These cysts develop further by sporogony—rapid meiotic divisions inside the cyst produce infective haploid spores called sporozoites.

The life cycles of apicomplexans may be very complex, involving several very different species of hosts—invertebrates and vertebrates. Many are bloodstream parasites. Many cause hypertrophy of the host cells in which they divide, causing a striking increase in the amount of host DNA, probably by polyploidization. That is, the spore infection leads to duplication of host chromatin so that the amount of host genetic material per chromosome is increased.

The coccidians are perhaps the best-known group of apicomplexans because many of them cause serious and even fatal diseases of their vertebrate and invertebrate hosts. *Isospora hominis* is the only coccidian that parasitizes man, but many others affect livestock and fowl. Because these parasites are generally acquired by being eaten and thus find their way into the digestive tract, the major symptoms of coccidian disease are diarrhea and dysentery.

An *Eimeria* infestation begins when an oocyst is eaten by a host animal. The oocyst germinates and produces sporozoites which escape from the oocyst and enter the host's epithelial cells, typically those of the gut lining. There they multiply by mitosis and escape to infect more host cells. Within the host cells, they develop into various forms called trophozoites, merozoites, and schizonts; this cycle, called the schizogony cycle, can be repeated many times. Eventually, some of the merozoites develop into microgametes, and others into macrogametes. Fertilization takes place inside host cells. The zygotes undergo meiosis and develop into oocysts. They typically pass out of the host's body with the feces. They survive in the soil until they are eaten by another host.

Some apicomplexans have played an important role in human history. The most famous are the malaria parasites, *Plasmodium* species, which are transmitted to human beings by the female *Anopheles* mosquito. Fertilization of *Plasmodium* takes place in the gut of the mosquito. The zygote, which is undulipodiated and motile, embeds itself in the gut wall, where it is transformed into a resistant oocyst. Within the oocyst, small infective cells are formed by meiosis and multiple fission, or sporogony. The sporozoites migrate to the salivary glands of the mosquito. With the bite of the mosquito, the sporozoites are injected into a human bloodstream, where they grow at the expense of the host—their diet requires iron, which they obtain from human hemoglobin. Inside a blood cell, a sporozoite develops into another form, called a trophozoite. Eventually, the trophozoite undergoes schizogony—it divides rapidly by mitosis to produce merozoites, small infective cells that escape from the ruined blood cell into the bloodstream. The merozoites attack and penetrate more blood cells, develop into trophozoites, and divide into more merozoites. After several such cycles, the merozoites differentiate into male and female gametes, which must be taken up by an *Anopheles* mosquito in order to complete the *Plasmodium* life cycle by fertilization. All the merozoites in a human host are produced and released more or less simultaneously—the pulse of formation and release of successive generations causes the characteristic periodic attacks of malarial fever.

There is a baffling variety of sexual life cycles and attack strategies in this group of advanced protoctists; an enormous and contradictory terminology has made their study an arcane delight for specialists—especially for those interested in veterinary medicine.

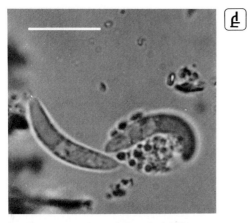

A Free sporozoites of *Eimeria falciformes.* LM, bar = 10 μm. [Courtesy of D. W. Duszynski.]

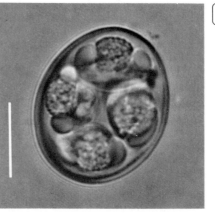

C Four sporocysts of *Eimeria nieschulzi* in sporulated oocyst. LM, bar = 10 μm. [Courtesy of D. W. Duszynski.]

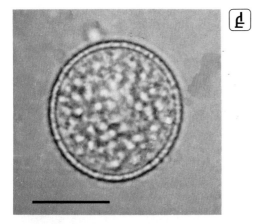

B Unsporulated oocyst of *Eimeria falciformes.* LM, bar = 10 μm. [Courtesy of T. Joseph (1974).]

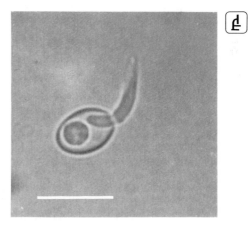

D Sporozoite of *Eimeria indianensis* excysting from oocyst. LM, bar = 10 μm. [Courtesy of T. Joseph (1974).]

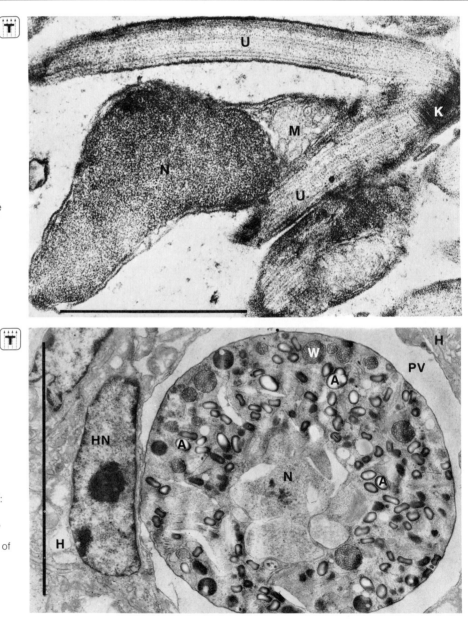

E Microgamete ("sperm") of *Eimeria labbeana,* an intracellular parasite of pigeons. N = nucleus; M = mitochondria; U = undulipodium; K = kinetosome. The structures above the nucleus are part of the apical complex. TEM, bar = 1 μm. [Courtesy of T. Varghese.]

F Macrogamete ("egg") of *Eimeria labbeana.* H = host cell; HN = host nucleus; PV = parasite vacuole in host cell: N = macrogamete nucleus; A = amylopectin granule; W = wall-forming bodies, which later coalesce to form the wall of the oocyst. TEM, bar = 5 μm. [Courtesy of T. Varghese.]

G The life cycle of *Eimeria* sp. The shaded part of the diagram represents the schizogony cycle, which may repeat itself many times before some of the merozoites differentiate into gametes. [Drawing by L. Meszoly.]

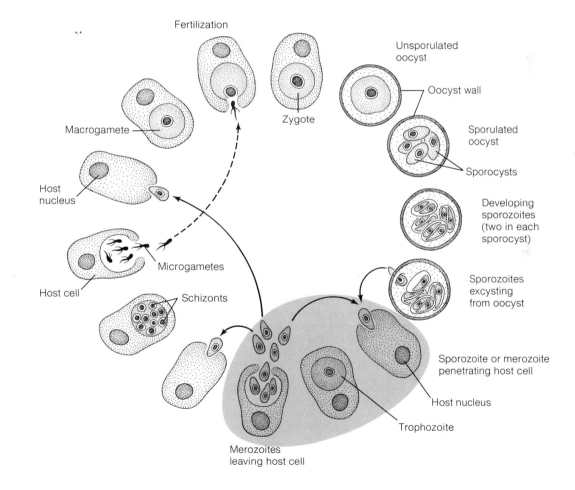

Fertilization

Unsporulated oocyst

Oocyst wall

Sporulated oocyst

Macrogamete

Zygote

Sporocysts

Developing sporozoites (two in each sporocyst)

Host nucleus

Sporozoites excysting from oocyst

Microgametes

Host cell

Schizonts

Sporozoite or merozoite penetrating host cell

Host nucleus

Trophozoite

Merozoites leaving host cell

Pr-20 Cnidosporidia

Greek *knide*, filament, nettle; Latin *spora*, spore

Ceratomyxa
Coccomyxa
Encephalitazoan
Glugea
Guyenotia
Helicosporidium
Ichthyosporidium
Myxidium
Myxobolus

Myxostoma
Nosema
Sphaeractinomyxon
Telomyxa
Triactinomyxon
Unicapsula

Cnidosporidians are heterotrophic microbes, all parasitic on animals. They all produce some sort of polar filament or thread. In the older literature, the cnidosporidians were classified with the organisms in our Phylum Apicomplexa (Pr-19)—all were called sporozoan parasites. However, they differ from apicomplexans in many ways—in particular, all cnidosporidians lack the apical complex. Apicomplexans have intricate sex lives, but whether cnidosporidians engage in any sex at all is still not known. Cnidosporidian spores are not infective; they are a resting stage in the life cycle and a modification for safe dissemination.

The phylum is a mixed bag. There are several distinct groups that are probably not related to each other except superficially. The polar thread itself differs completely in function and development among the classes of the phylum. The three classes listed here, Microsporida, Myxosporida, and Actinomyxida, it could be argued, ought to be phyla on their own.

Microsporidians are intracellular parasites, often of vertebrate hosts. They have a great reproductive capacity. A thick-walled chitinous spore contains the conspicuous polar filament and an infective body called the sporoplasm. When penetrating the host, the sporoplasm everts from the spore; it is squeezed through a narrow hollow tube (derived from the polar filament) and forced into the host cell. Thus, the microsporidians have independently evolved the injection needle.

Glugea stephani, illustrated here, is a microsporidian parasite of the starry flounder. Many microsporidians form large single-cell tumors in the tissue of their hosts. Many are extremely well integrated into their hosts, and do not cause them any harm at all. Others are severe pathogens; *Glugea*, for example, forms single-cell tumors on fish. *Ichthyosporidium* also grows on fish. Members of the genus *Encephalitazoan* are parasites of warm-blooded vertebrates. *Nosema* has caused devastating damage to the silk industry, as members of the genus are agents of pebrine, a disease of silkworm larvae.

Some microspiridians may be sexual, but the sexual life cycles that have been reported are not well documented. They lack mitochondria at all stages, and some, at least, have small ribosomes rather like those of prokaryotes. Microsporidians reproduce asexually inside the host cell by single or multiple fission. Their spores are multicellular, and 10–15 μm long. They may contain one or two nuclei per cell. The fish parasite *Glugea* develops a multinucleate plasmodium. The nuclei of the plasmodium fuse—without having formed undulipodiated gametes. The resulting diploid zygotes, called sporonts, undergo meiosis, the eventual products of which are filamented spores.

Myxosporidians are parasites of invertebrates and fishes, in which they can cause serious diseases. The polar filaments of their spores are enclosed in polar capsules, or cnidocysts; however, in this group the polar filaments are not used for injecting but for anchoring the spores to the host tissue. Myxosporidian spores are multinucleate. They have walls composed of two or more valves connected by sutures; the infective sporoplasm emerges from between the valves.

Myxosporidians penetrate the skin of nearly any part of a host's body and make their way to the intestine. There they release uninucleate amoeboid forms called amoebulinas. These penetrate the host's tissues and presumably are carried by the blood to target organs, where they often form large unsightly growths as much as several centimeters in diameter. They seem to favor hollow organs, such as urinary and gall bladders, but they also infest gills, muscles, intestines, liver, brain, bone, and skin. *Myxobolus pfeifferi* causes boil disease of the European barbel. The twist disease of salmon is caused by *Myxostoma cerebralis*, which lives in cartilage and forms tumors that exert pressure on the central nervous system.

The striking resemblance between the cnidocysts of myxosporidians and the nematocysts of coelenterates (Phylum A-3, Cnidaria) in morphology, developmental pattern, and function suggests that the myxosporidians may have evolved from coelenterate animals. However, their lack of a blastula and tissues prompts us to retain the myxosproridians in the protoctist kingdom.

The actinomyxids are parasites more like myxosporidians than like microsporidians, but they too are so distinctive that some favor raising them to phylum status. They are the most poorly known of these three classes. They are parasites of invertebrates, in particular, of annelids (Phylum A-23). Their spores are cells, called cnidocysts, that produce the filaments. *Sphaeractinomyxon gigas* is 40 μm long and lives in the coelom of the oligochete worm *Limnodrilus*. *Triactinomyxon* lives in the gut of *Tubifex*, also an oligochete. In both actinomyxid genera, the spore valves are elongated into large hooks for attachment to the host.

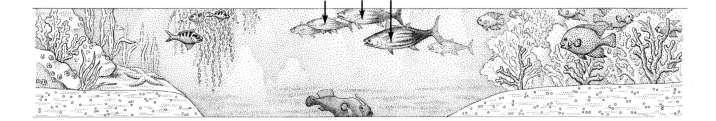

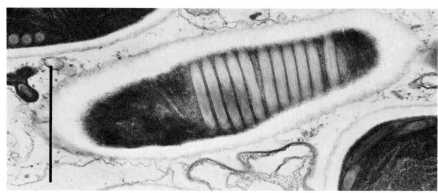

Glugea stephani, a parasite in the starry flounder, *Platiclothus stellatus.* Ultrastructure of a mature spore. TEM, bar = 1 μm. [Photograph courtesy of H. M. Jensen and S. R. Wellings, *Journal of Protozoology* 19:297–305; 1972. Drawings by R. Golder.]

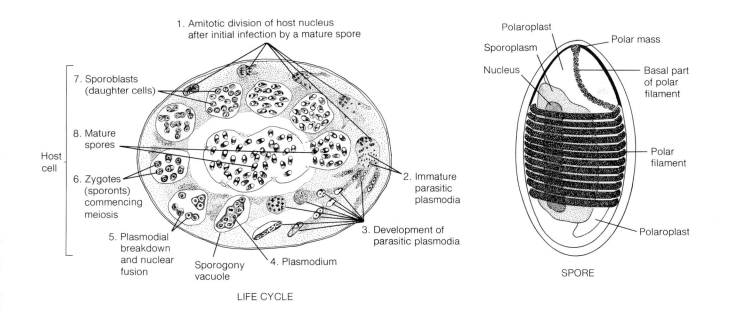

1. Amitotic division of host nucleus after initial infection by a mature spore

7. Sporoblasts (daughter cells)

8. Mature spores

6. Zygotes (sporonts) commencing meiosis

Host cell

5. Plasmodial breakdown and nuclear fusion

Sporogony vacuole

4. Plasmodium

3. Development of parasitic plasmodia

2. Immature parasitic plasmodia

LIFE CYCLE

Polaroplast

Sporoplasm

Nucleus

Polar mass

Basal part of polar filament

Polar filament

Polaroplast

SPORE

Pr-21   Labyrinthulamycota
(Slime nets)

Chlamydomyxa
Labyrinthorhiza
Labyrinthula
Pseudoplasmodium

Latin *labyrinthulum,* little labyrinth; Greek *mykes,* fungus

The slime nets, or labyrinthulids, are colonies of individual cells that move and grow entirely within a slime track of their own making. Only two genera have been studied, the marine forms *Labyrinthula* and *Labyrinthorhiza,* although freshwater and soil forms are known. In some texts, they have been called slime net amoebas, but there is nothing amoeboid about them.

The labyrinthulids form transparent colonies—they may be centimeters long. With the unaided eye they look like a slimy

A   Live cells of *Labyrinthula* sp. traveling in their slime net. LM, bar = 100 μm. [Courtesy of D. Porter.]

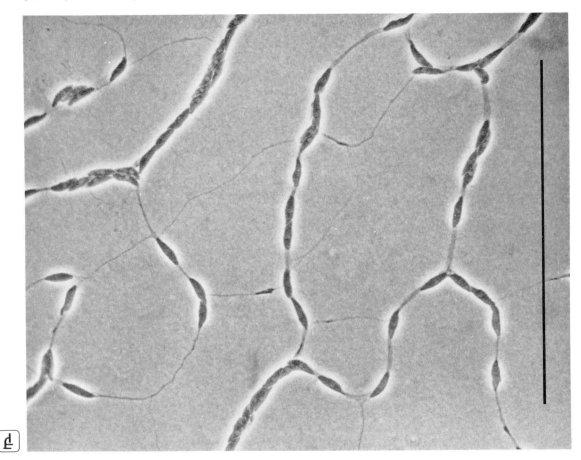

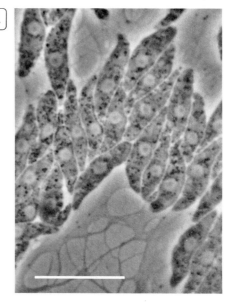

B Edge of a *Labyrinthula* colony on an agar plate. Bar = 1 mm. [Courtesy of D. Porter.]

C Live *Labyrinthula* cells in their slime net. LM, bar = 10 μm. [Courtesy of D. Porter.]

mass on marine grass. Under the microscope, spindle-shaped cells can be seen migrating back and forth in tunnels, like little cars in tracks. Labyrinthulid cells cannot move at all unless they are completely enclosed in the slime track. This track, probably mucopolysaccharides and actinlike proteins, is laid down in front of the cells. The mechanism of movement is unknown: it certainly does not involve undulipodia or pseudopods. Force seems to be generated between the external slime layer and the surface of the moving cells, which travel inside the slime at quite a clip, several micrometers per second.

The movements of trains of cells together back and forth in the track may look as if it is random. However, when a potential food source is sensed, such as a yeast colony, the cells of the slime net move toward the source. Extracellular digestive enzymes are produced and small food molecules diffuse through the slime to nourish the colony. Particulates are not ingested. The labyrin-

thulid colony changes shape and size as it goes about its business. Growth is by the mitotic division of cells inside the slime track, followed by separation of the offspring cells, all presumably capable of excreting slime.

When conditions become drier, labyrinthulids adopt a desiccation strategy. In the older parts of the colony, cells aggregate and form hardened, dark cystlike structures surrounded by tough membrane. These have no precise size or shape but are somewhat spherical. The cysts are tolerant; they wait until moisture and food become more available, then they rupture, liberating small spherical cells that grow again into spindle-shaped cells in a matrix.

Recently, undulipodiated stages of *Labyrinthula marina* have been found. They are the products of meiosis. These are isogametes (showing no difference between male and female gametes), which fuse to form a zygote. The zygote undergoes mitosis; its offspring apparently develop into the multicellular net, indicating that the labyrinthulid cells in the nets are probably diploid.

Labyrinthula is best known because it grows on the eel grass *Zostera marina* (Phylum Pl-9, Angiospermophyta) and on many algae, such as *Ulva* (Phylum Pr-15, Chlorophyta). *Zostera* is very important to the clam and oyster industries along Atlantic shores because it is the primary producer of ecosystems that include the clam and oyster beds. Sometimes, blooms of *Labyrinthula* cause a lethal disease of *Zostera* and thus have a drastic effect on the shellfish industry.

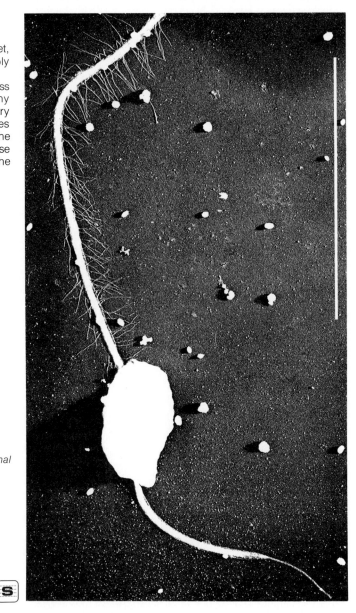

D   Zoospore of *Labyrinthula* sp. SEM, bar = 10 μm. [Courtesy of F. O. Perkins, from J. P. Amon and F. O. Perkins, *Journal of Protozoology* 15:543–546, 1968.]

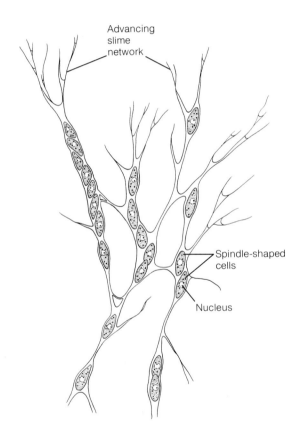

E *Labyrinthula* cells in a slime net.
[Drawing by R. Golder.]

Advancing
slime
network

Spindle-shaped
cells

Nucleus

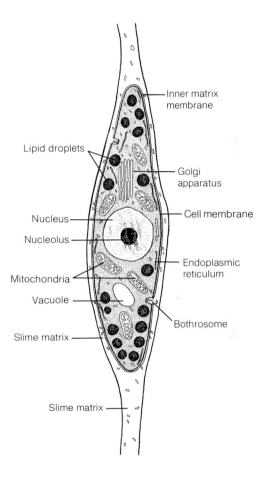

F Structure of a single *Labyrinthula* cell.
[Drawing by R. Golder.]

Inner matrix
membrane

Lipid droplets

Golgi
apparatus

Nucleus

Cell membrane

Nucleolus

Endoplasmic
reticulum

Mitochondria

Vacuole

Bothrosome

Slime matrix

Slime matrix

Pr-22 Acrasiomycota

(Cellular slime molds)

Greek *akrasia,* bad mixture; *mykes,* fungus

Acrasia
Acytostelium
Coenonia
Dictyostelium
Guttulina
Guttulinopsis
Pocheina
Polysphondylium

The cellular slime molds are heterotrophic protoctists found in fresh water, in damp soil, and on rotting vegetation, especially on fallen logs. In the course of their life cycle, independently feeding and dividing amoebas aggregate into a slimy mass or slug that eventually transforms itself into a spore-forming fruiting body; the scattered spores germinate into amoebas.

Because the slime molds have features that are animal (they move, they ingest whole food by phagocytosis, and they metamorphose), plant (they form spores on upright fruiting bodies), and fungal (their spores have tough cell walls and they germinate into colorless cells having absorptive nutrition—they live on dung and decaying plant material), the taxonomy of the group has always been contested. The zoologists have called them mycetozoa and classified them with the protozoa, and the mycologists have called them myxomycetes. In some classifications, four of our phyla—Pr-21 (Labyrinthulamycota), Pr-22 (Acrasiomycota), Pr-23 (Myxomycota), Pr-24 (Plasmodiophoromycota)—have been classified together as Gymnomycota or Gymnomyxa (naked fungi). As early as 1868, Ernst Haeckel considered them neither plants nor animals, but primitive forms that had not yet evolved into members of either of the two great kingdoms. Haeckel erected his new Kingdom Protoctista to accommodate such unruly organisms.

The phylum Acrasiomycota contains two classes, Acrasea and Dictyostelia. The members of both classes pass through a unicellular stage of amoeboid cells that feed on bacteria; they later form a multicellular fruiting structure that produces spores. Sexuality is rare or absent. In passing from the first stage to the second, the amoeboid cells aggregate to form a pseudoplasmodium. A true plasmodium, or syncytium, is a mass of protoplasm containing many nuclei formed by mitotic divisions but not separated by cell membranes. The acrasiomycote structure is called a pseudoplasmodium because its constituent cells retain their cell membranes. It only superficially resembles the plasmodium of the true plasmodial slime molds (Phylum Pr-23). Most acrasiomycotes will begin to aggregate if food is depleted and light is present. However, the exposure to light must be followed by a minimum period of darkness before development can continue.

The two classes differ in many ways and may not be directly related. In Acrasea, the stalk consists of live cells that are capable of germination and lack cellulosic walls; the dictyostelid stalk consists of a tube of cellulosic walls of dead cells. Dictyostelid amoeboid cells are aggregated by their attraction to cyclic adenosine monophosphate (cAMP); the acrasids do not respond to cAMP.

The feeding stage of the acrasids consists of amoeboid cells

A The development of a fruiting body from a slug of *Dictyostelium discoideum.* bar = 1 mm. [Courtesy of J. T. Bonner (© 1967), reprinted by permission of Princeton University Press.]

having broad rounded pseudopods. The families of this class are distinguished primarily by the structure of the sorocarp, or fruiting body. In some, the spore cells are different from the stalk cells; in others, the cells are all alike. Except for the newly discovered genus *Pocheina,* no acrasids have undulipodia. Still poorly known, pocheinas have been reported from the Soviet Union and North America; they live on conifer bark and lichenized dead wood. Their spores are borne in a swelling called a sorus either at the top of the stalk or just below the top. When the spores germinate, they divide into motile cells, each bearing two undulipodia of equal length.

The amoeboid cells of dictyostelids are usually uninucleate and haploid. However, cells having more than one nucleus and aneuploids, cells having an uneven number of chromosomes,

B Life cycle of the cellular slime mold *Dictyostelium discoideum.* [Drawing by R. Golder.]

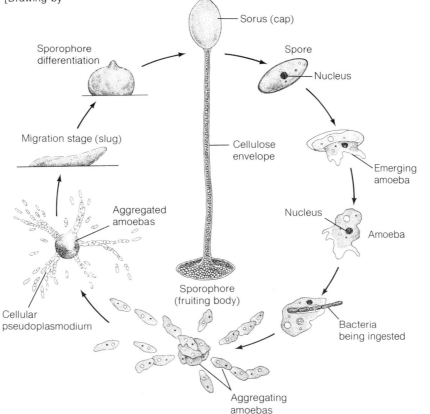

Sorus (cap)

Spore

Nucleus

Sporophore differentiation

Emerging amoeba

Cellulose envelope

Nucleus

Amoeba

Migration stage (slug)

Aggregated amoebas

Sporophore (fruiting body)

Cellular pseudoplasmodium

Bacteria being ingested

Aggregating amoebas

have been reported. Some strains consist of stable diploid cells. The amoebas have thin pseudopods and feed mainly on live bacteria. After the food supply is exhausted and the amoebal population has reached a certain density, the cells cease feeding and dividing. Because of a chemical attractant called acrasin (now known to be cAMP) secreted by the amoebas themselves, they begin to aggregate, streaming in toward aggregation centers. A thin slime sheath is produced around the mass of cells, the pseudoplasmodium, which takes on the appearance of a slug. The slug begins to wander, leaving behind a slime track derived from the sheath. As conditions become drier, the migration stops and

the differentiation of the fruiting structure begins. In a complicated developmental sequence involving differentiation, but not cell division, the fruiting structure, or sorocarp, forms with its cellulosic stalk.

Dictyostelids are far better known than acrasids. Four genera have been described: *Acytostelium, Dictyostelium, Polysphondylium,* and *Coenonia.* Each has a number of species, at least sixteen in *Dictyostelium. D. discoideum* is a favorite organism for the laboratory investigation of the mechanism of differentiation because it grows rapidly and its life cycle is easily manipulated.

Pr-23 Myxomycota

(Myxogastria, plasmodial slime molds)

Greek *myxa*, mucus; *mykes*, fungus

Arcyria	*Didymium*	*Metatrichia*
Barbeyella	*Echinostelium*	*Perichaena*
Ceratiomyxa	*Fuligo*	*Physarella*
Clastoderma	*Hemitrichia*	*Physarum*
Comatricha	*Leocarpus*	*Protostelium*
Dictydium	*Licea*	*Sappinia*
Diderma	*Lycogala*	*Stemonitis*

The myxomycotes have had many names: myxomycetes, mycetozoa, plasmodial slime molds, true slime molds, Myxomycotina, and others. Like the acrasiomycotes (Phylum Pr-22), they pass through an amoeboid stage that lacks cell walls and feeds by phagocytosis, they form fruiting structures that have stalks, and they are funguslike in appearance. However, the myxomycotes are sexually far more advanced than the acrasiomycotes. Like plants and foraminiferans, they show alternation of haploid and diploid generations.

Haploid cells bearing two anterior undulipodia of unequal lengths are formed; these cells readily convert into amoeboid cells called myxamoebas, and the myxamoebas reconvert just as readily into undulipodiated cells. Both the undulipodiated cells and the myxamoebas can differentiate into opposite mating types. Either two amoebas or two undulipodiated cells fuse to form a zygote, which divides repeatedly by mitosis to form a large mass of protoplasm, the plasmodium. Unlike the comparable structure of acrasiomycotes, the plasmodium of myxomycotes is not cellular—the nuclei are not separated by cell membranes.

Plasmodia are found as a slimy wet scum on fallen logs, bark, and many other substrates. Often the plasmodia are pigmented, usually orange or yellow, but none photosynthesize. They feed by engulfing decaying vegetation. (Undulipodiated cells and myxamoebas both feed on bacteria and absorb dissolved nutrients.) The size and shape of these slime molds is in no way predetermined; bits taken from a plasmodium can feed and grow independently. The plasmodium as a whole moves only by differential growth. However, it also shows a sort of internal shuttling movement, a pulsating back and forth of protoplasm that is very obvious, under the microscope, as an incessant intraplasmodial flow. The movement is thought to maintain an even distribution of metabolites and oxygen.

If its surroundings become drier, the plasmodial protoplasm may become concentrated into mounds, from which stalked sporangia, or sporocarps, grow. Meiosis takes place inside the maturing spores. At least in some cases, three of the meiotic products degenerate and the fourth develops into a mature spore, which can germinate into either an amoeba or an undulipodiated cell. Thus, most of the life cycle of these organisms is spent in the diploid stage. However, the development of amoebas and undulipodiates into plasmodia directly, without a fertilization step, is also known; thus, the ploidy of the slime-mold mass must be ascertained in each case.

Some four or five hundred species of myxomycotes are known. The color, shape, and size of the fruiting structure, the presence of a stalk, the presence of a sterile structure (the columella) at the top of the stalk, the presence of calcium carbonate ($CaCO_3$) granules or crystals in or on the fruiting body, and the structure of the spores are all of taxonomic importance in distinguishing the species. Furthermore, inside the young sporangium there develops a system of threads, which are sterile in that they do not give rise to spores. This thread system, called the capillitium, differs among various myxomycote groups and is used as a taxonomic marker. Five major subgroups are known, which have been listed as orders by other writers. In our classification, they become classes: Echinostelida, Trichida, Liceida, Stemontida, and Physarida.

The spores of members of the first three classes are pale, as a rule, and do not deposit $CaCO_3$. Echinostelid sporocarps are tiny (less than 1 mm high) and contain capillitia. The diploid feeding stage is a protoplasmodium—a microscopic amoebalike protoplast that lacks veins or reticulations. The echinostelids comprise two families: Echinosteliidae, having only the genus *Echinostelium,* and Clastodermidae, three tiny species belonging to the genera *Clastoderma* and *Barbeyella.*

The sporocarps of the trichids contain sculptured capillitia. The diploid feeding plasmodium is midway between an aphanoplasmodium (a thin, inconspicuous plasmodium consisting of a reticulum, or network, of veins having fan-shaped leading fronts) and a phaneroplasmodium (a thick, more conspicuous structure whose veins and leading fronts are visible to the unaided eye.) The genera include *Perichaena, Arcyria,* and *Metatrichia.*

The liceids may be either proto- or phaneroplasmodial. Their sporocarps are of diverse shapes and lack capillitia. Genera include *Licea* (with 19 species), *Lycogala,* and *Dictydium.*

The stemontids, which typically form aphanoplasmodia, and the physarids, which typically form phaneroplasmodia, both bear spores that are dark purplish to brown or black. These are the largest and best known slime molds. Stemontids, which include *Stemonitis* and *Comatricha,* form dark fingerlike upright sporocarps. The physarids often have conspicuous $CaCO_3$ deposits in the capillitium and other parts of the sporocarp. Some 85 species of the genus *Physarum* are known. Other genera are *Leocarpus, Fuligo, Didymium,* and *Diderma.*

Their very active protoplasmic streaming and the ease with which some myxomycotes can be grown in the laboratory have made them useful in the study of motile proteins. In particular, the plasmodia of *Physarum polycephalum* have yielded actinlike and myosinlike proteins somewhat similar to the actinomyosin complexes of vertebrate muscles.

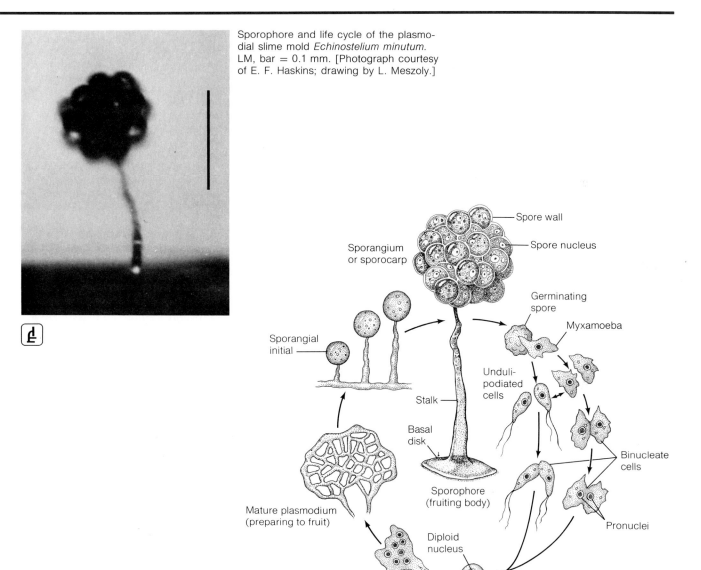

Sporophore and life cycle of the plasmodial slime mold *Echinostelium minutum*. LM, bar = 0.1 mm. [Photograph courtesy of E. F. Haskins; drawing by L. Meszoly.]

Pr-24 Plasmodiophoromycota

New Latin *plasmodium*, multinucleate mass of protoplasm not divided into cells; Greek *pherein*, to bear; *mykes*, fungus

The plasmodiophorans are all heterotrophic microbes and obligate parasites. Most species live inside plants. Because the feeding, or trophic, stage of plasmodiophorans is a multinucleate plasmodium that lacks cell walls, and because zoospores with two anterior undulipodia are formed, these organisms have sometimes been aligned with the acellular slime molds (Phylum Pr-23, Myxomycota). However, the two groups are quite different in life cycle and cell structure.

Plasmodiophoran cells form a mitotic spindle and from a genetic point of view they certainly have true mitosis. In some respects, however, their way of dividing is peculiar. The nuclear membrane persists during the mitotic stages. It constricts to form the two offspring nuclei, and during the constriction the nucleolus elongates and divides in half. At metaphase, the chromosomes form a folded ring inside the nucleus; this gives the nucleus in light-microscopic preparations a characteristic cruciform appearance. When duplicate chromosome sets separate, the two resultant rings of chromosomes pass to opposite ends of the nucleus along with the offspring nucleoli.

Although the details of the life cycle of most species are not entirely understood, it is thought that they can reproduce sexually. The life cycle of *Plasmodiophora brassicae* is the best known. A uninucleate cyst of this species that lands on an appropriate plant germinates into a primary zoospore bearing two smooth undulipodia of unequal lengths. When the zoospore swims, the short undulipodium is directed forward and the long tapered one is directed backward. This zoospore infects the host organism. Inside a host cell, each zoospore develops by nuclear divisions into a small primary plasmodium that forms a wall. This plasmodium cleaves into uninucleate segments called secondary sporangia, and each nucleus then divides by mitosis to produce at least four secondary zoospores resembling the primary ones. The secondary zoospores exit the plasmodium through an opening in its wall or through an exit tube formed by the plasmodium.

The secondary zoospores escape into the soil. It is thought that there they become gametes and fuse by pairs. In any case, after a time in the soil they reinfect the plant roots. Inside the infected plant cells, they develop by nuclear divisions into secondary plasmodia, which feed and grow. As a secondary plasmodium matures, its nuclei become less distinct, probably owing to changes in the function and morphology of the nucleoli. Following what have been purported to be two meiotic divisions of its nuclei, the plasmodium cleaves out uninucleate cysts; in each cyst, a spore forms. The cyst walls apparently lack cellulose, but the cysts can remain viable in soil for years before they land on an appropriate plant and begin the cycle again.

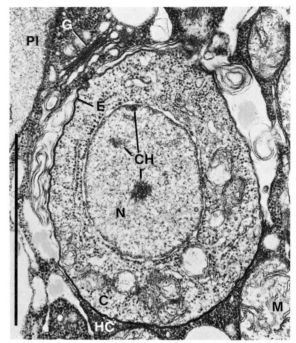

A Uninucleate stage (secondary sporangium) of *Plasmodiophora brassicae* in a cell of a cabbage root hair. Pl = plasmodium; HC = host cytoplasm; G = Golgi apparatus (of host); M = mitochondrion (of host); E = envelope of plasmodiophoran cell; C = cytoplasm of plasmodiophoran cell; N = nucleus of plasmodiophoran cell; CH = chromatin. TEM, bar = 0.5 μm. [Courtesy of P. H. Williams.]

Plasmodiophorans feed primarily by absorbing dissolved nutrients, but zoospores have also been seen to engulf particles. *Plasmodiophora brassicae* and *Spongospora subterranea* are of economic importance because they cause serious diseases of food plants: *P. brassicae* causes club root disease of cabbage and other cruciferous plants; *S. subterranea* causes the powdery scab of potatoes. However, most of the other plasmodiophorans

live as benign parasites or symbionts and do no obvious harm to their hosts.

About 35 species in nine genera of plasmodiophorans are known. The genera are distinguished by differences in such characters as the arrangement of the cysts in the host cells, the presence and morphology of cytosori (structures into which the cysts may be united), and the development of the plasmodia, their exit tubes, and papillae (raised bumps on the exit tubes).

C Life cycle of *Plasmodiophora brassicae* [Drawing by L. Meszoly.]

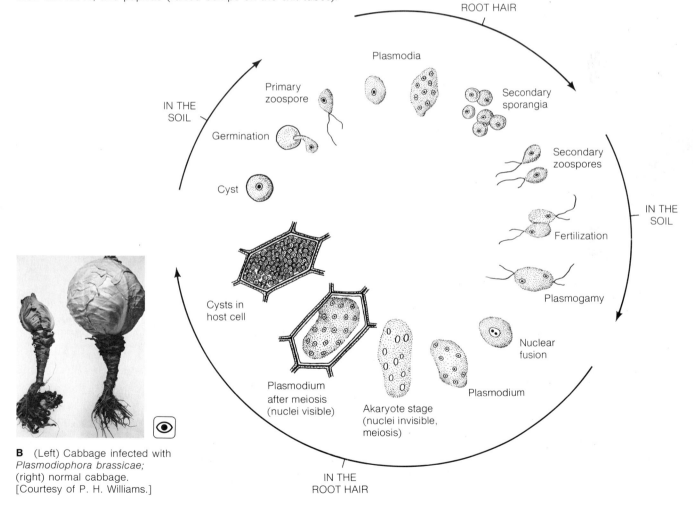

IN THE
ROOT HAIR

Plasmodia

IN THE
SOIL

Primary
zoospore

Secondary
sporangia

Germination

Secondary
zoospores

Cyst

IN THE
SOIL

Fertilization

Plasmogamy

Cysts in
host cell

Nuclear
fusion

Plasmodium
after meiosis
(nuclei visible)

Akaryote stage
(nuclei invisible,
meiosis)

Plasmodium

IN THE
ROOT HAIR

B (Left) Cabbage infected with *Plasmodiophora brassicae;* (right) normal cabbage. [Courtesy of P. H. Williams.]

129

Pr-25 Hyphochytridiomycota

Greek *hyphos*, web; *chytra*, little earthen cooking pot; *mykes*, fungus

Anisolpidium
Canteriomyces
Hyphochytrium
Latrostium
Rhizidiomyces

Hyphochytrids, chytrids (Phylum Pr-26), and oomycotes (Phylum Pr-27) have traditionally been considered fungi. These aquatic microbes do resemble fungi in their mode of nutrition, which is parasitic or saprobic: thin threads or filaments invade host tissue or dead organic debris, where they release digestive enzymes and absorb the resulting nutrients. However, they differ from fungi in that they produce undulipodiated cells.

Hyphochytrids live in fresh water; they either are parasites on algae and fungi or live on insect carcasses and plant debris. The body, or thallus, of a hyphochytrid may be holocarpic (the entire thallus converts into a reproductive structure) or eucarpic (a part of the thallus develops into a reproductive structure while the remaining part continues its somatic function). In the holocarpic species, the thallus is formed inside the tissues of the host. In holocarpic species, the thallus consists of only a single reproductive organ bearing branched rhizoids (rootlike tubes that penetrate the substrate) or hyphae (feeding tubes that grow out of the substrate), in which cross walls, or septa, may or may not develop. In eucarpic species, the thallus may be on the surface rather than inside the host tissues.

In any case, from the reproductive organ, or zoosporangium, zoospores emerge through discharge tubes. Hyphochytrid zoospores are very active swimmers. Each bears one mastigoneme ("hairy") anterior undulipodium, which can move in helical waves as well as in whiplash fashion. The zoospores swim to new hosts or food sources. Each zoospore can develop into a thallus again. All reproduction is asexual, by zoospores; no reports either of sexuality or of resistant spores have been confirmed.

Hyphochytrids have been isolated primarily from soil and tropical fresh waters. However, they are probably distributed all over the world wherever their hosts are found. The fifteen known species have been assigned to six or seven genera, grouped here into three classes: Anisolpidia, Rhizidiomycetae, and Hyphochytria.

The best known hyphochytrid, *Rhizidiomyces apophysatus,* is parasitic on water molds, such as *Saprolegnia* (Phylum Pr-27, Oomycota). After *Rhizidiomyces* zoospores come to rest on their hosts, they round up. They do not divide but withdraw their undulipodia and germinate, extending a germ tube into the host. With continued growth, the tube ramifies into a branching system of rhizoids in the host tissue. Between the rhizoids and the surface of the host, a swelling develops into a baglike structure called the apophysis of the sporangium. The sporangium itself grows at the outer end of the apophysis; it enlarges and forms an exit papilla, a raised bump with a hole in it, which becomes a discharge tube. Nuclear divisions take place in the sporangium,

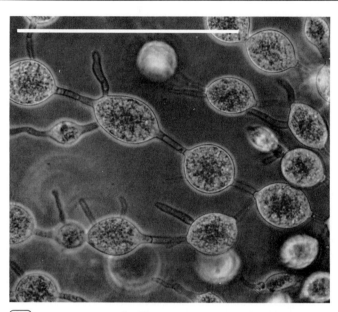

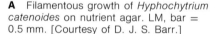

A Filamentous growth of *Hyphochytrium catenoides* on nutrient agar. LM, bar = 0.5 mm. [Courtesy of D. J. S. Barr.]

and a plasmodial mass is formed. This mass of multinucleate protoplasm passes through the discharge tube and emerges from it. It then cleaves into a mass of individual zoospores, which develop outward-directed undulipodia. The zoospores swim away to begin the cycle again.

The details of the life cycles of most hyphochytrids are not known, partly because they are difficult to observe. *Hyphochytrium catenoides,* for example, grows mainly on pollen of conifers; the thallus stages develop inside empty pollen grains. It is possible that the group is polyphyletic—derived from more than one ancestral line. The cell walls of all hyphochytrids that have been studied are composed of chitin, but in some species they contain cellulose as well. In some species, lysine is synthesized by the diaminopimelic acid pathway (as in chytrids, Phylum Pr-26; monerans; and plants); in others, lysine is synthesized by the aminoadipic pathway, as in fungi and animals.

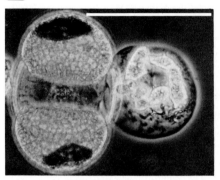

B Sporangium of *Hyphochytrium catenoides* on a ruptured pine pollen grain. Grain at left, sporangium at right. LM, bar = 0.5 mm. [Courtesy of D. J. S. Barr.]

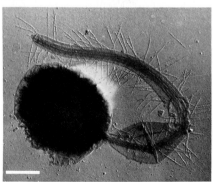

C Zoospore of *Rhizidiomyces apophysatum*. TEM (negative stain), bar = 1 μm. [Courtesy of M. S. Fuller.]

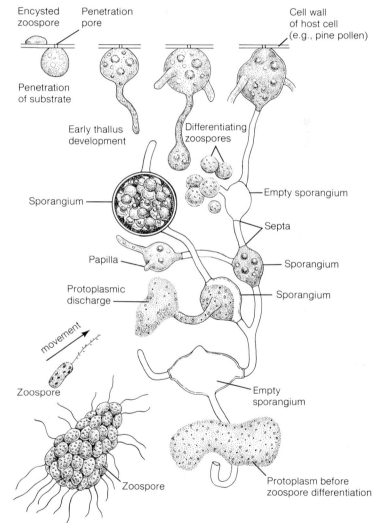

D Life Cycle of *Hyphochytrium* sp. [Drawing by R. Golder.]

Pr-26 Chytridiomycota

Greek *chytra*, little earthen cooking pot; *mykes*, fungus

Allomyces
Blastocladiella
Chytria
Cladochytrium
Coelomomyces
Monoblepharis
Olpidium
Physoderma
Rhizophydium
Synchytrium

Like the hyphochytrids (Phylum Pr-25) and the oomycotes (Phylum Pr-27), chytrids are parasitic or saprobic microbes that live in fresh water or soil. Like fungi, they grow and feed by extending threadlike hyphae or rhizoids into living hosts or dead organic debris, where they secrete digestive enzymes and absorb the resulting nutrients. Some chytrids are pathogens in higher plants; *Physoderma zea-maydis,* for example, causes the brown spot disease of corn.

The simplest chytrids grow and develop entirely within the cells of their host. The more complex produce reproductive structures on the host's surface, even though the vegetative and feeding parts of the chytrid body, or thallus, are sunk deep in the host's tissues.

The cell walls of all chytrids are composed of chitin; some contain cellulose as well. The chytrid thallus is coenocytic; that is, its many nuclei are not separated by cell walls. However, a septum, a solid plate composed of cell-wall material separates each reproductive organ from the thallus. In addition, the hyphae of many chytrids have pseudosepta, regular partitions composed of substances different from the outer chytrid wall. Pseudosepta are rather more pluglike than platelike, and typically, are not complete partitions.

Unlike the hyphochytrids, the chytrids are sexual. The gametes that fuse to form a zygote may be identical to each other or may differ; at least one is motile. A motile chytrid gamete has one smooth posterior undulipodium. The zygote converts itself into a resting structure. In some species, this structure eventually releases resistant spores that germinate into new chytrid thalli; in

Development of the ordinary colorless sporangium of *Blastocladiella emersonii.* The hours are time elapsed after water was added to an initial small dry sporangium. After 18 hours, there is swelling and a proliferation of rhizoids. After 36 hours, the protoplasm has migrated into the anterior cell that becomes the sporangium. After 83 hours, the sporangium has thickened and zoospores have begun to differentiate from the coenocytic nuclei inside. LM, bar = 1 mm. [Photographs courtesy of E. C. Cantino and J. S. Lovett (1966); drawings by R. Golder.]

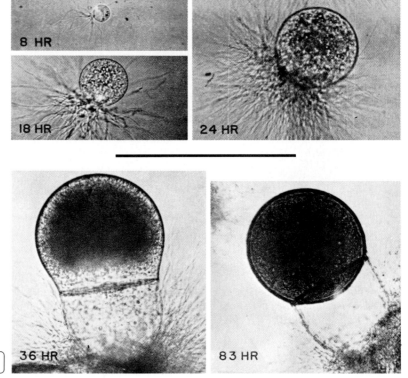

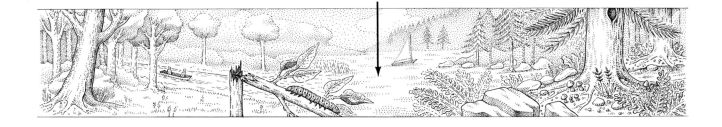

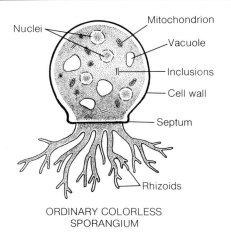

ORDINARY COLORLESS
SPORANGIUM

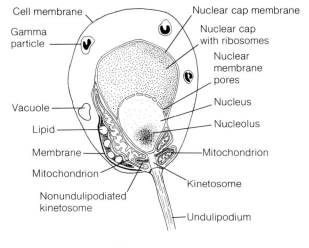

ZOOSPORE

others, it germinates directly and grows into a new thallus.

There are four classes: Chytridia, Blastocladia, Monoblepharida, and Harpochytridia.

The chytridias are unicellular. They have no well-developed mycelium. Some species do produce a rhizomycelium, a system of branched hyphae that, like rhizoids, emerge from the posterior of the chytrid body. Other species produce no mycelium at all. Sexual reproduction is by the fusion of equal zoosporelike gametes to form a zygote.

The blastocladias have well-developed branching mycelia. Many of them have complex life cycles with several alternative developmental pathways. *Blastocladiella emersonii,* for example, produces zoospores that have three distinct developmental options: a zoospore can form an ordinary colorless thallus, a stiff and resistant dark thallus, or a tiny thallus that releases only a single zoospore. Which option is taken depends on the quantity of food, moisture, and carbon dioxide in the medium. These factors, in turn, are related to the degree of crowding. Blastocladias also reproduce sexually, by the fusion of undulipodiated cells called planogametes (Greek *planos,* wandering), which look like the asexual zoospores. The planogametes are formed inside a thick-walled sporangium.

Blastocladiella zoospores have an unusual feature that can be seen by light microscopy. Virtually all of the ribosomes of the cell are packed near the nucleus in a membrane-bounded structure called the nuclear cap. The significance of this arrangement is not known.

In the monoblepharids, male planogametes are released from a specialized part of the thallus called the male gametangium, or antheridium. Other specialized hyphae of the thallus produce a walled female gametangium, or oogonium. Inside the oogonium, the protoplasm differentiates into a uninucleate oosphere, a fancy name for what is really an egg: it is a large, nonundulipodiated, nearly nonmotile gamete that is fertilized by a motile gamete. The zygote hardens over to form a structure traditionally misnamed a zoospore, in which meiosis takes place. The zoospore germinates to begin growing the hyphae of a mycelium; presumably, the nuclei in the mycelium are haploid. The vegetative mass grows; eventually, either it produces sporangia that release asexual zoospores or it differentiates male and female gametangia. In *Monoblepharis polymorpha,* the same thallus that produces asexual sporangia at certain temperatures is capable of producing the male and female gametangia at slightly higher temperatures.

Like all members of our Kingdom Fungi and some hypochytrids (Phylum Pr-25), chytrids synthesize the amino acid lysine by the aminoadipic pathway. Other funguslike protoctists (such as other hypochytrids and the oomycotes, Phylum Pr-27), use the diaminopimelic-acid pathway.

Pr-27 Oomycota

Greek *oion*, egg; *mykes*, fungus

Achlya
Albugo
Aphanomyces
Apodachlya
Dictyuchus
Isoachlya
Lagenidium
Myzocytium
Peronospora

Phytophthora
Plasmopara
Pythium
Rhipidium
Saprolegnia

The oomycotes (oomycetes) include organisms called water molds, white rusts, and downy mildews. Like the hyphochytrids (Phylum Pr-25) and chytrids (Phylum Pr-26), they are parasitic or saprobic. They feed by extending funguslike threads, or hyphae, into their host's tissues, where they release digestive enzymes and absorb the resulting nutrients. They are all coenocytic, lacking even the partial septa of the chytrids; normally, only the reproductive organs are separated from the other hyphae by septa. (However, other septa may form in response to injury.) Their cell walls are made of cellulose. The mycelia of oomycotes show less regularity in their branching patterns than those of chytrids. Most live in fresh water or moist soil; a few are parasites of terrestrial plants and rely on air currents to disperse their offspring.

The oomycotes are distinguished from the other funguslike protoctists by their zoospores and the nature of their sexual life cycle. The zoospores bear two undulipodia of unequal length. One is directed forward during swimming and is mastigonemate ("hairy"), while the other trails behind and is smooth. The zoospores are produced and released by a sporangium, an asexual reproductive organ that differentiates at the tip of a vegetative hypha. Either immediately or after some transformations, a zoospore can germinate to start growing a new oomycote thallus, or body.

Although sexuality is well developed in this phylum, it never involves undulipodiated gametes. Instead, the tips of vegetative hyphae produce specialized male and female structures, the antheridia and oogonia. An antheridium in contact with an oogonium produces fertilization tubes, which penetrate the oogonial tissue and through which the male nuclei migrate.

Fertilization takes place inside the walled oogonium, which usually contains a few large spheres called oospheres. Terminology notwithstanding, these are really egg cells. Most oospheres are uninucleate; those that are multinucleate are called compound oospheres. By fertilization, the oospheres become zygotes; they may develop thick chitinous walls and become resistant to starvation and desiccation. Such developed zygotes are called oospores or, sometimes, zygospores, after an analogous structure of zygomycotes (Phylum F-1). The walls of the oospores are usually dark and sculptured in patterns peculiar to each species.

In some oomycotes, such as *Saprolegnia,* an oospore can germinate to form a new thallus directly. Alternatively, meiosis may take place inside the oospore, so that, eventually, haploid zoo-

Saprolegnia ferax, an oomycote from freshwater ponds.

A Oogonium. LM, bar = 50 μm.

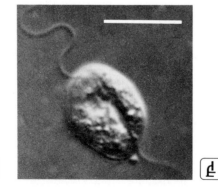

B Zoospore. LM, bar = 10 μm.

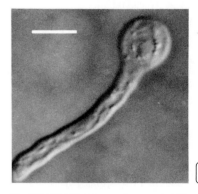

C Germinating secondary cyst. LM, bar = 10 μm.

spores are released; these zoospores then grow into thalli. This would imply that the haploid phase of the life cycle is dominant.

There are hundreds of species. The simplest produce merely a unicellular thallus and the characteristic reproductive structures. Other species produce a highly branched, profuse, and abundantly growing thallus. The most complex of the oomycotes live within specific hosts and depend on wind to disseminate the oospores or the sporangia full of zoospores. Even wind-disseminated oomycotes produce zoospores at some stage in their life cycle, indicating their affinities with free-living and less complex oomycotes. Some species have two types of zoospores. *Saprolegnia ferax,* for example, has ovoid primary zoospores bearing two undulipodia at their apex, and pear-shaped secondary zoospores bearing oppositely directed undulipodia at their side. The two types are produced in succession, not simultaneously. Some oomycotes are eucarpic (only part of the thallus differentiates for the reproductive function); others are holocarpic (the entire thallus becomes a reproductive sporangium). In some, the zoospores are formed within the sporangium; in others, within an evanescent vesicle that is produced by the sporangium.

Four classes (orders in other classification schemes) have been described. These are (with some representative genera) Saprolegnia (*Saprolegnia, Achlya, Isoachlya, Dictyuchus*), Leptomita (*Apodachlya, Rhipidium*), Lagenidia (*Lagenidium, Myzocytium*), and Peronospora (*Pythium, Albugo, Peronospora*). Members of the last class have been of utmost importance in history. The most infamous of its genera, *Phytophthora infestans,* causes the late blight of potatoes. In the nineteenth century, this oomycote destroyed the entire potato harvest of Ireland and Germany and caused mass migrations of people from their homelands. *Plasmopara viticola,* another dangerous peronosporan, causes mildew of grapes.

The saprolegnias are widely distributed in freshwater environments. The many species of the genus *Saprolegnia* include *Saprolegnia parasitica,* an oomycote that attacks fish and their eggs. It is seen very commonly in freshwater aquaria as a white fuzz on fish fins. Other species of the genus have been cultured in the laboratory and are useful in the study of developmental and mitotic processes. The hyphe of saprolegnias are profusely branched. At the hyphal tips, sporangia form, inside of which zoospores develop. Some saprolegnias have an additional type of asexual reproduction: they produce chlamydospores or gemmae. These irregular bits of protoplasm separate from the mycelium (mass of hyphae) and grow by forming germ tubes; these grow into hyphae bearing typical sporangia.

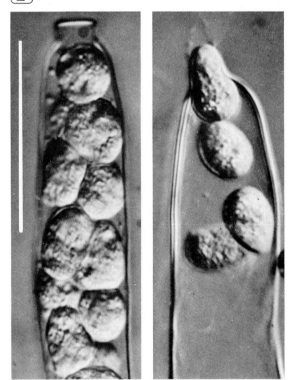

D Zoospores in zoosporangium (left) and their release (right). LM, bar = 50 μm. [Photographs courtesy of I. B. Heath.]

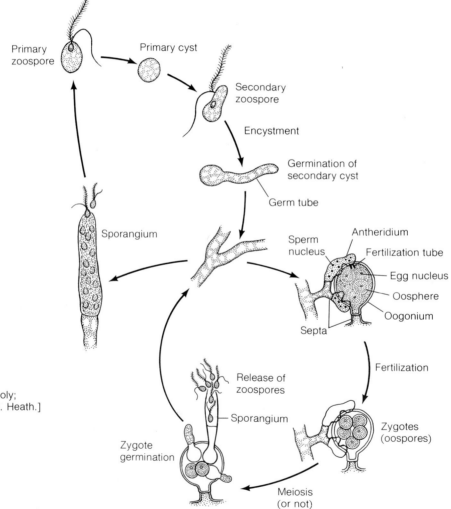

E Life cycle.
[Drawing by L. Meszoly; information from I. B. Heath.]

Primary zoospore

Primary cyst

Secondary zoospore

Encystment

Germination of secondary cyst

Germ tube

Sporangium

Sperm nucleus

Antheridium

Fertilization tube

Egg nucleus

Oosphere

Oogonium

Septa

Fertilization

Release of zoospores

Sporangium

Zygote germination

Zygotes (oospores)

Meiosis (or not)

Bibliography

General

Copeland, H. F., *The classification of lower organisms.* Pacific Books; Palo Alto, California; 1956.

Fritsch, F. E., *Structure and reproduction of the algae,* 2 vols., 2nd ed. Cambridge University Press; Cambridge and New York; 1961.

Grell, K. G., *Protozoology,* 2nd ed. Springer-Verlag; New York, Heidelberg, and Berlin; 1973.

Hogg, J., "On the distinctions of a plant and an animal, and on a fourth kingdom of nature." *The Edinburgh New Philosophical Journal* 12 (new series): 216–225; 1861.

Leedale, G. F., "How many are the kingdoms of organisms?" *Taxon* 23:261–270; 1974.

Pr-1   Caryoblastea

Daniels, E. W., and E. P. Breyer, "Ultrastructure of the giant amoeba *Pelomyxa palustris.*" *Journal of Protozoology* 14:167–179; 1967.

Grell, K. G., *Protozoology,* 2nd ed. Springer-Verlag; New York, Heidelberg, and Berlin; 1973.

Jeon, K. W., ed., *The biology of amoeba.* Academic Press; New York and London; 1973.

Whatley, J. M., "Bacteria and nuclei in *Pelomyxa palustris:* Comments on the theory of serial endosymbiosis." *New Phytologist* 76:111–118; 1976.

Pr-2   Dinoflagellata

Fritsch, F. E., *Structure and reproduction of the algae,* vol. 1, 2nd ed. Cambridge University Press; Cambridge and New York; 1961.

Hutner, S. H., and J. J. A. McLaughlin, "Poisonous tides." *Scientific American* 199:92–98; August 1958.

Medcof, J. C., A. H. Leim, A. B. Needler, A. W. H. Needler, J. Gibbard, and J. Naubert, "Paralytic shellfish poisoning on the Canadian Atlantic coast." *Bulletin of the Fisheries Research Board of Canada* 75:1–32; 1947.

Scagel, R. F., R. J. Bandoni, G. E. Rouse, W. B. Schofield, J. R. Stein, and T. M. C. Taylor, *An evolutionary survey of the plant kingdom.* Wadsworth; Belmont, California; 1966.

Taylor, F. J. R., "On dinoflagellate evolution." *BioSystems* 13:65–108; 1980.

Pr-3   Rhizopoda

Grell, K. G., *Protozoology,* 2nd ed. Springer-Verlag; New York, Heidelberg, and Berlin; 1973.

Hall, R. P., *Protozoology.* Prentice-Hall; New York; 1953.

Jeon, K. W., ed., *The biology of amoeba.* Academic Press; New York and London; 1973.

Page, F. C., "A revised classification of the gymnamoebia (Protozoa: Sarcodina)." *Zoological Journal of the Linnaean Society* 58:61–77; 1976.

Pr-4   Chrysophyta

Bouck, G. B., "The structure, origin, isolation, and composition of the tubular mastigonemes of the *Ochromonas* flagellum." *Journal of Cell Biology* 50:362–384; 1971.

Fritsch, F. E. "Chrysophyta." In G. M. Smith, ed., *Manual of phycology.* Ronald Press; New York; 1951.

Fritsch, F. E., *Structure and reproduction of the algae,* vol. 1, 2nd ed. Cambridge University Press; Cambridge and New York; 1961.

Hibberd, D. J., "The ultrastructure and taxonomy of the Chrysophyceae and Prymnesiophyceae (Haptophyceae): A survey with some new observations on the ultrastructure of the Chrysophyceae." *Botanical Journal of the Linnaean Society* 72:55–80; 1976.

Kudo, R. R., *Protozoology,* 5th ed. Charles C. Thomas; Springfield, Illinois; 1977.

Pr-5   Haptophyta

Fritsch, F. E. "Chrysophyta." In G. M. Smith, ed., *Manual of phycology.* Ronald Press; New York; 1951.

Hibberd, D. J., "The ultrastructure and taxonomy of the Chrysophyceae and Prymnesiophyceae (Haptophyceae): A survey with some new observations on the ultrastructure of the Chrysophyceae." *Botanical Journal of the Linnaean Society* 72:55–80; 1976.

Manton, I., and G. F. Leedale, "Observations on the fine structure of *Prymnesium parvum* Carter." *Archiv für Mikrobiologie* 45:285–303; 1963.

Prescott, G. W., *The algae: A review.* Houghton Mifflin; Boston; 1968.

Pr-6 Euglenophyta

Bold, H. C., *Morphology of plants,* 3rd ed. Harper & Row; New York, Evanston, San Francisco, and London; 1973.

Fritsch, F. E., *Structure and reproduction of the algae,* vol. 1., 2nd ed. Cambridge University Press; Cambridge and New York; 1961.

Leedale, G. F., "*Euglena:* A new look with the electron microscope." *Advancement of Science* 23(107):22–37; 1966.

Leedale, G. F., *Euglenoid flagellates.* Prentice-Hall; Englewood Cliffs, New Jersey; 1967.

Leedale, G. F., B. J. D. Meeuse, and E. G. Pringsheim, "Structure and physiology of *Euglena spirogyra,*" *Archiv für Mikrobiologie* 50:133–155; 1965.

Pr-7 Cryptophyta

Fritsch, F. E., *Structure and reproduction of the algae,* vol. 1, 2nd ed. Cambridge University Press; Cambridge and New York; 1961.

Graham, H. W., "Cryptophyceae." In G. M. Smith, ed., *Manual of phycology.* Ronald Press; New York; 1951.

Grell, K. G., *Protozoology,* 2nd ed. Springer-Verlag; New York, Heidelberg, and Berlin; 1973.

Meglitsch, P. A., *Invertebrate zoology,* 2nd ed. Oxford University Press; New York, London, and Toronto; 1972.

Pringsheim, E. G., "Some aspects of taxonomy in the Cryptophyceae," *New Phytologist* 43:143–150; 1944.

Pr-8 Zoomastigina

Corliss, J. O., "The opalinid infusorians: Flagellates or ciliates?" *Journal of Protozoology* 2:107–114; 1955.

Grell, K. G., *Protozoology,* 2nd ed. Springer-Verlag; New York, Heidelberg, and Berlin; 1973.

Hall, R. P., *Protozoology.* Prentice-Hall; New York; 1953.

Hickman, C. P., *Biology of the invertebrates,* 2nd ed. C. V. Mosby; St. Louis, Missouri; 1973.

Hollande, A., and J. Valentin, "Appareil de Golgi, pinocytose, lysomes, mitochondries, bactéries symbiontiques, attractophores et pleuromitose chez les hypermastigines du genre *Joenia.* Affinités entre Joeniides et Trichomonadines." *Protistologica* 5:39–86; 1969.

Pr-9 Xanthophyta

Bold, H. C., *Morphology of plants,* 3rd ed. Harper & Row; New York, Evanston, San Francisco, and London; 1973.

Fritsch, F. E., *Structure and function of the algae,* vol. 1, 2nd ed. Cambridge University Press; Cambridge and New York; 1961.

Fritsch, F. E., "Chrysophyta," In G. M. Smith, ed., *Manual of phycology.* Ronald Press; New York; 1951.

Hibberd, D. J., and G. F. Leedale, "Cytology and ultrastructure of the Xanthophyceae. II. The zoospore and vegetative cell of coccoid forms, with special reference to *Ophiocytium majus* Naegeli." *British Phycological Journal* 6:1–23; 1971.

Pr-10 Eustigmatophyta

Bold, H. C., *Morphology of plants,* 3rd ed. Harper & Row; New York, Evanston, San Francisco, and London; 1973.

Hibberd, D. J., and G. F. Leedale, "Eustigmatophyceae—A new algal class with unique organisation of the motile cell." *Nature* 225:758–760; 1970.

Hibberd, D. J., and G. F. Leedale, "Observations on the cytology and ultrastructure of the new algal class Eustigmatophyceae." *Annals of Botany* 36:49–71; 1972.

Pr-11 Bacillariophyta

Fritsch, F. E., *Structure and reproduction of the algae,* vol. 1, 2nd ed. Cambridge University Press; Cambridge and New York; 1961.

Lewin, J. C., and R. R. L. Guillard, "Diatoms." *Annual Review of Microbiology* 17:373–414; 1963.

Patrick, R., "A discussion of natural and abnormal diatom communities." In D. F. Jackson, ed., *Algae and man.* Plenum; New York; 1964.

Patrick, R., and C. W. Reimer, *The diatoms of the United States.* Academy of Natural Sciences; Philadelphia; 1966.

Werner, D., *The biology of diatoms.* University of California Press; Berkeley; 1977.

Pr-12 Phaeophyta

Alexopoulos, C. J., and H. C. Bold, *Algae and fungi.* Macmillan; New York; 1967.

Dawson, E. Y., *Marine botany: An introduction.* Holt, Rinehart and Winston; New York, Chicago, San Francisco, Toronto, and London; 1966.

Manton, I. and B. Clarke, "An electron microscope study of the spermatozoid of *Fucus serratus.*" *Annals of Botany* 15:461–471; 1951.

Papenfuss, G. F., "Phaeophyta." In G. M. Smith, ed., *Manual of phycology.* Ronald Press; New York; 1951.

Pr-13   Rhodophyta

Dixon, P. S., "The Rhodophyta: Some aspects of their biology." *Oceanography and Marine Biology Annual Review* 1:177–196; 1963.

Drew, K. M., "Rhodophyta." In G. M. Smith, ed., *Manual of phycology.* Ronald Press; New York; 1951.

Fritsch, F. E., *The structure and reproduction of the algae,* vol. 2, 2nd ed. Cambridge University Press; Cambridge and New York; 1961.

Prescott, G. W., *The algae: A review.* Houghton Mifflin; Boston; 1968.

Scagel, R. F., R. J. Bandoni, G. E. Rouse, W. B. Schofield, J. R. Stein, and T. M. C. Taylor, *An evolutionary survey of the plant kingdom.* Wadsworth; Belmont, California; 1966.

Pr-14   Gamophyta

Bold, H. C., *Morphology of plants,* 3rd ed. Harper & Row; New York, Evanston, San Francisco, and London; 1973.

Fritsch, F. E., *Structure and reproduction of the algae,* vol. 1, 2nd ed. Cambridge University Press; Cambridge and New York; 1961.

Pickett-Heaps, J. D., *Green algae.* Sinauer; Sunderland, Massachusetts; 1975.

Randhawa, M. S., *Zygnemaceae.* Indian Council of Agricultural Research; New Delhi; 1959.

Pr-15   Chlorophyta

Johnson, U. G., and K. R. Porter, "Fine structure of cell division in *Chlamydomonas reinhardi:* Basal bodies and microtubules." *Journal of Cell Biology* 38:403–425; 1968.

Pickett-Heaps, J. D., *Green algae.* Sinauer; Sunderland, Massachusetts; 1975.

Puiseux-Dao, S., *Acetabularia and cell biology.* Springer-Verlag; New York, Heidelberg, and Berlin; 1970.

Raven, P. H., R. F. Evert, and H. Curtis, *Biology of plants,* 2nd ed. Worth; New York; 1976.

Robbins, W. W., T. E. Weier, and C. R. Stocking, *Botany: An introduction to plant science,* 3rd ed. John Wiley and Sons; New York, London, and Sydney; 1964.

Pr-16   Actinopoda

Grell, K. G., *Protozoology,* 2nd ed. Springer-Verlag; New York, Heidelberg, and Berlin; 1973.

Hall, R. P., *Protozoology.* Prentice-Hall; New York; 1953.

Hollande, A., J. Cachon, and M. Cachon-Enjumet, "L'infrastructure des axopodes chez les Radiolaires Sphaerellaires periaxoplastidies," *Comptes Rendu Hebdomedaire Seances Academie des Science* (Séries D) 261: 1388–1391, 1965.

Hollande, A., and M. Enjumet, "Cytologie, évolution et systématique des Sphaeroidés (Radiolaires)." *Archives de Musée National de Histoire Naturelle, Paris* (7th Séries)7:1–134, 1960.

Meglitsch, P. A., *Invertebrate zoology,* 2nd ed. Oxford University Press; New York, London, and Toronto; 1972.

Pr-17   Foraminifera

Cushman, J. A., *Foraminifera: Their classification and economic use,* 4th ed. Harvard University Press; Cambridge, Massachusetts; 1948.

Grell, K. G., *Protozoology,* 2nd ed. Springer-Verlag; New York, Heidelberg, and Berlin; 1973.

Hedley, R. H., "The biology of foraminifera." *International Review of General and Experimental Zoology* 1:1–45; 1964.

Hedley, R. H., and C. G. Adams, eds., *Foraminifera,* 3 vols. Academic Press; New York; 1974–1978.

McEnery, M. E., and J. J. Lee, "*Allogromia laticollaris:* A foraminiferan with an unusual apogamic metagenic life cycle." *Journal of Protozoology.* 23(1):94–108; 1976.

Pr-18   Ciliophora

Corliss, J. O., *The ciliated protozoa: Characterization, classification, and guide to the literature,* 2nd ed. Pergamon Press; London and New York; 1979.

Grell, K. G., *Protozoology,* 2nd ed. Springer-Verlag; New York, Heidelberg, and Berlin; 1973.

Olive, L. S., "Sorocarp development by a newly discovered ciliate." *Science* 202:530–532; 1978.

Olive, L. S., and R. L. Blanton, "Aerial sorocarp development by the aggregative ciliate, *Sorogena stoianovitchae.*" *Journal of Protozoology* 27:293–299; 1980.

Pr-19   Apicomplexa

Hammond, D. M., and P. L. Long, *The coccidia:* Eimeria, Isospora, Toxoplasma, *and related genera.* University Park Press; Baltimore, Maryland; 1973.

Joseph, T., "*Eimeria indianesis* sp. n. and an *Isospora* sp. from the opossum *Didelphis virginiana* (Kerr)." *Journal of Protozoology* 21(1):12–15; 1974.

Kreier, J. P., ed., *Parasitic protozoa,* vol. 3, *Gregarines, coccidians and plasmodia.* Academic Press; New York; 1977.

Noble, E. R., and G. A., Noble, *Parasitology,* 4th ed. Lea and Febiger; Philadelphia; 1976.

Pr-20 Cnidosporidia

Kreier, J. P., ed., *Parasitic protozoa,* vol. 4, Babesia, Theileria, *Myxosporida, Microsporida, Bartonellaceae, Anaplasmataceae,* Ehrlichia, *and* Pneumocystis. Academic Press; New York, San Francisco, and London; 1977.

Kudo, R. R., "A biologic and taxonomic study of the Microsporidia," *Illinois Biological Monographs* 9:77–344; 1924.

Kudo, R. R., and E. W. Daniels, "An electron microscope study of the spore of a microsporidian, *Thelohania californica.*" *Journal of Protozoology* 10:112–120; 1963.

Noble, E. R., and G. A. Noble, *Parasitology,* 4th ed. Lea and Febiger; Philadelphia; 1976.

Sprague, V., and S. H. Vernick, "Light and electron microscope study of a new species of *Glugea* (Microsporida, Nosematidae) in the 4-spined stickleback *Apeltes quadracus,*" *Journal of Protozoology* 15:547–571; 1968.

Pr-21 Labyrinthulamycota

Alexopoulos, C. J., *Introductory mycology,* 2nd ed. John Wiley and Sons; New York, London, and Sydney; 1962.

Bonner, J. T., *The cellular slime molds,* 2nd ed. Princeton University Press; Princeton, New Jersey; 1967.

Hollande, A., and M. Enjumet, "Sur l'évolution et la systématique des Labyrinthulidae. Étude de *Labyrinthula algeriensis* nov. sp." *Annales des Sciences Naturelles, Zoologie et Biologie Animale* (11e sér.) 17:357–368; 1955.

Pokorny, K. S., *"Labyrinthula."* *Journal of Protozoology* 14(4):697–708; 1967.

Young, E. L., "Studies on *Labyrinthula.* The etiologic agent of the wasting disease of eel grass." *American Journal of Botany* 30:586–593; 1943.

Pr-22 Acrasiomycota

Bonner, J. T., *The cellular slime molds,* 2nd ed. Princeton University Press; Princeton, New Jersey; 1967.

Erdos, G. W., K. B. Raper, and L. K. Vogen, "Mating types and macrocyst formation in *Dictyostelium discoideum.*" *Proceedings of the National Academy of Sciences* (USA) 70:1828–1830; 1973.

Olive, L. S., *The mycetozoans.* Academic Press; New York, San Francisco, and London; 1975.

Wolf, F. A., and F. T. Wolf, *The fungi,* vol. 1. John Wiley and Sons; New York; 1947. (Reprinted by Hafner; New York; 1969.)

Pr-23 Myxomycota

Alexopoulos, C. J., "Morphology and laboratory cultivation of *Echinostelium minutum.*" *American Journal of Botany* 47:37–43; 1960.

Hagelstein, R., *The mycetozoa of North America.* Hagelstein; Mineola, New York; 1944.

Olive, L. S., *The mycetozoans.* Academic Press; New York, San Francisco, and London; 1975.

Poindexter, J. S., *Microbiology: An introduction to protists.* Macmillan; New York; 1971.

Pr-24 Plasmodiophoromycota

Aist, J. R., and P. H. Williams, "The cytology and kinetics of cabbage root hair penetration by *Plasmodiophora brassicae.*" *Canadian Journal of Botany* 49: 2023–2034; 1971.

Cook, W. R. I., "A monograph of the Plasmodiophorales." *Archiv für Protistenkunde* 80:179–254; 1933.

Karling, J. S., *The Plasmodiophorales,* 2nd rev. ed. Hafner; New York; 1968.

Lechevalier, H. A., and D. Pramer, *The microbes.* J. B. Lippincott; Philadelphia and Toronto; 1971.

Olive, L. S., *The mycetozoans.* Academic Press; New York, San Francisco, and London; 1975.

Sparrow, F. K. Jr., *Aquatic phycomycetes,* 2nd ed. University of Michigan Press; Ann Arbor; 1960.

Pr-25 Hyphochytridiomycota

Alexopoulos, C. J., *Introductory mycology,* 2nd ed. John Wiley and Sons; New York, London, and Sydney; 1962.

Barr, D. J. S., "*Hyphochytrium catenoides:* A morphological and physiological study of North American isolates." *Mycologia* 62:492–503; 1970.

Karling, J. S., "A new fungus with anteriorly uniciliate zoospores: *Hyphochytrium catenoides.*" *American Journal of Botany* 26:512–519; 1939.

Moore-Landecker, E., *Fundamentals of the fungi.* Prentice-Hall; Englewood Cliffs, New Jersey; 1972.

Sparrow, F. K. Jr., *Aquatic phycomycetes,* 2nd ed. University of Michigan Press; Ann Arbor; 1960.

Pr-26 Chytridiomycota

Cantino, E. C., "Morphogenesis in aquatic fungi." In G. C. Ainsworth and A. S. Sussman, eds., *The fungi: An advanced treatise,* vol. 2. Academic Press; New York and London; 1966.

Lovett, J. S., "Growth and differentiation of the water mold *Blastocladiella emersonii:* Cytodifferentiation and the role of RNA and protein synthesis." *Bacteriological Reviews* 39:345–404; 1975.

Sparrow, F. K. Jr., *Aquatic phycomycetes,* 2nd ed. University of Michigan Press; Ann Arbor; 1960.

Webster, J., *Introduction to fungi.* Cambridge University Press; New York and London; 1970.

Pr-27 Oomycota

Alexopoulos, C. J., *Introductory mycology,* 2nd ed. John Wiley and Sons; New York, London, and Sydney; 1962.

Alexopoulos, C. J., and H. C. Bold, *Algae and fungi.* Macmillan; New York; 1967.

Lechevalier, H. A., and D. Pramer, *The microbes.* J. B. Lippincott; Philadelphia and Toronto; 1971.

Webster, J., *Introduction to fungi.* Cambridge University Press; New York and London; 1970.

FUNGI

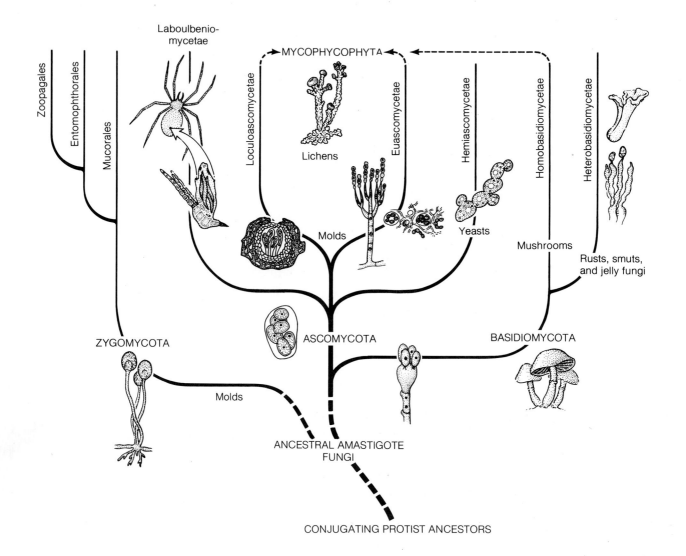

Zoopagales

Entomophthorales

Mucorales

Laboulbenio-
mycetae

Loculoascomycetae

MYCOPHYCOPHYTA

Lichens

Euascomycetae

Hemiascomycetae

Homobasidiomycetae

Heterobasidiomycetae

Molds

Yeasts

Mushrooms

Rusts, smuts,
and jelly fungi

ZYGOMYCOTA

ASCOMYCOTA

BASIDIOMYCOTA

Molds

ANCESTRAL AMASTIGOTE
FUNGI

CONJUGATING PROTIST ANCESTORS

FUNGI

Latin *fungus,* probably from Greek *sp(h)ongos,* sponge

Kingdom Fungi, as defined in this book, is limited to eukaryotes that form spores and are amastigote (lack undulipodia) at all stages of their life cycle.* It is estimated that there are 100,000 species of fungi, mostly terrestrial, although a few truly marine species are known. Because fungi often differ only in subtle characteristics, such as the pigments and complex organic compounds they produce, it is likely that many yet unknown species exist.

Fungal spores can germinate to grow slender tubes called *hyphae* (singular, *hypha*), which are divided into ''cells'' by cross walls called *septa* (singular, *septum*). Each such cell may contain more than one nucleus— the exact number depends on the species of fungus. The septa seldom separate the cells completely; thus, cytoplasm can flow more or less freely through the hyphae. In fact, the hyphae of some fungi have no septa at all.

A large mass of hyphae is called a *mycelium* (plural, *mycelia*), which is the vegetative form of most fungi. From time to time, reproductive structures are formed, also made of hyphae. Such structures are commonly noticed as molds, morels, and mushrooms. The largest and most complex are the large mushrooms and shelf fungi, some of which arise from mycelia meters in diameter. Many others are microscopic.

Fungi are sexual by conjugation, in which hyphae of different mating types come together and fuse (see the photographs illustrating Phylum F-3, Basidiomycota). Even after conjugation, the hyphal nuclei, which are always haploid, do not immediately fuse. Instead, each parental nucleus grows and divides within the hyphae, often for long periods of time. The offspring nuclei remain in pairs, one nucleus descended from each parent. A hypha containing paired haploid nuclei, whether or not they have been shown to come from separate parents, is called *dikaryotic.* A mycelium of such

---

* Traditional classifications often include funguslike microbes in the fungi (such as chytrids and oomycotes) all in the Plantae as Subkingdom Fungi. We have classified these that have undulipodia as protoctists (Phyla Pr-25, 26, and 27).

hyphae is called a *dikaryon.* If the nuclei of each pair are demonstrably different, the mycelium is called a *heterokaryon.* If the hyphae contain only single, unpaired nuclei, the mycelium is called a *monokaryon.*

In most fungal phyla, the dikaryotic state is eventually followed by fusion of nuclei to form diploid zygotes, but diploidy is transient. The zygote immediately undergoes meiosis to reestablish the haploid state. Meiotic cell division results in the formation of spores. All fungi form some sort of spores. The deuteromycotes, which lack sexual stages, produce only asexual spores. In all cases, the spores are haploid and are capable of germinating into haploid hyphae. In most fungi, hyphae of different mating types fuse later in the life cycle, and the dikaryotic or sexual stage follows. In asexual species, the hyphae are permanently haploid.

Fungi lack embryological development: the spores develop directly into hyphae or, in some cases, single vegetative cells. In many species, spores are formed in special structures called sporangia, asci, or basidia. Most fungi, even those that have sexual stages, can form spores vegetatively, without sex. In fact, most reproduce asexually more often than they do sexually. Vegetative spores called conidia form at the tips of hyphae. Most conidia are dispersed by the wind and can endure conditions of heat, cold, and desiccation unfavorable to the growth of fungi. Under favorable conditions, the conidia grow into hyphae and form mycelia. This is the general reproductive pattern, although there are many variations.

Nearly all fungi are aerobes, and all of them are heterotrophs. Fungi do not ingest their food but absorb it. They excrete powerful enzymes that break down food into molecules outside the fungus; these molecules are then transported in through the fungal membrane. The various fungal strategies for survival include the production of complex organic compounds, such as the ergot and amanita alkaloids, which can induce hallucinations or even death in mammals.

Fungi are tenacious, resisting severe desiccation and other insults. Their cell walls, composed of the nitrogenous polysaccharide chitin, are hard and stiff and resist the loss of water. Some grow in acid, others survive in environments that contain nearly no nitrogen. They are the most resilient of the eukaryotes.

Many fungi cause diseases, especially in plants. However, many more form important and constructive associations with plants. These associations can be quite intimate. Most orchid seeds, for example, require specific fungal partners in order to germinate, and fungi inhabiting the roots of forest trees are apparently responsible for transporting nutrients from the soil to the plants. Some fungi are important sources of antibiotics: *Penicillium chrysogenum* (Phylum F-4, Deuteromycota) produces penicillin. Molds and yeasts are used in the production of cheese and beer.

The ancestry of fungi is not well understood. The fungal way of life has evolved many times and in many groups—slime molds and nets, chytrids, and oomycotes (Phyla Pr-21 to Pr-27). It is possible that the true fungi descended from conjugating protists and thus share an ancestor with the gamophytes (Phylum Pr-14) or the rhodophytes (Phylum Pr-15). The ascomycotes and basidiomycotes seem related to each other and probably did descend from a zygomycote ancestor; the deuteromycotes clearly descended from either the ascomycotes or the basidiomycotes by loss of sexual stages. At any rate, fungi differ from animals and plants in life cycle, in mode of nutrition, in pattern of development, and in many other ways. Thus, many mycologists feel that it is high time fungi were raised to kingdom status.

The oldest fossil fungi, dating from the Devonian, are intimately associated with fossil plant tissue. It has been suggested that plant-fungus associations made it possible for plants to become truly terrestrial: the fungi could have transported nutrients to plants and prevented them from drying out. In any case, the association between plants and fungi has persisted for at least 300 million years.

F-1 Zygomycota

(Zygomycetes)

Greek *zygon*, pair; *mykes*, fungus

Blakeslea
Chaetocladium
Cochlonema
Conidiobolus
Endocochlus
Endogone
Kickxella
Mortierella
Mucor

Phycomyces
Pilobolus
Rhizopus
Stylopage

Members of this phylum lack cross walls or septa, except for separations between reproductive structures and the rest of the mycelium. There are about 600 species of zygomycetous fungi: they live on land throughout the world. Many are saprobes, living on decaying vegetation; some are highly specialized parasites, living on animals or even on each other.

Two modes of reproduction characterize the group. One is the asexual formation of resistant spores called conidia. These develop inside sporangia, which are often borne aloft on special hyphae called sporangiophores. The hyphae that develop from these spores are haploid. The other reproductive mode is conjugation. Special hyphae, called gametangia, of apparently opposite mating types, grow toward each other until they touch. The ends of the hyphae swell and the two cytoplasms intermingle. Nuclei from both parents enter the joined swellings, which develop into a thick-walled zygospore. In this organ, karyogamy (nuclear fusion) takes place, to be followed by meiosis and eventually, dispersion of haploid spores. Zygomycote matings are more like orgies than simple couplings because so many nuclei are brought together, fuse (probably two at a time), and undergo meiosis.

There are three classes of zygomycotes, which in some classification schemes are considered orders: Mucorales, Entomophthorales, and Zoopagales.

The Mucorales are mainly saprobic organisms; some are parasitic on plants or vertebrates. They form sporangia both asexually and by conjugation. Many of these sporangia, which contain one or many spores, have conspicuous walls that break when desiccated to release their contents. The best-known genera in this class are the common black bread mold *Rhizopus stolonifer, Mucor,* and *Phycomyces.* The sensitivity of members of the genus *Phycomyces* to light is truly astounding—only a few photons are required to initiate the development of the fruiting body. The photoreceptor, which absorbs blue light, seems to be the common B vitamin riboflavin.

Most of the Entomophthorales are parasites on animals, mainly on insects. They reproduce asexually by sporangia that do not release spores while they are attached to the fungus, but are forceably discharged as a unit. For example, *Pilobolus* (the "hat thrower"), a genus that produces its mycelium on horse dung, is able to shoot its sporangia into the air to a height of two meters.

The Zoopagales comprise about 65 species in ten genera, all of which parasitize amoebas, nematodes, and other protists and small animals. Most of them produce a mycelium that sends penetrating hyphae into their hosts. Some produce short, thick,

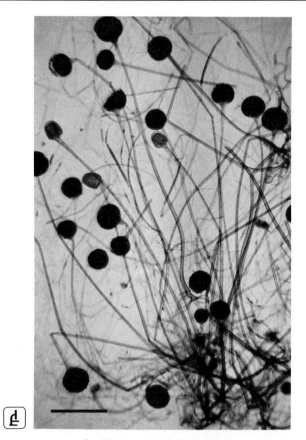

A *Rhizopus stolonifer* hyphae, sporangiophores, and sporangia, LM, bar = 1 mm. [Courtesy of W. Ormerod.]

spirally wound hyphae that form a coil inside the parasitized host. The coils are released with a violent discharge, debilitating the host. Club-shaped gametangia (in this case, special hyphal cells that act as gametes) emerge from the remains of the host, fuse, and form zygospores. Some genera are *Cochlonema* (*C. verrucosum* is a parasite of amoebas), *Endocochlus,* and *Stylopage.*

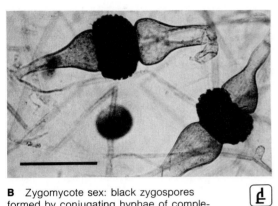

B Zygomycote sex: black zygospores formed by conjugating hyphae of complementary mating types. LM, bar = 1 mm. [Courtesy of W. Ormerod.]

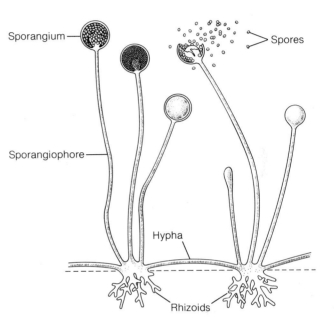

D *Rhizopus* sp. [Drawing by R. Golder.]

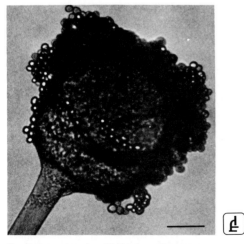

C Sporangium bearing asexual spores. LM, bar = 0.1 mm. [Courtesy of G. Cope. Reproduced by permission of Elementary Science Study of Education Development Corporation, Inc.]

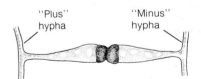

E Conjugation in *Rhizopus.* [Drawing by R. Golder.]

F-2 Ascomycota

Greek *askos,* bladder; *mykes,* fungus

Amorphomyces
Cerotocystis
Chaetomium
Claviceps
Elsinoe
Morchella
Mycosphaella
Neurospora
Rhizomyces

Saccharomyces
Sarcocypha
Sordaria
Tuber

Familiar as yeasts, bread molds, morels, and truffles, the ascomycotes are a large, diverse, and economically important group of fungi comprising tens of thousands of species. Ascomycotes are distinguished from other fungi by their possession of an ascus, a microscopic reproductive structure. In some species, the asci are so numerous that their organized mass forms a visible fruiting structure, or ascocarp. The hyphae of ascomycotes are long, slender, branched tubes that form a visible cottony mass, the mycelium.

An ascus is a capsule formed when two hyphae of compatible mating types conjugate. In the ascus, the partners' nuclei fuse, mingling their sets of chromosomes temporarily in one zygote nucleus. This transient diploid nucleus undergoes reduction division by meiosis, producing four new haploid nuclei with the same number (but a different combination) of chromosomes as in the nuclei of the parent fungi. These newly formed nuclei then undergo one or two mitotic divisions, and a protective spore wall is laid down around each new nucleus and some cytoplasm. The walled cells formed by meiosis are called ascospores. Nestled in the ascus like peas in a pod and released when mature, these ascospores may travel long distances, usually on wind currents. If they land in an appropriate nutrient-rich place, they germinate and send out hyphae of their own.

The formation of ascospores by the fusion of sexually different hyphae is not the sole means of ascomycote propagation. In fact, many of them have lost all sexual processes. However, asexual reproduction is universal in ascomycotes, even in those that can reproduce sexually. One individual's hyphae simply segment into huge numbers of genetically identical conidiospores, which, like ascospores, are dispersed by wind, water, or animals, often insects, to germinate elsewhere.

Ascomycotes are nourished by dead or living plant and animal material; they secrete digestive enzymes into their immediate environment and take in the dissolved nutrients thus formed. Ascomycotes play an essential ecological role by attacking and digesting resistant plant and animal molecules such as cellulose, lignin, and collagen. Valuable biological building blocks—compounds of carbon, nitrogen, phosphorus, among others—locked in such macromolecules are thus recycled. Ascomycotous pathogens have nearly eliminated such trees as the chestnut and elm from North America; however, other members of the phylum form necessary and healthy mycorrhizal (fungus and root) association with almost all trees. The transfer of inorganic nutrients from soil to tree is probably increased greatly by such connections. The prevalence of mycorrhizal associations in modern terrestrial ecosystems may indicate that trees had such fungal as-

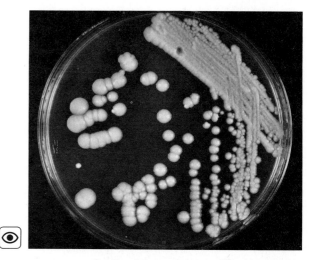

A *Saccharomyces cerevisiae.* Yeast colonies on nutrient agar Petri dish. [Courtesy of P. B. Moens.]

sistance when their ancestors successfully colonized the land more than three hundred million years ago.

There are four classes: Hemiascomycetae, Euascomycetae, Loculoascomycetae, and Laboulbeniomycetae.

The hemiascomycetes are morphologically most simple. They have short mycelia or even none at all; the ascus is formed directly, rather than on ascogenous hyphae that grow from the conjugated cells; and they lack ascocarps. Probably the best known of the hemiascomycetes are the yeasts, represented here by *Saccharomyces cerevisiae.* Yeasts generally do not grow hyphae. Many ascomycotes have both single-celled and mycelial stages, but yeasts have apparently reverted to a single-celled way of life. However, this resemblance to protoctists is only superficial. Haploid cells conjugate to form a diploid zygote, which undergoes meiosis and forms a true ascus and ascospores in a tetrahedral arrangement as shown here. The ascospores germinate by simply budding off a few cells. Yeasts ferment sugars such as glucose and sucrose to ethyl alcohol; this ability is utilized in the making of wine and beer. In the presence of gaseous oxygen, yeasts oxidize sugars to carbon dioxide, seen as gas bubbles in bread making. Brewer's and baker's yeasts have been cultivated for thousands of years.

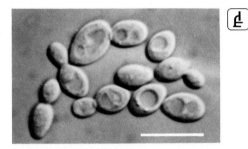

B Budding yeast cells after a day's growth. LM, bar = 0.1 mm. [Courtesy of P. B. Moens.]

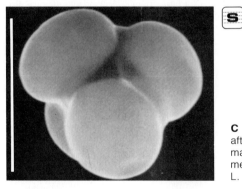

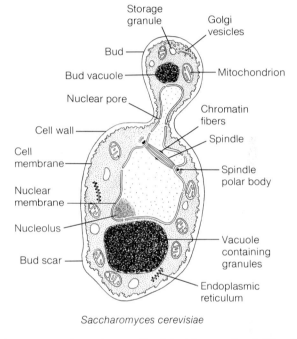

Saccharomyces cerevisiae

C Tetrad of yeast ascospores formed after fertilization. Cells of complementary mating types have fused and undergone meiosis. SEM, bar = 0.1 mm. [Courtesy of L. Bulla.]

D A budding yeast cell. [Drawing by R. Golder.]

The euascomycetes are the largest and best known class. Morels, truffles, and most fungal partners in lichens (Phylum F-5) are euascomycetes. The asci generally develop from ascogenous hyphae, which, in most cases, are enclosed in a true ascocarp. The asci are usually unitunicate—the inner and outer ascus walls are more or less rigid and do not separate when spores are ejected. The genus *Neurospora* is widely used in genetic research. In each ascus, the four products of meiosis divide once to form eight cells that remain fixed in a row in the order that they were formed. Each ascospore in an ascus can be picked up in order and grown to determine its genetic constitution. The information thus obtained reveals the behavior of chromosomes during a single meiosis and the position of genes on the chromosomes.

The loculoascomycetes have bitunicate asci—the inner wall is elastic and expands greatly beyond the outer wall when spores are released. Ascocarps form in a mass of supporting tissue, the stroma. The genus *Mycosphaella* has more than a thousand species, many parasitic on economically important food plants. The genus *Elsinoe* contains many pathogenic species that cause diseases of citrus, raspberry, and avocado, among others.

Laboulbeniomycetes are all minute parasites of insects. They are highly host specific—some will parasitize only one sex of the host species or only one body part, such as the legs or the wings. Their ascospores germinate directly into fruiting structures between 0.1 mm and 1 mm in diameter, whose number of cells is fixed for each species. Some genera are *Rhizomyces* and *Amorphomyces*.

F-3 Basidiomycota

Greek *basidion*, small base; *mykes*, fungus

Agaricus
Amanita
Boletus
Calvatia
Cantharellus
Clavaria
Cyathus
Fomes
Geastrum

Lepiota
Phallus
Polyporus
Puccinia
Schizophyllum
Tremella
Ustilago

Basidiomycotes include the smuts, rusts, jelly fungi, mushrooms, puffballs, and stinkhorns. They can be distinguished from all other fungi by the basidium—a microscopic clublike reproductive structure from which their name is derived. Each basidium bears several haploid spores, usually four. Called basidiospores, they are produced by sex and meiosis. There are about 25,000 species of basidiomycotes, several of which are parasites on agricultural crops and forest trees. Others, such as the domesticated mushroom *Agaricus*, are popular items in the human diet.

Most basidiomycotes go through three stages of development. A basidiospore germinates and grows into a primary mycelium, which is monokaryotic and lacks septa. Eventually, septa form and divide the hyphae into uninucleate segments. Two segments of compatible hyphae (of opposite mating types) conjugate. Often, the hyphae have grown from the same basidiospore. This self-fertilizing ability is called homothallism.

From the fused hyphae, a secondary mycelium develops. This constitutes the dikaryotic stage. Dikaryosis, the state of having two paired nuclei in each hyphal segment, is extremely common in this phylum. If the nuclei come from two different parents (grown from different basidiospores), the state is called heterokaryosis. The condition is analogous to the period of delay between the fusion of animal egg and sperm cytoplasm and the fusion of their nuclei.

Dikaryotic and heterokaryotic mycelia grow by the simultaneous division of the two nuclei in a hyphal segment and the formation of a new septum. To ensure that each new segment contains a nucleus descended from each parent, a clamp connection is formed. This is a short hyphal loop through which, before the new septum forms, one of the four nuclei moves to another place in the segment—the result is to "shuffle" the order of the nuclei. When the new septum forms, there are nuclei of two different parental origins on each side.

A tertiary mycelium is represented by the complex tissues that the dikaryon forms during basidiosporogenesis. These may be complex fruiting bodies such as mushrooms and jelly structures, or simpler fruiting protrusions that are seen often in plants infected with smut or rust disease.

There are two classes of basidiomycotes. The Heterobasidiomycetae include the jelly fungi, rusts, and smuts, which typically have more than one type of basidiospore-producing structure. Some (for example, the ergot fungi which cause a disease of rye flowers and are poisonous to human beings) have enormously complex life cycles that are linked to seasonal conditions and the developmental biology of their host plants.

A Fruiting bodies of *Boletus chrysenteron,* a yellow polypore bolete mushroom. Bar = 5 cm. [Courtesy of W. Ormerod.]

The Homobasidiomycetae all produce nonseptate basidia, usually containing four basidiospores. There are two subclasses. The Hymenomycetes include the common mushrooms, shelf fungi, and coral fungi. They all bear their basidia in a well-developed structure, the hymenium, that is exposed while the basidiospores develop. In the gilled mushrooms, basidia are borne on the lower surface of the gills; in the pore mushrooms, such as the *Boletus* shown here, the basidia line the inside of tubes that lead to exterior pores. The Gasteromycetes include the puffballs, earthstars, stinkhorns, and bird's-nest fungi. Their spores mature inside an enclosing structure called a basidiocarp, from which they are liberated only when the fruiting body decays or is ruptured.

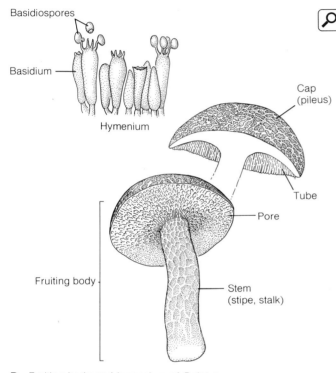

Basidiospores

Basidium

Hymenium

Cap
(pileus)

Tube

Pore

Fruiting body

Stem
(stipe, stalk)

B Fruiting body and hymenium of *Boletus chrysenteron.* [Drawing by L. Meszoly.]

C "Pores" of *Boletus chrysenteron.* Bar = 1 mm. [Courtesy of W. Ormerod.]

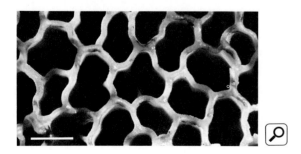

D Underside of *Schizophyllum Commune,* showing the gills. The white double lines of the gills bear the basidia in rows. Bar = 1 mm. [Courtesy of W. Ormerod.]

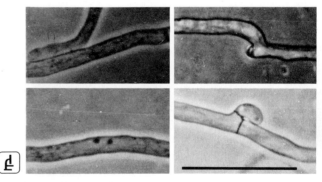

E Conjugation of basidiomycote hyphae of complementary mating types. (Left top) Approach of hyphae. (Right top) Incipient fusion. (Left bottom) Fused hyphae, two nucleoli apparent. (Right bottom) Clamp connection. The species is *Schizophyllum commune.* LM, bar = 40 μm. [Courtesy of W. Ormerod.]

F-4   Deuteromycota

Greek *deuteros*, second; *mykes*, fungus

Alternaria
Aspergillus
Candida (Monilia)
Clypeoseptoria
Cryptococcus
Cryptosporium
Fusarium
Geotrichum
Histoplasma

Penicillium
Rhizoctonia
Trichophyton
Verticillium

Because they do not represent a natural grouping, the deuteromycotes, or Fungi Imperfecti, are often called a form phylum or form class. The word *perfect* is an old botanical term that denotes a complete set of sexual reproductive structures. Thus, the deuteromycotes are fungi that lack organs for sexual reproduction. However, like the ascomycotes and basidiomycotes, they develop from spores, or conidia, into mycelia whose hyphae are divided by septa.

Although they lack standard meiotic sexuality, some deuteromycotes exhibit a parasexual cycle; that is, in the laboratory, one can form recombinant mycelia having different genetic traits by fusing the hyphae from two different organisms. From these, by processes not understood, new true-breeding offspring appear and persist. The parasexual process does not involve specialized mycelia.

The deuteromycotes are thought to be ascomycotes or basidiomycotes that have lost their potential to differentiate asci or basidia. Most of them are thought to be derived from ascomycotes because of the resemblance of the mycelium and the conidia to those of known ascomycote genera. Sexual structures have been discovered in some species originally classified as deuteromycotes, enabling the reclassification of these species with the ascomycotes or basidiomycotes.

There are about 25,000 species of these fungi, including some of great economic and medical importance. In our classification scheme, the phylum is divided into four major classes.

The Sphaeropsida reproduce by means of structures called pycnidia, hollow fruiting bodies lined on the inside with conidiophores, spore-bearing hyphae that are well known in the ascomycotes. *Clypeoseptoria aparothospermi* grows inside leaf tissue and forms needle-shaped spores.

The Melanconia reproduce asexually by conidia that are borne on short, closely packed conidiophores that form a mat. These mats, called acervuli, can often be seen as flat, disc-shaped cushions on host plants. They are also characteristic of an ascomycote order called the Melanconiales, from which these particular deuteromycotes are thought to have descended. The melanconia *Cryptosporium lunasporum* produces crescent-shaped conidiospores.

The Monilia comprise more than 10,000 species. In this class

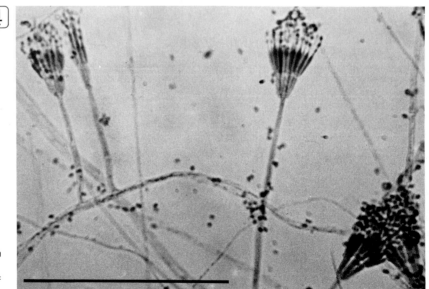

A Hyphae of a *Penicillium* species with several conidiophores bearing conidia (asexual spores) at their tips. LM, bar = 0.1 mm.

are the pathogenic and other yeasts that do not form asci or basidia. They reproduce by means of oedia, thin-walled cells that break off from the tips of ordinary hyphae. *Penicillium* belongs to this class. Some monilias reproduce by budding. Moniliasis, a common vaginal infection is caused by the proliferation of the monilia *Candida albicans.*

The Mycelia Sterilia include those genera that have no special reproductive structures at all. The mycelia simply grow without any visible differentiation. Of the two dozen or so genera belonging to this conglomerate class, the best known is *Rhizoctonia,* a common soil fungus that causes damping off and root rot of plants, including cultivated ones of economic importance.

B Colony of *Penicillium* derived from a single conidium growing on nutrient agar in a dish. Pigmented conidia form from the center, in the older portions of the colony; only unpigmented hyphae are at the outer edge. Bar = 1 cm. [Courtesy of W. Ormerod.]

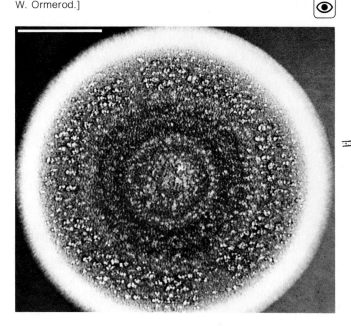

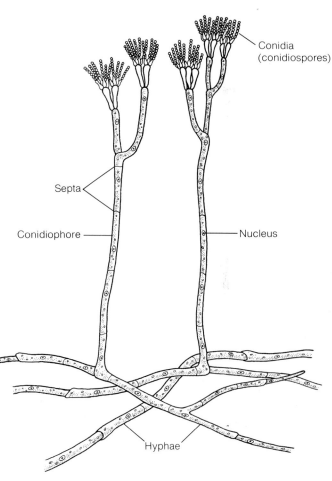

C *Penicillium* sp. [Drawing by R. Golder.]

F-5 Mycophycophyta
(Lichens)

Greek *mykes,* fungus; *phykos,* seaweed, alga; *phyton,* plant

Calicium
Cladonia
Collema
Coniocybe
Lepraria
Lichina
Lichenothrix
Ochrolechia

Parmelia
Umbilicaria
Usnea

Because they often grow on bare rock and are the first vegetation to cover burnt-out or newly exposed volcanic regions, lichens have been called "pioneer plants." Although it is true that lichens are pioneers, they are not plants. A lichen is a symbiotic partnership between a fungus, usually an ascomycote (Phylum F-2), and a chlorophyte alga (Phylum Pr-15) or a cyanobacterium (Phylum M-7). Thus, like the deuteromycotes (Phylum F-4), lichens constitute a form phylum. There are some 25,000 species.

In the laboratory, some lichen partners have been separated and grown by themselves. At least 26 genera of photosynthetic partners, including one xanthophyte (Phylum Pr-9), have been reported. In most lichens, the algae are either the chlorophytes *Trebouxia* or *Pseudotrebouxia* or cyanobacteria, such as *Nostoc.* A far greater number of fungal species have entered lichen partnerships. Thus, in saying that there are 25,000 species of lichens, we are referring to the number of distinct fungal species.

The lichen partners are quite different from their free-living counterparts. It has been known for almost a century that symbiosis is a crucial mechanism in the morphology, development, and evolution of lichens. In the last two decades, it has been found that lichens can synthesize compounds, such as the lichenic acids and pigments, that are lacking in the individual algae and fungi when grown alone.

Lichens are most abundant in the Antarctic, the Arctic tundra, high mountains, the tropics, and northern forests; they commonly grow on the bark of trees or on rocks. Another common habitat is the supratidal zone of rocky coasts, where lichens become highly diversified in the sea-spray zone. Lichens are well known for their resistance to desiccation; what is less well known is that they *must* have alternating dry and wet periods, because continuous drought or continuous dampness kills them. Their slow growth is legendary. Many studies of lichens on gravestones and other dated monuments indicate that they grow but a few millimeters in a century.

Lichens are very important initiators of ecological change. By slowly wearing away and dissolving the rocks on which they are established, they prepare the surfaces for the germination of seeds and the formation of rooted plant communities. They thus accelerate weathering and initiate the formation of soils. Despite their hardiness, lichens are very sensitive to certain gases (for example, sulfur dioxide and volatile metal compounds) that are released when coal is burned. Thus, their presence and state of health are used as pollutant indicators.

Naturalists group lichens according to their external appearance. Depending on their growth habit, lichens are called crustose (low and crusty), foliose (leafy), or fruticose (bushy—see our illustration of *Cladonia cristatella*). The thallus—crusty or leafy—is the vegetative part of the organism. The sexual part usually has a distinct appearance. Reproduction often involves the formation of ascospores by meiosis. Perhaps the most common form of lichen propagation is the release of soredia, small fragments consisting of at least one algal cell surrounded by fungal hyphae. Soredia are easily dispersed by air currents; in a suitable environment, they develop into new lichens.

The fact that lichens are partnerships has caused dissension about their classification. However, most lichenologists concede that, if possible, lichens ought to be classified with the ascomycote or basidiomycote groups to which their fungal partners belong. There is a precise analogy between this notion and the concept that the photosynthetic protoctists ought to be classified according to their life cycles and the structure of the nonplastid part of their cells. This is because they probably arose polyphyletically—different heterotrophic protoctists may independently have acquired the same photosynthetic partners, which developed into plastids. It is likely that the lichens also originated polyphyletically. Thus, like some other large taxa (Deuteromycota and Zoomastigina, Phyla F-4 and Pr-8), the lichens are a group that we maintain intact only because of their common features and easy recognition in nature.

We have divided the lichens into three classes depending on whether the fungal partner is an ascomycote, a basidiomycote, or a deuteromycote. Lichens containing ascomycotes typically produce open fruiting bodies. They include *Lichina, Collema,* and *Cladonia.* Common in the northeastern part of North America, *Cladonia* species are eaten by many animals (in emergencies, by people). They are at the base of the Arctic food chains that include caribou and reindeer. Most lichenized basidiomycotes belong to the homobasidiomycete families Clavariaceae and Coraceae. In deuteromycote-containing lichens (for example, *Lepraria* and *Lichenothrix*), no fruiting structures are known.

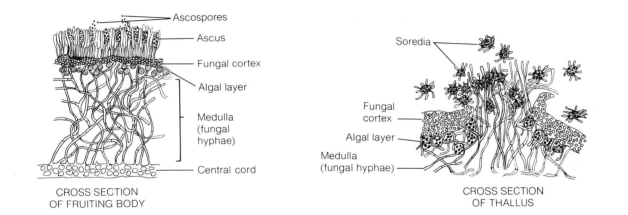

Ascospores

Ascus

Fungal cortex

Algal layer

Medulla
(fungal
hyphae)

Central cord

CROSS SECTION
OF FRUITING BODY

Soredia

Fungal
cortex

Algal layer

Medulla
(fungal hyphae)

CROSS SECTION
OF THALLUS

Fruiting body

THALLUS WITH FRUITING BODIES

Cladonia cristatella, the British soldier
lichen of New England woodlands. Bar =
1 mm. [Photograph courtesy of J. G.
Schaadt; drawings by E. Hoffman.]

Bibliography

General

Ainsworth, G. C., and A. S. Sussman, eds., *The fungi,* 4 vols. Academic Press; New York; 1965–1973.

Brightman, F. H., *Oxford book of flowerless plants.* Oxford University Press; New York; 1966.

Large, E. C., *The advance of the fungi.* Dover Publications; New York; 1962.

Moore-Landecker, E., *Fundamentals of the fungi.* Prentice-Hall; Englewood Cliffs, New Jersey; 1972.

F-1   Zygomycota

Alexopoulos, C. J., *Introductory mycology.* 2nd ed. John Wiley and Sons; New York; 1962.

Moore-Landecker, E., *Fundamentals of the fungi.* Prentice-Hall; Englewood Cliffs, New Jersey; 1972.

F-2   Ascomycota

Bold, H. C., *Morphology of plants,* 3rd ed. Harper & Row; New York; 1973.

Fincham, J. R. S., and P. R. Day, *Fungal genetics,* 2nd ed. F. A. Davis; Philadelphia; 1965.

Phaff, H. J., M. W. Miller, and E. M. Mrak, *The life of yeasts,* rev. ed. Harvard University Press; Cambridge, Massachusetts; 1978.

F-3   Basidiomycota

Alexopoulos, C. J., *Introductory mycology,* 2nd ed. John Wiley and Sons; New York; 1962.

Smith, A. H., *The mushroom hunter's field guide.* University of Michigan Press; Ann Arbor; 1963.

F-4   Deuteromycota

Alexopoulos, C. J., *Introductory mycology,* 2nd ed. John Wiley and Sons; New York; 1962.

Barnett, H. L., and B. B. Hunter, *Illustrated genera of imperfect fungi,* 3rd ed. Burgess; Minneapolis; 1972.

F-5   Mycophycophyta

Ahmadjian, V., *Lichen symbiosis.* Blaisdell; Waltham, Massachusetts; 1967.

Ahmadjian, V., and M. E. Hale, eds., *The lichens.* Academic Press; New York; 1974.

Hale, M. E., *How to know the lichens,* 2nd ed. Wm. C. Brown; Dubuque, Iowa; 1979.

Richardson, D. H. S., *The vanishing lichens: Their history, biology and importance.* Hafner; New York; 1975.

ANIMALIA

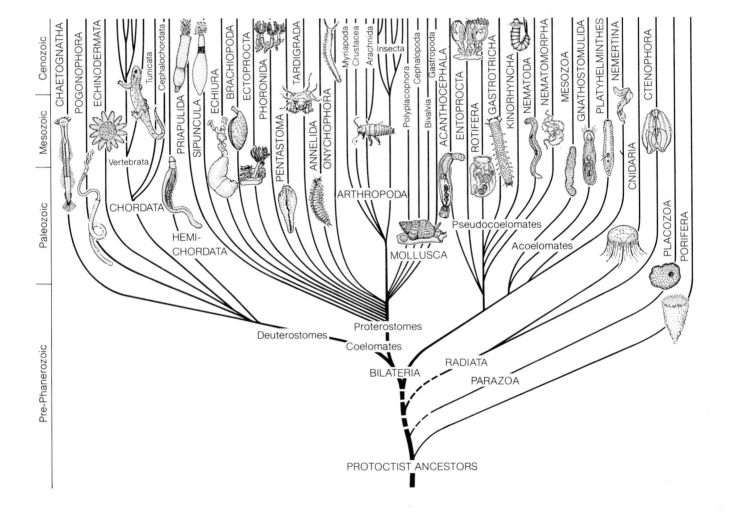

CHAETOGNATHA
POGONOPHORA
ECHINODERMATA
Tunicata
Cephalochordata
Vertebrata
CHORDATA
PRIAPULIDA
SIPUNCULA
ECHIURA
BRACHIOPODA
ECTOPROCTA
PHORONIDA
HEMI-
CHORDATA
PENTASTOMA
TARDIGRADA
ANNELIDA
ONYCHOPHORA
Myriapoda
Crustacea
Arachnida
Insecta
ARTHROPODA
Polyplacophora
Bivalvia
Cephalopoda
Gastropoda
ACANTHOCEPHALA
ENTOPROCTA
ROTIFERA
GASTROTRICHA
KINORHYNCHA
NEMATODA
NEMATOMORPHA
MESOZOA
GNATHOSTOMULIDA
PLATYHELMINTHES
NEMERTINA
CTENOPHORA
CNIDARIA
PLACOZOA
PORIFERA
MOLLUSCA
Pseudocoelomates
Acoelomates

Proterostomes
Deuterostomes
Coelomates
BILATERIA
RADIATA
PARAZOA

PROTOCTIST ANCESTORS

Cenozoic
Mesozoic
Paleozoic
Pre-Phanerozoic

160

ANIMALIA

Latin *anima,* breath, soul

In traditional two-kingdom systems, the multicellular animals were referred to broadly as *metazoa* to distinguish them from one-celled ''animals,'' the *protozoa.* In our classification system, the traditional protozoa belong to Kingdom Protoctista; in our system, animals may be defined as multicellular, heterotrophic, diploid organisms that develop anisogamously—from two different haploid gametes, a large egg and a smaller sperm. The product of fertilization of the egg by the sperm is a diploid zygote that develops by a sequence of mitotic cell divisions. These mitoses result in first a solid ball of cells and then a hollow ball of cells called a *blastula.* All animals (as defined in this book) develop from a blastula. In most animals, the blastula invaginates, folds inward at a point, to form a *gastrula,* a hollow sac having an opening at one end. Further growth and movement of cells produce a hollow digestive system called an *enteron* if it is open at only one end, and a *gut* or *intestine* if it has developed a second opening.

The details of further embryonic development differ widely from phylum to phylum but are fairly constant within each phylum. Such developmental details provide very important criteria for determining relationships between the phyla. In many phyla, developmental details are known for very few species; in some, not for any. In all cases they cannot be summarized in a few words. For this reason, concise and precise definitions of the phyla cannot always be given here; our descriptions are rather more informal.

Although multicellularity is found in all the kingdoms, it has developed most impressively in the animals—their cells are joined by complex junctions into tissues. Such elaborate joints—desmosomes, gap junctions, and septate junctions, for example—ensure and control communication and the flow of materials between cells. These junctions—and there are more kinds than those listed here—can be seen with an electron microscope. Indeed, the study of cells in tissues, and of tissues in organs, is a science in itself— histology.

Most animals have ingestive nutrition: they take food into their bodies and then either engulf particles or droplets of it into digestive cells by the process of phagocytosis (''cell eating'') or pinocytosis (''cell drinking'') or absorb food molecules through cell membranes. Although behavior of various kinds (attraction to light and avoidance of noxious chemicals, sensing of dissolved gases, and so forth) can be found in members of all five kingdoms, the animals have elaborated upon this theme too, far more than members of the other kingdoms. Mammalian behavior is perhaps the most complex.

The animals are the most diverse in form of all the kingdoms. The smallest are microscopic—smaller than many protists—and the largest today are whales, sea mammals in our own class (Mammalia) and phylum (Chordata). The members of most phyla are found in shallow waters. Truly land-dwelling forms are found only in two phyla, Arthropoda (Phylum A-27) and Chordata (Phylum A-32). Several phyla contain species that live on land in the soil (for example, earthworms), but these require constant moisture and have not really freed themselves from an aqueous environment throughout their life cycle. In fact, most animal phyla are aquatic worms of one kind or another. Most species of animals are extinct—these form the subject of another science: paleontology. Only living forms have been included in this book.

Of all organisms on Earth, only the animals have succeeded in actively invading the atmosphere. Representatives of all five kingdoms that spend significant fractions of their life cycle in the atmosphere can be found (for example, spores of bacteria, fungi, and plants). However, none in any kingdom spend their entire life cycle in the atmosphere, and only animals fly. Active locomotion of animals through the air has been independently achieved several but not many times, and only in two phyla: Arthropoda, class Insecta, and Chordata, classes Aves (birds), Mammalia (bats and *Homo sapiens* only), and Reptilia (several extinct flying dinosaurs).

For many years (and even today), biologists divided the animals, protozoans and metazoans together, into two large groups: the invertebrates, those without backbones, and the vertebrates, those with. In fact, all animals except the Craniata, a subphylum of Phylum Chordata, belong to the invertebrate group. This invertebrate/vertebrate dichotomy amply represents the skewed perspective we have as members of Phylum Chordata. Our pets, beasts of burden, sources of food, leather, and bone—that is, the animals closest to our size and best known to us—are members of our own phylum. We now realize that, from a less species-centered point of view, characteristics other than backbones are more basic and reflect much earlier evolutionary divergences.

The animal phyla are described here in approximate order of increasingly morphological complexity. Two phyla of animals, set apart as Subkingdom Parazoa, lack tissues organized into organs and have an indeterminate shape. These are the newly discovered Placozoa (Phylum A-1) and the well-known sponges (Phylum A-2, Porifera). The other thirty phyla, constituting Subkingdom Eumetazoa (true metazoans), have tissues organized into organs and organ systems.

There are two branches of the Eumetazoa. One consists of radially symmetrical organisms, the coelenterates (Phylum A-3) and the comb jellies (Phylum A-4, Ctenophora). These animals are planktonic and thus face a uniform environment on all sides; their radial symmetry is both internal and external. All the rest of the 28 phyla show bilateral symmetry, at least internally.

The bilaterally symmetrical phyla may be divided into three groups, or *grades:* those that lack a coelom (Acoelomata, Phyla A-5 through A-8), those that have a body cavity but lack a true coelom (Pseudocoelomata, Phyla A-9 through A-15), and those that develop a true coelom (Coelomata, Phyla A-16 through A-32). What is the coelom? The process of gastrulation leads to the development of three tissue layers in all animals more complex than the coelenterates and ctenophores. These tissue layers, called the endoderm, mesoderm, and ectoderm (listed from the inside out), are the masses of cells from which the organ systems of animals develop. In general, the intestine and other digestive organs develop from endoderm, the muscle and skeletal materials from mesoderm, and the nervous tissue and outer integument from the ectoderm. In the coelomates, the mesodermal tissues open to contain a space that widens and eventually forms a body cavity in which digestive and reproductive organs, among others, develop and are suspended. This true body cavity is called the coelom. A pseudocoelom is an internal space that does not develop from a space surrounded by mesoderm.

Two groups, called *series,* of coelomate animals are distinguished according to the fate of an early developmental feature called the *blastopore.* The invagination of the blastula, the hollow ball of cells into which the animal zygote develops, is the blastopore. This embryonic structure enlarges as cells divide, grow, and move over each other. In animals of Series Protostoma (Phyla A-16 through A-27), the blastopore eventually becomes the mouth of the adult. In Series Deuterostoma (Phyla A-28 through A-32), the blastopore becomes the anus, the rear end of the intestine; the mouth forms as a secondary opening at the end of the animal opposite from the anus. The five deuterostome phyla are thought to have common ancestors more recent than the ones they have with any protostome phyla. However, this divergence probably occurred more than 680 million years ago, judging from the putative presence of both protostomes and deuterostomes in the Ediacaran fauna.

Virtually all biologists agree that animals evolved from protoctists. However, which protoctists, when, and in what sort of environments are questions that are still actively debated. E. D. Hanson has amassed a great deal of information on the protoctist-animal connection but admits the problem has not been solved for the Eumetazoa.* The Parazoa, or at least the Porifera (Phylum A-2), are thought to have evolved from the choanoflagellates (Phylum Pr-8, Zoomastigina). This is deduced from the details of fine structure of the cells. It is possible, in fact likely, that the other animal phyla, especially the eumetazoans, had different ancestors among the protoctists.

---

*E. D. Hanson, *Origin and early evolution of animals.* Wesleyan University Press; Middletown, Connecticut; 1977.

A-1 Placozoa

Greek *plakos,* flat; *zoion,* animal

Only a single species, *Trichoplax adhaerens,* is known in this phylum. Like the other Parazoa, it lacks tissues, organs and organ systems, head and tail, and left and right. Soft bodied and inconspicuous, it is the simplest of animals. *Trichoplax* looks like a very large amoeba, it is barely visible to the naked eye. One can see under higher magnification that the "amoeba" is really an animal composed of a few thousand cells. Because most of the surface cells bear undulipodia (cilia), the entire surface of the animal is ciliated.

The grayish white body of *Trichoplax* changes shape continuously; the periphery is the most mobile region. The dorsal, or back, side of the animal is composed of flattened cells bearing scattered cilia and lipid droplets, whereas the ventral, or belly, surface is composed of more columnar cells evenly covered with cilia. Between the dorsal and ventral layers of cells, there is a middle layer of fluid that harbors scattered cells containing unidentified lumpy bodies. Although *Trichoplax* has a dorsal side and a ventral side, one end does not differ from the other. Neither do left and right. The animal travels along in any direction, ventral side down and dorsal side up.

Both asexual and sexual modes of replication are known. Large trichoplaxes simply divide by "fission" into two animals, each multicellular and having two layers of cells. Cut trichoplaxes heal rapidly, and a complete animal regenerates from any cutaway portion. *Trichoplax* is also known to develop from eggs that probably come from the layer of cells on the ventral surface. Eggs will form in profusion if the population of animals becomes dense. Spermlike cells have been seen in aquaria containing *Trichoplax,* but they have not yet been caught in the act of fertilizing eggs. A membrane has been seen to rise up and coat the eggs; this is assumed to be a manifestation of fertilization.

Cleavage has been observed: the two-cell stage becomes the four-cell and then the eight-cell stage, and so on until a blastula-like ball of cells is formed. However, because blastulas that have been watched have always stopped developing, the later stages are not known. In the nuclei of body cells are twelve chromosomes. Cells of the middle layer have twice as much DNA as epithelial cells, but the extra DNA may belong to intracellular bacteria. The best assumption is that *Trichoplax* is a diploid animal that develops from eggs fertilized by sperm.

Only very small trichoplaxes—probably just recently developed from eggs—can swim. Large adults use their undulipodia to crawl. No evidence of nerve or muscle cells has been found, although the fibrous cells in the middle fluid layer are thought to contract the body. How the animals feed is a mystery: they have no mouths at all. Sometimes, on the ventral side of the body, a broad pocket is formed over organic debris and algae thought to be food. It is possible that *Trichoplax* secretes enzymes into this pocket after it settles over the area it will feed on, thus forming a temporary "stomach" from which food is taken in by phagocytosis. But this is still conjecture.

Trichoplax is so obscure that there is no common name either for it or for the phylum to which it belongs. It was discovered in the sea water aquarium of the Graz Zoological Institute in Austria in 1883, but it has been seldom seen since except by ardent zoologists who have searched for it. *Trichoplax* has always been found in sea water—from the Red Sea, the Plymouth Marine Station on the southern coast of England, the Rosenstiel marine station at Miami, Florida, and on the walls of the marine aquarium at Temple University in Philadelphia. This distribution undoubtedly better reflects the distribution of interested zoologists than the distribution of the little trichoplaxes.

A *Trichoplax* cell has very little DNA—of the order of 10^{10} daltons (molecular-weight units), which is more like a bacterial nucleoid or the nucleus of some small protists than like other animal nuclei. Its chromosomes are very small, from 0.6 μm to 1 μm in length, each one the size of a bacterial cell, a little dot even at high power under the light microscope.

Because the organization of the body of *Trichoplax* is superficially quite like the planula larvae of coelenterates (Phylum A-3, Cnidaria), they were misidentified before careful studies showed them to be adults of their own unique group. They may have evolved from planulalike ancestors of the coelenterates; but because there is no fossil record for Placozoa and so little is known about *Trichoplax* and its life cycle, it is hard to tell.

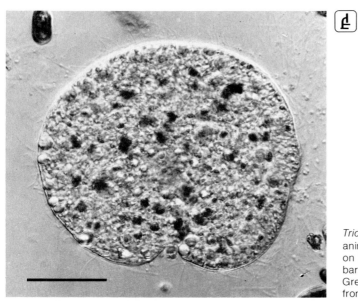

Trichoplax adhaerens, the simplest of all animals, found adhering to and crawling on the walls of marine aquaria. LM, bar = 0.1 mm. [Photograph courtesy of K. Grell; drawing by L. Meszoly, information from R. Miller.]

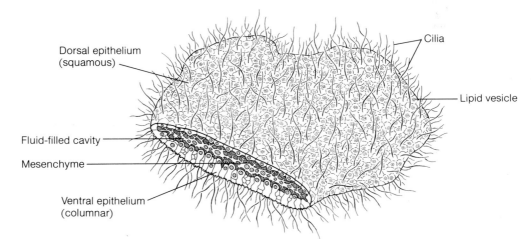

Dorsal epithelium (squamous)

Cilia

Lipid vesicle

Fluid-filled cavity

Mesenchyme

Ventral epithelium (columnar)

CUTAWAY VIEW

A-2 Porifera

Latin *porus*, pore; *ferre*, to bear

Euplectella
Gelliodes
Grantia
Halisarca
Leucosolenia
Microciona
Scolymastra
Spongia
Spongilla
Stromatospongia

Poriferans, or sponges, are named for their thousands of pores. All sponges have these holes, many of which open into canals through which water flows. Like placozoans (Phylum A-1), sponges lack symmetry. Because they have no tissues or organs, they belong to the Subkingdom Parazoa. All sponges are aquatic. About 10,000 species are known, but only 150 live in fresh water. They take fantastic forms—fans, cups, crusts, and tubes—and range in size from a few millimeters wide to more than one meter tall (*Scolymastra joubini,* a barrel-like glass sponge of the Antarctic).

Sponges are made of two layers of cells; between the layers lies a gelatinous layer, the mesenchyme, which contains amoeboid cells and spicules (skeletal needles) or fibers. Although the inner and outer cell layers contain some specialized cells, the sponge body lacks the cell cohesiveness and coordination typical of true metazoans. Nearly all sponges are sessile, permanently attached. Their size and shape tend to be determined by such factors as the degree of water turbulence and the amount of available space. Many coat logs and rocks and are so shapeless that they are not recognized as animals.

Sponges lack mouths, intestines, and respiratory and circulatory organs. Oxygen diffuses through their body walls. Food is filtered from copious quantities of water that flow through them. Plankton, such as dinoflagellates (Phylum Pr-2), makes up about 20% of their food; another 80% consists of detrital organic particles. Choanocytes, very distinctive collared cells having beating undulipodia, move water into the pores and along incoming waterways called inhalent canals. Some food particles stick to the collars first; others are directly engulfed by phagocytotic cells lining the canals. Waste particles and fluids either diffuse out of the sponge directly through the body wall or flow out through a single large opening called the excurrent opening.

Most sponges are hermaphroditic: mature individuals bear both eggs and sperm. In a few species, the sexes are separate. Cells floating in the inner mesenchyme develop into eggs and sperm. Clouds of sperm are released through the excurrent opening. Sperm enter other sponges with food and water. Amoeboid cells carry sperm to the eggs, which remain in the mesenchyme. In the mesenchyme, fertilization occurs and zygotes form. The zygotes develop into multicellular, free-swimming, ciliated larvae. Some sponges release their developing larvae into the water; others retain them for some time. The larvae metamorphose into adult-type sponges by turning inside out, bringing their cilia inside.

Some sponges reproduce without benefit of sex. Fragments may break off and be moved away by water currents; such fragments continue growing as individual sponges. Many freshwater and a few marine sponges also reproduce asexually by means of gemmules, particles composed of nutrient-laden amoeboid cells surrounded by a layer of epithelial cells. These disperse and grow to form new sponges.

Many sponges have symbiotic algae that provide food and probably also oxygen, remove waste, and screen the sunlight. These may be blue-green (Phylum M-7), red (Phylum Pr-13), green (Phylum Pr-15), or brown (Phylum Pr-12) algae. In some species, symbionts are transmitted from adult sponges to their offspring by adhering to the gemmules. Because of their algal symbionts and their own pigments, sponges may be white, red, orange, yellow, green, blue, purple, or brown. Some are even bioluminescent.

There are four classes in this phylum. The Calcarea, calcareous sponges, have spicules composed of calcium carbonate in the form of calcite. Sponges in other classes have siliceous spicules fashioned from opaline silica (SiO_2). The SiO_2 is extracted from sea water, concentrated, and laid down in delicate patterns inside special cells. The detailed process of spicule formation is not known. The Desmospongiae have a skeletal network of an organic material called spongin, a class of proteins related to the keratins of hair and nails; depending on species, the network may contain siliceous spicules. The Sclerospongiae also have a spongin network; it contains calcium carbonate in the form of aragonite. Some species also contain siliceous spicules. The Hexactinellida, or glass sponges, contain siliceous spicules with six rays pointed along three axes at right angles to each other.

Parasitic sponges are not known, and most animals prefer other things as food. Apparently, sponge spicules are obnoxious tasting and a successful deterrent to predators, with the exception of snails and nudibranchs (Phylum A-19), a few starfish (Phylum A-29), and a few fish (Phylum A-32). Broken and maimed sponges show remarkable powers of regeneration.

The oldest sponges in the fossil record date from the early Cambrian, about 550 million years ago. Sponges seem to form a cohesive group not ancestral to any other metazoans. Of all the animals, they are easiest to relate to their protist ancestors, complex advanced colonial choanoflagellates (in Phylum Pr-8, Zoomastigina). This is deduced from the remarkable similarity between sponge choanocytes, or collar cells, and the nearly identical free-living choanoflagellates and craspedomonads.

Gelliodes digitalis, one of the simpler sponges, is here shown live in its marine habitat. It has a single excurrent opening, the osculum, on top. Bar = 10 cm. [Photograph courtesy of W. Sacco; drawing by L. Meszoly, information from W. Hartman.]

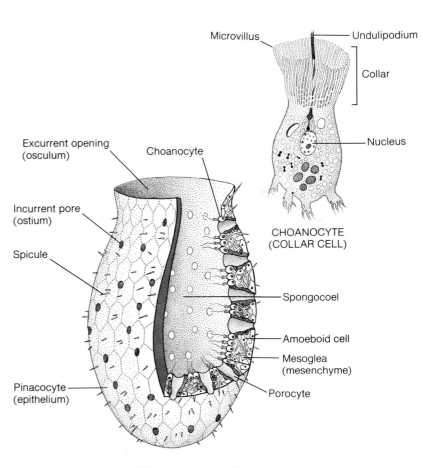

CHOANOCYTE (COLLAR CELL)

CUTAWAY VIEW OF WHOLE SPONGE

A-3 Cnidaria

(Coelenterates)

Greek *knide*, nettle; *koilos*, hollow; *enteron*, intestine

Antipathes	*Metridium*
Atolla	*Millepora*
Branchioceranthus	*Obelia*
Corallium	*Physalia*
Craspedacusta	*Renilla*
Cyanea	*Tubipora*
Haliclystus	*Tubularia*
Heliopora	*Velella*
Hydra	

Coelenterates are nearly all marine. They are radially symmetrical animals, the least morphologically complex of the true metazoa. They form tissues and organs. They also produce stinging cells called nematocysts on tentacles, which typically are arranged in a ring around the mouth of the animal. These nematocysts forcibly discharge their stings if the nearby undulipodia, which are also abundant on the surface of coelenterate tentacles, are mechanically or chemically stimulated.

They are found as two basic forms: the polyp and the medusa. Polyps are cylindrical animals. Some are sessile and attached; others move by gliding or somersaulting. Their tentacle-ringed mouth end faces upward. Medusae are shaped like umbrellas or cubes and are usually free swimming. The tentacles of medusae resemble the snaky locks of the mythical Medusa; medusae face down and the tentacles float outward. Some species of coelenterates are always found as polyps, others always as medusae, and still others alternate between the two forms.

Coelenterates range in size from microscopic polyps to *Branchioceranthus*, a hydrozoan polyp that may reach 2 meters in length. The sea blubber *Cyanea*, largest of the jellyfish, is more than 3.6 meters wide and its tentacles are more than 30 meters long.

Coelenterates are all carnivores; they do not pursue their prey, but sting when they contact it. They attack worms, crustaceans, fish, comb jellies, and diatoms and other protists. Unlike sponges (Phylum A-2), coelenterates have a single body cavity, the gastrovascular cavity, which opens through the mouth. The one opening serves as "anus" too, because waste also leaves by it. There are no blood vessels. Between an outer layer of cells, the ectoderm, and an inner layer, the gastrodermis or endoderm, lies a jellylike layer called the mesoglea, containing loose cells.

Coelenterates have nerve nets, but not centralized nervous systems. Their nerve fibers are not coated with membranes or with other cells—they have the only truly naked nerve fibers in the animal kingdom. Some of their nerve junctions transmit impulses in both directions; others, like those of the higher animals, transmit impulses in only one direction.

Three types of rigidifying structures are found in coelenterates: the gut itself, stiffened by fluid pressure, the mesoglea, and the carbonate or organic horny exoskeleton, formed only by polyps.

The three major classes of coelenterates are the Hydrozoa (hydras), the Scyphozoa (true jellyfish), and the Anthozoa (most corals and sea anemones).

Hydrozoans, of which there are about 3100 species, include the freshwater hydras and colonial hydroids, and some corals

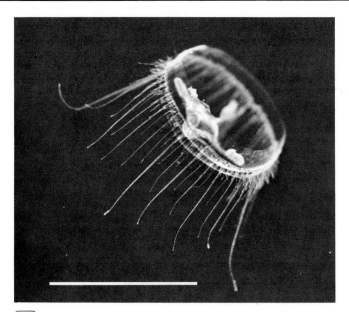

A *Craspedacusta sowerbii*, the living medusa of a freshwater coelenterate. This species was first described from individuals taken from the waterlily tank in Regent's Park, London. Bar = 10 mm. [Courtesy of C. M. Flaten and C. F. Lytle.]

called fire corals. Hydrozoans usually reproduce by asexual budding: polyps bud off daughter polyps that resemble their parent. The *Craspedacusta* shown in our photograph is produced asexually from a tiny solitary polyp. Hydrozoans may also reproduce sexually: the *Craspedacusta* medusa produces eggs and sperm. The zygote formed by fertilization of the egg develops into a free-swimming, ciliated, and mouthless larva called a planula. These larvae metamorphose into polyps.

There are some 200 species of Scyphozoans; most are free-swimming medusae. After fertilization, these jellyfish develop into planulae that become polyps. By budding, these may produce

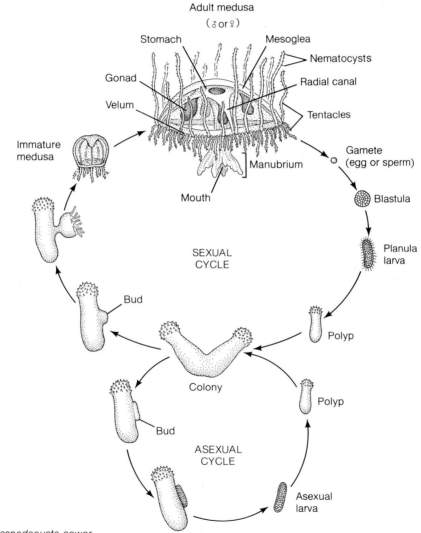

Adult medusa
(♂ or ♀)

Stomach

Mesoglea

Nematocysts

Gonad

Radial canal

Velum

Tentacles

Immature medusa

Gamete
(egg or sperm)

Manubrium

Blastula

Mouth

SEXUAL
CYCLE

Planula
larva

Bud

Polyp

Colony

Polyp

Bud

ASEXUAL
CYCLE

Asexual
larva

B The life cycle of *Craspedacusta sowerbii.* [Drawing by L. Meszoly, information from C. F. Lytle.]

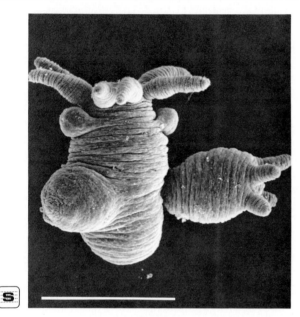

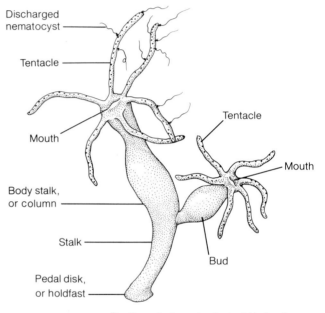

C   A sexually mature *Hydra viridis* (Ohio strain). The tentacles are at the top, two spermaries are located near the tentacles, a large swollen ovary is at the left of the picture, and an asexual bud at the right. *H. viridis* is normally about 3 mm long when extended, but this one shrank by about 1 mm when it was prepared for this photograph. SEM, bar = 1 mm.

D   Overall view of a typical *Hydra,* between 0.5 mm and 10 mm long, depending on species. [Drawing by L. M. Reeves.]

male and female medusae for years. The planulae of some open-ocean scyphozoans metamorphose directly into medusae, bypassing the polyp stage.

Anthozoans include sea anemones, sea pens, sea fans, sea pansies, and stony and soft corals. There are some 6200 species. They are solitary or colonial marine animals. Some species are hermaphroditic, while others have separate sexes. Fertilized anthozoan eggs usually develop into planula larvae that settle, attach, and then turn into polyps. Anthozoans are never found as

medusae. Many anemones are viviparous: developed offspring of sexually mature polyps are "born." Anemones may also reproduce asexually by splitting into two animals.

Nearly all shallow-water corals, horny corals, sea anemones, and many medusae harbor intracellular, photosynthetically active dinoflagellates (Phylum Pr-2). Many freshwater hydras harbor chlorophyte symbionts (Phylum Pr-15). Although most coelenterates can survive without their symbionts, they grow faster in sunlight with them. Furthermore, skeletalization is far more

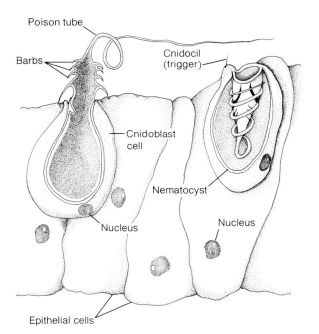

Poison tube

Barbs

Cnidocil
(trigger)

Cnidoblast
cell

Nematocyst

Nucleus

Nucleus

Epithelial cells

E Discharged and undischarged nematocysts. The undischarged nematocyst is
about 100 μm long. [Drawing by L. M.
Reeves.]

active in coelenterates that have symbionts than in those that
lack them. Symbionts provide oxygen, remove carbon dioxide,
and supply food in nutrient-poor waters.

Coral reefs are underwater limestone ridges near the surface
of the sea. They are usually formed by combined secretions of
several species of coelenterates and other carbonate-precipitat-
ing organisms, such as chlorophytes (*Halimeda, Penicillus,* and
Acetabularia, Phylum Pr-15) and rhodophytes (*Lithothamnion,*
Phylum Pr-13). Marine algae also encrust and shield reefs from
wave destruction. Reef corals are restricted to warm shallow
seas. Soft corals predominate in Atlantic reefs; hard corals are
more important in the Pacific. Below a depth of 60 meters, corals
are incapable of reef formation because the shortage of light
limits photosynthesis by their symbionts. Corals that lack algae
can live down to a depth of about 80 meters.

Reefs encroach on the sea; as they grow out, they change the
patterns of wave action and turbulence. Coral reefs have produced much dry land, such as Bermuda, the Bahamas, and the
Fiji Islands; they provide a haven for hundreds of fish, invertebrate, and protoctist species. Jewelry and other decorations
have been carved from the internal limestone skeletons of the
red coral *Corallium rubrum* since pre-Roman times, from black
coral *Antipathes,* and from the blue coral *Heliopora.*

Coelenterates are among the oldest fossil metazoans. *Ediacara,* a famous South Australian fossil found as a sandstone im-
print, has been interpreted to be a hydrozoan—it is nearly 700
million years old. Scyphozoan, hydrozoan, and medusan fossils
are found in Cambrian rocks 500 million years old. Fossil corals
abound from the Ordovician through the Devonian (from 500
million to 430 million years ago).

Although they differ in tentacle morphology and pattern of
development, coelenterates and ctenophores (Phylum A-4) are
believed to derive from the same evolutionary line. They share
radial symmetry, nematocysts (one species of comb jelly), and
mesoglea. No other phylum of animals is thought to have
evolved from them.

A-4 Ctenophora

Greek *kteis*, comb; *pherein*, to bear

Beroe
Bolinopsis
Cestum
Coeloplana
Ctenoplana
Euchlora
Lampea (= *Gastrodes*)
Mertensia
Mnemiopsis

Pleurobrachia
Tjalfiella
Velamen

Ctenophores are called comb jellies because of their comb plates, short rows of external cilia. In each comb plate, the cilia are fused together at their bases. Eight bands of comb plates stretch along the length of a ctenophore. With synchronized sweeps, they propel the animal's mouth end forward through the sea.

About 90 species of these transparent animals have been described. Most have two retractable tentacles, which are arranged in a bilateral symmetry—the comb bands form a radial symmetry. These tentacles are far longer than the bodies—by as much as twenty times—and are usually extended into handsomely curved fishing nets. They catch prey but play no role in locomotion.

Although ctenophoran bodies look glassy, they are flexible and motile, of the consistency of soft jelly in a membrane bag. Most are pelagic and are carried along by ocean currents. Ctenophores differ from all other macroscopic animals in that they retain cilia as locomotor organs throughout life. Both internal and external surfaces are ciliated. Some ctenophores, such as *Ctenoplana* (from Sumatra, New Guinea, and Papua) and *Coeloplana* (from Japan, Florida, and the Red Sea), creep instead of swimming weakly by waving their comb bands. The large comb jelly *Cestum veneris,* Venus's girdle, undulates through the water by muscular contractions of its ribbonlike body.

Some comb jellies live at depths of several kilometers. All of them usually live solitary lives, but tides and currents can concentrate them in vast numbers. They and jellyfish (coelenterates, Phylum A-3) are often responsible for phosphorescent patches in the night ocean. Some tropical comb jellies are delicately colored—violet, rose, yellow, or brown—by symbiotic algae. Sea walnuts, sea gooseberries, and cat's eyes are comb jellies often found stranded on the beach or caught in tow nets, where they collapse under their own weight and quickly die. *Lampea pancerina* (= *Gastrodes parasiticum*) is the only parasitic ctenophore so far described. It dwells in the mantle cavity of the floating tunicate *Salpa* (Phylum A-32).

As in the coelenterates, the nervous system of comb jellies is a diffuse net. They lack any special respiratory, circulatory, or excretory organs besides two small anal pores. An aboral sense organ acts as a statocyst, determining the orientation of the animal. Tilting the comb jelly brings several hundred calcareous particles, between 5 μm and 10 μm in diameter, to bear on one of four fused groups of S-shaped cilia, which support the particles as on a spring. This stimulus is transmitted by a ciliated groove to a corresponding comb band; the comb band responds with movements that restore the animal to an upright position.

Special cells, named after lassos, stud the tentacles. They entangle live fish, crustaceans, and other zooplankters. The sticky head of the lasso cell is attached to two filaments, one of which spirals around the other and acts as a spring. This design prevents struggling prey from tearing away. Comb jellies wipe their tentacles across their mouths to unload captive food. Whatever the food—live larvae, eggs, tiny fish (including economically important ones), arrow worms (Phylum A-30), copepods—it is all carried by cilia and mucus into the digestive cavity.

Food is then broken down by the secretion of digestive enzymes into this cavity. The resulting nutrient fluid is distributed along a canal system that runs through the mesenchyme, a thick jelly layer lying between epidermis and digestive cavity. This canal system ends blindly beneath the comb plates. Undigestible matter is voided through the mouth and through two small anal pores beside the statocyst.

The mesenchyme layer also contains amoeboid cells and smooth muscle strands, and it provides buoyancy.

Some comb jellies, such as *Bolinopsis,* have reduced tentacles. They may gather prey with their oral lobes instead of with tentacles, and a pair of ciliated flaps, called auricles, guide food into the mouth. The North Sea thimble jelly *Beroe gracilis* has no tentacles at any stage in its development. This organism is highly specialized—it feeds exclusively on *Pleurobrachia pileus,* another comb jelly. The food is sucked by mobile lips ringed with membranelles—these look like large cilia but, under the electron microscope, can be seen to be made of about 3000 interconnected normal-size cilia.

Ctenophores reproduce only sexually. In most, a fertilized egg develops into a free-swimming larva, which grows to become the sexually mature adult. Reproduction in temperate regions takes place once a year in late summer or in autumn. Ctenophores are hermaphroditic: each individual can form both female and male organs. The ovaries and testes are many on each animal—they develop along each digestive canal under the comb rows. Both eggs and sperm are shed through the mouth. In almost all ctenophores, fertilization occurs externally, in the sea. *Tjalfiella,* a viviparous comb jelly, is an exception. The young are brooded in special pouches, then released as a free-swimming stage. *Tjalfiella* adults lose their comb plates.

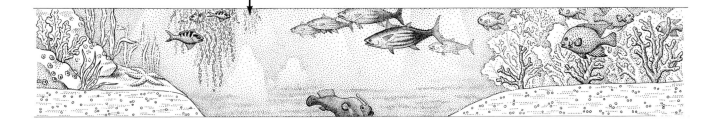

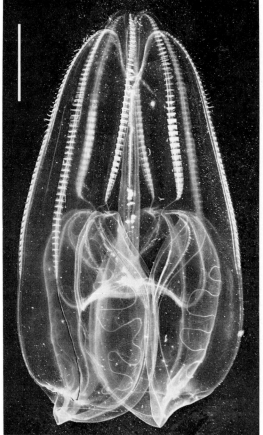

A *Bolinopsis infundibulum* (alive), a comb jelly, swims with its mouth end forward and gathers prey with its extended ciliated oral lobes. Bar = 1 cm. [Courtesy of M. S. Laverack.]

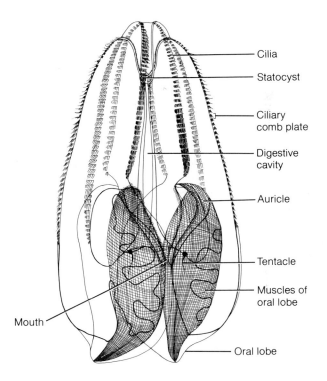

Cilia

Statocyst

Ciliary comb plate

Digestive cavity

Auricle

Tentacle

Muscles of oral lobe

Oral lobe

Mouth

B *Bolinopsis infundibulum.* [Drawing by I. Atema, information from M. S. Laverack.]

Ctenophora have such soft and fragile bodies that their potential for preservation is very low. No fossils have ever been found, so evolution of the group must be inferred from living forms. Some have suggested ctenophores evolved from medusa-shaped coelenterates (Phylum A-3), but the similarity of medusas to ctenophorans may be only superficial. On the other hand, it is more generally thought that comb jellies and hydroid coelenterates share a common ancestor. Coelenterates and ctenophores have evolved roughly parallel adaptations: a radial body plan with a branched body cavity, mesoglea or mesenchyme interposed between ectoderm and endoderm, carnivory, luminescence, a netlike nervous system, and the ability to regenerate broken-off body parts. (Both combs and lasso cells re-form regularly.) However, the phyla differ in two fundamental ways: except for one species, *Euchlora*, ctenophores lack the nematocysts so characteristic of coelenterates; they have true muscles (derived from mesoderm) so conspicuously lacking in coelenterates.

C Alive comb jelly, *Beroe cucumis*, which naturally lacks tentacles. Highly phosphorescent *Beroe* is common in plankton from Arctic to Antarctic seas. Bar = 1 cm. [Courtesy of M. S. Laverack.]

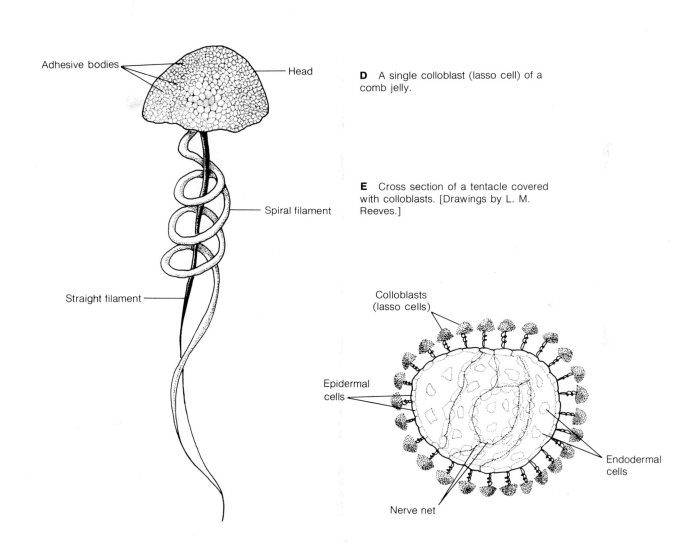

Adhesive bodies

Head

Spiral filament

Straight filament

D A single colloblast (lasso cell) of a comb jelly.

E Cross section of a tentacle covered with colloblasts. [Drawings by L. M. Reeves.]

Colloblasts (lasso cells)

Epidermal cells

Endodermal cells

Nerve net

A-5 Mesozoa

Greek *mesos*, middle; *zoion*, animal

Conocyema
Dicyema
Dicyemennea
Microcyema
Pseudicyema
Rhopalura
Stoecharthrum

These small wormlike organisms are bilaterally symmetrical, but have two instead of three tissue layers and lack organs and organ systems—with one exception, the gonad. The name of the phylum indicates that they are possibly intermediate between the protoctists, which lack tissues, and the more complex metazoans, which contain millions of cells connected into tissues, organs, and organ systems.

Adult mesozoans, which range from 0.5 mm to about 8 mm in length, lack body cavity, circulatory, respiratory, skeletal, muscular, nervous system, execretory and digestive organs. They have only one organ: a gonad. They must absorb nutrients directly from the urine of a host. The outer layer consists of from 20 to 30 "jacket" cells, a constant number in each species; these cells enclose a long, cylindrical axial cell. Within the axial cell are from one to more than a hundred cells called axoblasts, each containing a polyploid nucleus. The function of the large specialized axial cell is solely to give rise to young.

The two mesozoan classes, dicyemids and orthonectids, are probably not related directly, and two phyla probably will supplant this one eventually. However, both classes show an alternation of sexual and asexual generations and both live inside marine invertebrates. Dicyemids live in the kidneys of octopus, squid, and other cephalopod mollusks (Phylum A-19). Their habitat is the interface between urine and mucus-covered epithelial kidney tissue. Orthonectids live in flatworms (Phylum A-6), nemertines (Phylum A-7), polychete annelids (Phylum A-23), bivalve mollusks (Phylum A-19), and echinoderms, such as brittle stars (Phylum A-29). Because they are frequently found in hosts that are widespread in shallow seas, mesozoans are common.

The life cycle of dicyemids is not completely known. Free-swimming larvae, called infusoriform larvae, develop from eggs. Soon after they hatch, the larvae are acquired somehow by young bottom-dwelling cephalopods. The larvae enter the kidneys of their host and attach lightly by their anterior cells. They develop into the adult form on the inner surface of the kidney.

Adult dicyemids have morphologically similar but functionally distinct reproductive phases called nematogen and rhombogen. Nematogen adults reproduce asexually; they release immature vermiform (wormlike) larvae into the urine of the young mollusks, where they attach to spongy tissue of the kidney and grow. As long as the cephalopod is immature, new generations of nematogens are produced. Rhombogens are sexual adults. When the population of adult dicyemids becomes dense, they develop either hermaphroditic gonads (producing both eggs and tailed sperm) or separate male and female gonads. Within the axial

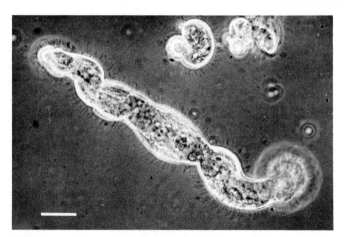

A An extended adult *Dicyema truncatum*, with a small contracted one above. LM, bar = 10 μm. [Courtesy of H. Morowitz.]

cells of the rhombogen parent, zygotes develop into infusoriform larvae. They escape from the parent into host urine and are shed into the sea. Their fate there is unknown; it is possible that intermediate hosts transmit mesozoans to other cephalopods.

In orthonectids, tiny free-swimming adult males and females comprise the sexual generation. In each individual, the outer cell layer encloses either eggs or sperm. Sperm are released into the ocean; they enter the tiny female's body, where fertilization takes place. The zygotes develop into ciliated larvae, which penetrate and infest tissue spaces of host organisms. Once inside, the larvae lose their outer ciliated cells. The inner cell mass becomes syncitial by nuclear mitoses that are not succeeded by cell division. These syncytia develop into adult males and females of the sexual generation.

The dicyemid infusoriform larva is spherical or top shaped, and is about 0.04 mm long. It consists of 28 cells; each of the four interior cells contains another cell, like Chinese boxes. The larva grows by differentiation and enlargement of existing cells rather than by cell division. It is weighted to the sea bottom by two cells filled with a high-density substance, magnesium inositol hexaphosphate. The orthonectid larva is amoeboid and multinucleate. It does not attach to the kidney, but grows through host tissues and may even cause castration.

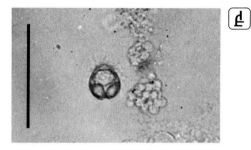

B *Dicyema truncatum* larva found in the kidneys of cephalopod mollusks. LM, bar = 100 μm. [Courtesy of H. Morowitz.]

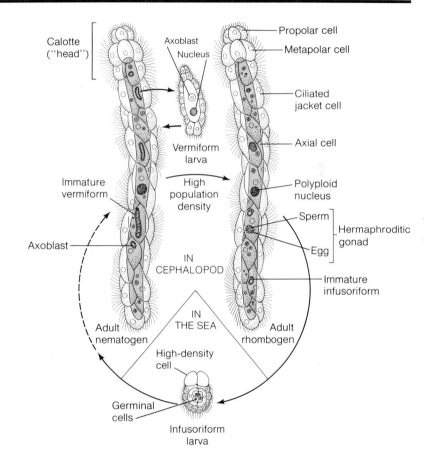

C *Dicyema truncatum* life cycle. [Drawing by L. Meszoly, information from E. Lapan.]

Except for placozoans (Phylum A-1), mesozoans are the least complex of the animals. However, some scientists have suggested that they have evolved by degeneration from flatworms (Phylum A-6). Evidence that mesozoans are not degenerate flatworms includes their unique cell-within-a-cell arrangement, the intracellular development of their embryos, their ciliated larvae and adults, the polyploid nucleus of their axial cell, and their alternation of asexual and sexual generations. Furthermore, the percentage of combined guanine and cytosine in the DNA of flatworms (35%–50%) is considerably higher than that measured in mesozoans (*Dicyemennea* has 23%).

A-6 Platyhelminthes

Greek *platys*, flat; *helmis*, worm

Dipylidium
Dugesia
Echinococcus
Fasciola
Hymenolepis
Opisthorchis
Planaria
Procotyla
Schistosoma
Taenia

Platyhelminthes are the flatworms, ribbon shaped and soft bodied. Of all the animals that have heads, they are the least complex: they have a mouth—an opening into a gut—but no rear exit or anus. They are acoelomates: no mesodermal coelom develops. However, platyhelminths are more complex than coelenterates (Phylum A-3) because they are bilaterally symmetrical and have organs composed of tissues and organized into systems.

Platyhelminths are a highly successful phylum of uncertain origin. They are adapted to a great variety of habitats. Some live in bat guano; others in the mantle fold of limpets. Members of nearly every other phylum—and certainly an enormous number of vertebrate species—play host to the ubiquitous flatworm.

Some species of free-living flatworms are marine; others dwell in moist soil or fresh water. Soil flatworms are mainly tropical, while aquatic ones are more abundant in temperate than in tropical waters. Taken as a whole, the phylum tolerates an immense temperature range—from $-50°C$ to $47°C$. Many flatworms frequent certain submerged plants and are easily collected by baiting the plants with meat. Some secrete mucus and glide slowly over it on cilia. They are often seen gliding over surface films on water, soil, and plants.

There are about 15,000 species of flatworms. Some species are white; others are richly colored. Some are bright green because they harbor symbiotic algae called zoochlorellae. The largest platyhelminths are the tapeworms, which reach a length of more than 10 meters; the smallest are barely visible to the naked eye. The flattened body of platyhelminths ensures large surface area for exchange of oxygen, carbon dioxide, and ammonia. Gases diffuse into the tissues from the host digestive system or from the water. Flatworms entirely lack a blood vascular system. Because they lack an anus, they void waste through their mouth as coelenterates do. Ciliated flame cells regulate water content by wafting liquid through protonephridia, systems of tubes that collect dissolved wastes and release them from the body through pores.

The nervous system is very simple. It is composed of pigment-cup eyespots (either single or in groups), a cluster of nerve cells in the head, and longitudinal nerve cords. Free-living flatworms detect food, chemicals, objects, and currents with the sides of the head, which in some bears sensory pits or tentacles. If they wander away from a source of scent, they turn more frequently, so that eventually they arrive at the source.

Flatworms are capable of extensive regeneration. Slices of *Dugesia* regenerate to form entire worms. In some flatworm species, an individual multiplies asexually by pinching itself into two parts—each fragment becomes a complete new animal. Most free-living flatworm species are hermaphroditic. Each individual bears eggs in ovaries and sperm in testes in the posterior region of the body. However, self-fertilization is rare. Sperm are deposited in a copulatory bursa or injected into the body as a mating pair mutually copulates. Fertilized eggs are laid in cocoons strung in ribbons—in freshwater species, they are often attached to stones. In most free-living flatworms, the eggs hatch into miniature adults, but many parasitic flatworms have complex reproductive cycles, with a succession of larval stages. Some flatworms are parthenogenetic: females asexually produce females.

There are three classes: Turbellaria (free-living flatworms), Trematoda (flukes), and Cestoda (tapeworms).

Turbellarians are carnivores and scavengers. They eat insects, crustaceans, other worms, bacteria, flagellates (Phylum Pr-8), ciliates (Phylum Pr-18), and diatoms (Phylum Pr-11). Using a tubular, muscular pharynx, which may project through the mouth, on the ventral side, they suck out the soft parts of their prey. Food is digested by enzymes secreted into the gut; intestinal cells also engulf whole food by forming food vacuoles. Primitive turbellarians probably were the ancestors of other platyhelminths, nematodes (Phylum A-14), and gastrotrichs (Phylum A-9).

Flukes are internal or external parasites of animal hosts, to which they attach by means of hooks and suckers. Tapeworms are internal parasites of animals—they are very common in the alimentary tract of vertebrates. Their larvae often develop in one or more intermediate hosts, animals of different species. Tapeworms, which are obligatory parasites, lack a gut altogether. Both fluke and tapeworm surfaces are covered with microvilli, small tissue projections that absorb nutrients (amino acids and sugars) from the parasitized host by active transport. Some flukes also feed through oral suckers. The suckers of tapeworms, however, serve only for attachment to the host.

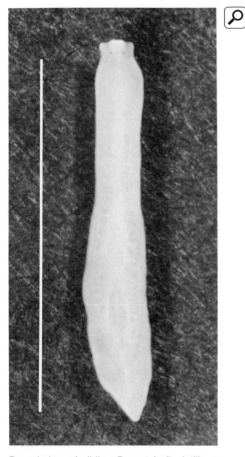

Dorsal view of gliding *Procotyla fluviatilis,* a live freshwater turbellarian flatworm from Great Falls, Virginia. Its pharynx is visible through its translucent body. Bar = 1 cm. [Photograph courtesy of R. Kenk; drawing by L. Meszoly, information from R. Kenk.]

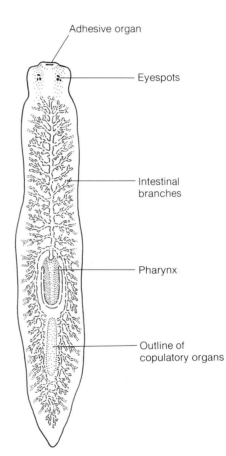

Adhesive organ

Eyespots

Intestinal branches

Pharynx

Outline of copulatory organs

A-7 Nemertina

Greek *Nemertes,* a sea nymph

Amphiporus
Cephalothrix
Cerebratulus
Emplectonema
Geonemertes
Lineus
Malacobdella
Nectonemertes

Paranemertes
Prostoma
Tubulanus

Nemertines are also called ribbon worms. Their unsegmented bodies are velvety, fragile, and flat. Their characteristic structure is a long, sensitive anterior proboscis. This unusual organ is not connected with the digestive tract, but resides in a special body cavity, from which it can be extended to as much as three times the length of the body of the worm. In some worms, it is branched. It is used to explore the environment and to capture prey.

Unlike flatworms (Phylum A-6), nemertines have an anus and a blood vascular system. The blood is pumped in all directions through the body. Depending on the species, the blood may be colorless, red, yellow, or green. Nemertines respire through their outer epidermis. The excretory system of these worms resembles that of flatworms: they have surface pores and tubes lined with ciliated flame cells, which regulate salts, water, and, possibly, waste. The nervous system of nemertines is also like that of flatworms: they have eyespots (as many as several hundred), a lobed brain, and nerves that run longitudinally down the body.

Most ribbon worms live in the sea. Some species live in the intertidal zone; others live down to 4000 meters. *Tubulanus* lives in a mucus tube that it secretes around itself. *Prostoma* is a freshwater nemertine that lives on aquatic plants in quiet pools in the United States and Europe. Some ribbon worms, such as *Geonemertes,* live in the soil of subtropical forests and, when introduced, thrive in greenhouse soils.

Although nemertines are abundant in the intertidal zone, they are rarely seen because they hide during the day and are active at night. They burrow by elongating to shove their anterior end into the mud; then they shorten and swell, forcing an opening in the mud. On tidal mudflats they may be found among algae, mussels, and tube-dwelling annelids. They will sometimes creep out from among seaweed placed in dishes of sea water.

About 750 nemertine species are known. They range from less than 0.5 mm to 30 meters in length. For example, *Lineus longissimus,* the iridescent bootlace worm, about 30 meters long, is one of the longest invertebrates known. *Cerebratulus lacteus* can extend itself from about 1 meter to 10 meters long. The members of the genus *Emplectonema* are bioluminescent, but no other luminescent nemertines are known. Most nemertines, especially the bottom dwellers, are pale. A few are striped, speckled, or marbled multicolor ribbons.

Pelagic species of nemertines swim with lateral undulations. Others move by means of cilia over a slime track, or by muscular contractions. Some may attach themselves to an object with their proboscis and pull themselves forward.

Nemertines feed on a wide variety of prey: protists, diatoms, annelids, crustaceans, flatworms, mollusks, roundworms, and

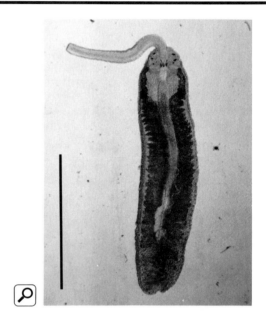

even small fish. Muscular pressure on the fluid in the proboscis chamber forces explosive eversion of the proboscis. The proboscis of some nemertines, such as *Prostoma,* has a venomous stylet with which the worm stabs its prey repeatedly, paralyzing it. The proboscis is very accurate. Live prey are sucked whole into the mouth or juices are sucked out. Cilia move food along the gut. Phagocytosis and intracellular digestion occur in the intestine, which has numerous pouches. The proboscis also may be shot out in self-defense. Irritated ribbon worms release their proboscis, which then regenerate. Nemertines are the prey of crustaceans, annelids, and other invertebrates. Along the Atlantic coast of North America they serve as fish bait, but they are not edible by human beings.

Nemertines are capable of both sexual and asexual reproduction. In most species, the sexes are separate. The worms reproduce asexually, by fragmentation, only in summer; a complete worm grows from each fragment. In sexual reproduction, numerous temporary gonads (ovaries or spermaries) form in mesenchyme tissue between intestinal pouches. Each gonad has its own opening, through a surface pore, to the outside. Eggs are laid in gel strings, and fertilization typically takes place in the

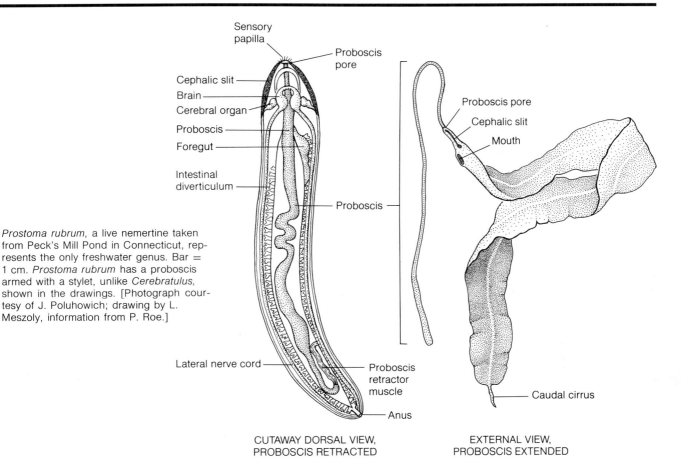

Sensory
papilla

Proboscis
pore

Cephalic slit

Brain

Cerebral organ

Proboscis

Foregut

Intestinal
diverticulum

Proboscis

Proboscis pore

Cephalic slit

Mouth

Prostoma rubrum, a live nemertine taken from Peck's Mill Pond in Connecticut, represents the only freshwater genus. Bar = 1 cm. *Prostoma rubrum* has a proboscis armed with a stylet, unlike *Cerebratulus,* shown in the drawings. [Photograph courtesy of J. Poluhowich; drawing by L. Meszoly, information from P. Roe.]

Lateral nerve cord

Proboscis
retractor
muscle

Anus

Caudal cirrus

CUTAWAY DORSAL VIEW,
PROBOSCIS RETRACTED

EXTERNAL VIEW,
PROBOSCIS EXTENDED

water. The eggs develop either directly into adults or first into a dipilidium larva, which looks like a ciliated cap with ear flaps and a topknot of undulipodia. Some species are hermaphrodites capable of self-fertilization. Members of the hermaphroditic terrestrial genus *Geonemertes* bear live young. In *Nectonemertes,* a genus of active swimmers, males clasp females with special attachment organs.

The fossil record of nemertines is sparse. The Cambrian *Amiskwia* may be a nemertine fossil, but it is also claimed to be a chaetognath (Phylum A-30). Nemertines probably evolved from flatworm ancestors. Evidence of the common ancestry of flat-worms (Phylum A-6) and nemertines is ciliated epidermis, absence of a coelom and of respiratory organs, similarity of excretory and sensory systems, and the presence of multiple reproductive organs along the body. Some flatworms also have an eversible proboscis. Flatworms, sipunculans (Phylum A-21), mollusks (Phylum A-19), and arthropods (Phylum A-27) all resemble nemertines in the development of the fertile egg—they all undergo spiral cleavage. Nemertines, the most complex of the acoelomate invertebrate protostomes, have been proposed to have common ancestors with deuterostomes (Phyla A-28 through A-32).

A-8 Gnathostomulida

Greek *gnathos,* jaw; *stoma,* mouth

Austrognatharia
Gnathostomula
Haplognathia
Nanognathia
Problognathia
Semaeognathia

Gnathostomulids are microscopic acoelomate marine worms that have a characteristic feeding apparatus made of a hard noncellular material. Gnathostomulids are found mainly in shallow salty waters and in the spaces between sand grains. The worms are tiny—they can flow through pores 900 μm^2 in cross-sectional area. Species are known from Maine, the Florida Keys, Bimini, the Caribbean, Bodega Bay (California), the Indo-Pacific, and the White Sea.

About 80 species in 18 genera of gnathostomulids have been described so far, and there are probably more than a thousand species alive today. They live in the sulfuretum—deep, black, fine sand smelling of hydrogen sulfide produced by bacteria—and also on the thallus (leafy parts) of algae or on leaves of marine plants, such as the eel grass *Zostera* and the marsh grass *Spartina* (Pl-9). In certain sediments, gnathostomulids may outnumber even nematodes (Phylum A-14): there can be more than 6000 individuals per liter of sediment. Despite their enormous numbers, gnathostomulids remained unrecognized and undescribed for so long because they stick to sand and other particles and because they degenerate so quickly after death. It requires sophisticated extraction techniques to pull them off these surfaces alive.

Gnathostomulids are hermaphrodites. The single ovary produces large eggs, which mature one at a time. Posterior to the ovary is at least one testis. They do cross-fertilize—some, like the *Problognathia* shown here, have a penis stiffened by a stylet. One gnathostomulid injects sperm packets between the skin and gut of his mate. The undulipodiated sperm swim to a storage sac called a bursa. Fertilization takes place internally, after which the fertilized egg ruptures the body wall and is released. Development is direct, from egg to adult hermaphrodite worm; there are no larval stages. However, at least in some species, a nonsexual feeding stage may alternate with a distinct nonfeeding sexual stage—taking a year to complete the cycle.

Gnathostomulids have transparent bodies 0.3 mm to 1.0 mm long; a slight constriction separates the head from the trunk, and there is no external cuticle. They have no circulatory or respiratory organs and no skeleton. There is a modest nervous system in the outer epithelial layer. In some species, the head bears well-developed sense organs—stiff bundles of cilia and pits lined with cilia, collectively called the sensorium. On the belly side of the head are the feeding structures: a hard comblike basal plate (used for grazing on films of bacteria, protists, and fungi), two lips, a mouth, and a pair of toothed lateral jaws within a muscular pharynx. The jaws work rapidly, snapping open and closed in about a quarter of a second. Particles of food are passed into the gut, a tubular sac having no anus—undigested food leaves by the mouth. A pair of excretory organs, each made of two or three cyrtocytes (curved cells), is located on the sides of the bursa.

Gnathostomulids move by using their body cilia and by rapid muscle contractions. They nod their heads from side to side, swim, glide, constrict, and twist by using three or four paired longitudinal muscles. The cells of the external epithelium bear only one cilium each, and the ciliary propulsion can reverse direction—these traits distinguish gnathostomulids from flatworms (Phylum A-6), which they otherwise resemble. Both phyla have body cilia, lack a coelom and anus, and are hermaphrodites. In fact, gnathostomulids have been recognized as an independent phylum only since 1969. Evidence against a close relationship of the two phyla is that sperm tails of gnathostomulids have a typical 9+2 cross section (see Figure 2 in the Introduction to this book), whereas those of flatworms have a 9+1 cross section. Their jaw structure may relate gnathostomulids to rotifers (Phylum A-10); on the other hand, monociliated epithelium has been found only in gastrotrichs (Phylum A-9) and gnathostomulids.

It may be that some famous fossils called conodonts are remains of the hard parts of ancient gnathostomulids. They are toothlike fossils found in black, silty shale deposited under anaerobic conditions from the late Cambrian to late Triassic, from about 500 million to 200 million years ago. It is surmised that conodonts were used as "teeth" to tear apart fungal hyphae and algal mats. However, the basal plates of modern gnathostomulids differ from conodonts in that they are made of tough organic matter rather than of calcium phosphate.

An adult *Problognathia minima*. It glides between sand grains in the intertidal zone and shallow waters off Bermuda. LM (phase contrast), bar = 0.1 mm. [Photograph courtesy of W. Sterrer (*Trans. Am. Micro. Soc. 93*:357–367); drawing by L. Meszoly, information from W. Sterrer.]

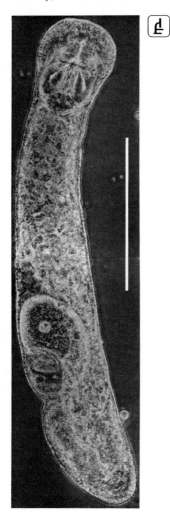

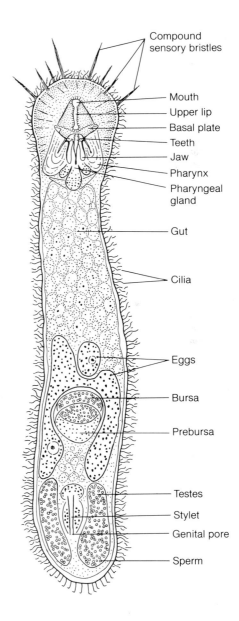

Compound sensory bristles

Mouth
Upper lip
Basal plate
Teeth
Jaw
Pharynx
Pharyngeal gland

Gut

Cilia

Eggs

Bursa

Prebursa

Testes
Stylet
Genital pore
Sperm

A-9 Gastrotricha

Greek *gaster*, stomach; *thrix*, hair

Chaetonotus
Dactylopodola
Lepidodermella
Macrodasys
Tetranchyroderma
Turbanella
Urodasys

Gastrotrichs are free-living, bilaterally symmetrical, transparent wormlike animals having lobed heads. From their sides, primarily near their posterior ends, project from 2 to 250 adhesive tubes that are used for clinging to vegetation or other surfaces. Their bodies are flat and unsegmented, ranging from about 0.05 mm to 4 mm in length. Their backs and sides are usually spiny, bristly, or scaly.

About 400 species of gastrotrichs are known. They are abundant in aquatic environments all over the world. Marine species favor the surfaces of coral or intertidal and subtidal sands. Freshwater gastrotrichs prefer bogs, mossy pools, and plant-choked ponds rather than clean running water. They may be seen crawling over water lily leaves or on the underside of duckweed.

Gastrotrichs are pseudocoelomate animals lacking circulatory, respiratory, and skeletal organs. Fluid fills the pseudocoelom, a body cavity that both acts as a hydraulic skeleton and provides the medium for the exchange of oxygen, carbon dioxide, food, and waste. The body is covered by a thin cuticle that lacks chitin. The ventral surface bears cilia in distinctive patterns that are important for classification. Currents induced by the movements of these cilia bring food—organic debris, algae, foraminiferans (Phylum Pr-17), diatoms (Phylum Pr-11)—to the mouth. Food is moved farther into the body by cilia waving in the muscular, pumping pharynx.

Some gastrotrichs have a pair of protonephridia, rudimentary excretory tubules having blind inner ends. In the blind ends, flame cells bearing quivering cilia help to move collected waste out through a midventral excretory pore. Marine gastrotrichs lack flame cells. Muscles that encircle the body move the adhesive tubes and the bristles; longitudinal muscles steer the animal as it swims. The nervous system consists of a large brain (which surrounds the pharynx), a pair of longitudinal nerve cords, and, on the head and trunk, sensory bristles with tufts of sensory cilia. Red spots on both sides of the brain may be photoreceptors.

Marine gastrotrichs are hermaphrodites. In most species, individuals produce eggs and sperm alternately but not from the same gonad; that is, they are protandric. In *Dactylopodola* and *Urodasys,* for example, spermatophores (packets of sperm) are transferred from individuals that are behaving as males to those that are behaving as females. The end of the sperm duct serves as a penis for sperm transfer. Freshwater species of gastrotrichs lack males (except *Lepidodermella squammata*); the females reproduce parthenogenetically. Rodlike sperm are found in some *Lepidodermella squammata* individuals that also contain ova. Female gastrotrichs are not very prolific: they lay from one to five large eggs in a lifetime, depositing them on algae, discarded skeletons, or pebbles. When water touches the egg, an egg shell forms. Freshwater gastrotrichs produce two egg types. One type is modified to resist harsh conditions—it must dry, freeze, or be exposed to high temperatures before it will develop. The other type, thin-walled eggs, cleave as soon as they are laid. There is no larval stage; the young resemble the adults.

Gastrotrichs are of only indirect importance to humans. They serve as food for amoebas (Phylum Pr-3), hydras (Phylum A-3), turbellarian flatworms (Phylum A-6), insects and crustaceans (Phylum A-27), and annelids (Phylum A-23) in aquatic food webs.

Because gastrotrichs retain the embryonic blastocoel as the adult body cavity, they are thought to be related to other pseudocoelomate invertebrates (Phyla A-9 through A-15). Their body form, musculature, and protonephridia with flame cells are like those of rotifers (Phylum A-10); however, unlike rotifers, gastrotrichs have no crown of cilia. Gastrotrichs are probably most closely related to nematodes (Phylum A-14). Both have ornamented cuticles, complete digestive systems, a muscular pharynx, and adhesive tubes. In both phyla, transverse sections of gut are triangular in shape. Turbellarian flatworms also resemble gastrotrichs. Both glide on cilia, are hermaphroditic, digest food intracellularly, have protonephridia, and have sense organs that are ciliated pits on the sides of the head.

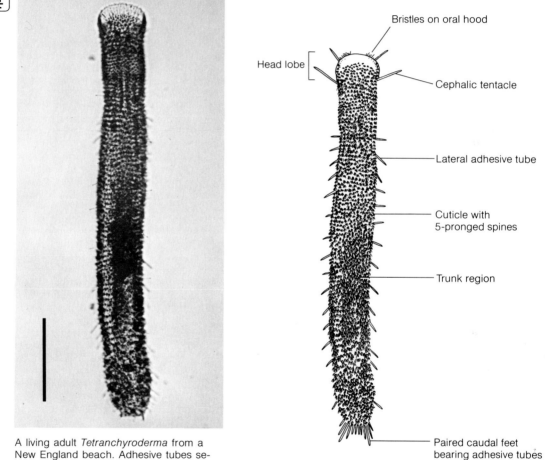

A living adult *Tetranchyroderma* from a New England beach. Adhesive tubes secrete materials that anchor it temporarily to sand in the intertidal zone. LM, bar = 0.1 mm. [Photograph courtesy of W. Hummon; drawing by I. Atema, information from W. Hummon.]

Bristles on oral hood

Head lobe

Cephalic tentacle

Lateral adhesive tube

Cuticle with 5-pronged spines

Trunk region

Paired caudal feet bearing adhesive tubes

A-10 Rotifera

Latin *rota*, wheel; *ferre*, to bear

Albertia
Brachionus
Chromogaster
Conochilus
Cupelopagis
Embata
Euchlanis
Floscularia
Notommata

Philodina
Proales
Seison
Synchaeta

Rotifers are common little aquatic animals named for the crown of cilia at their heads. The cilia in motion give the crown the appearance of a rapidly rotating wheel. Rotifers have complex jaws, which, with only a hand lens, can be seen to grind rapidly. Their bodies are bilaterally symmetrical and covered with an external layer of chitin, called a lorica; in many species, the lorica can collapse down like a telescope. Their shapes range from trumpetlike to spherical. Free-swimming rotifers, such as *Brachionus,* have spines that enhance flotation and stalks on which their eggs are carried externally. In some rotifers, the wheel cilia are fused into tentacles or platelike structures, or even are lacking altogether.

Rotifers are usually transparent, but some appear green, orange, red, or brown from colored food in their gut, which can be seen right through the body wall. They are among the smallest of the animals, ranging from 0.04 mm to 2.0 mm in length.

Rotifers entirely lack a blood circulatory system and respire through their body surface. Posterior to the jaws and mouth is the mastax, a muscular pharynx containing several hard parts called trophi that, in different species, are designed for pumping, grabbing, or grinding. Food passes through the mastax into the stomach, into which gastric glands secrete digestive enzymes. Dissolved wastes are collected within a pair of protonephridia, convoluted tubules that drain to a bladder or to the intestine. The flow is maintained by the rapidly beating cilia of flame cells. Rotifer muscles are organized into scattered bundles rather than into layers. Dorsal to the mastax is the small brain, from which paired nerves extend toward the posterior. Rotifers have eyespots. *Asplanchna brightwelli* has photoreceptors containing membranes layered like cabbage leaves, but they are not formed from cilia, as those of vertebrates are. *Euchlanis* photoreceptors actually contain a lens.

About two thousand rotifer species have been reported; only about fifty are marine. Most are free-swimming individuals, although *Conochilus* swims as a revolving colony of about a hundred individuals anchored in a jelly sphere. Others are sedentary, secreting jelly cases of, like *Floscularia,* building exquisite tubes. Rotifers are perhaps the most common, abundant, and cosmopolitan of the freshwater zooplankton. They live in bogs, sandy beaches, lakes, rivers, and glacial muds. They also inhabit gutters, ditches, moss and lichen pads, rocks, and tree bark; they resist desiccation by secreting a protective envelope of gel around their bodies.

Some rotifers are symbiotic or parasitic. *Seison* moves leech-like on crustacean gills, nourished both by eating its host's eggs and by food in the water current. *Proales* lives on *Daphnia,* the common freshwater flea (Phylum A-27), in snail eggs, in the heliozoan *Acanthocystis* (Pr-16), in *Vaucheria* filament tips (Phylum Pr-9) and in *Volvox* (Phylum Pr-15)—all of these hosts are regular and abundant inhabitants of ponds and lakes. The rotifer *Albertia* is a wormlike obligatory parasite in the coelom of annelids (Phylum A-23).

Free-living rotifers typically have eclectic diets feeding on bacteria, protists, other rotifers and small animals, and suspended organic matter. The sessile rotifer *Cupelopagis* traps protists by means of a retractile funnel. *Chromogaster* specializes in grabbing dinoflagellates (Phylum Pr-2) with its pincer-like mastax. It drills through the test of the prey and sucks it dry.

Many rotifer populations contain only females. These reproduce parthenogenetically, producing eggs that develop only into females. In most, the young are born live: the eggs mature in a germinarium (ovary) and the embryos develop inside a structure called the vitellarium from which they are released from their mothers. Some rotifers, such as *Brachionus,* carry their eggs on external stalks. Some species produce two kinds of eggs. Diploid eggs, which hatch unfertilized into female adults, are the rule; however, when the home ponds dry up, haploid eggs are formed. If they are not fertilized, these eggs hatch into small, degenerate males. These males, although incapable of even feeding themselves, may produce sperm and fertilize other haploid eggs. They produce two distinct types of sperm: ordinary ones, which fertilize the egg after they penetrate the female body wall, and rod-shaped bodies thought to assist the regular sperm. The products of fertilization are heavy-shelled ''resting'' or ''winter'' eggs. In spring, they hatch into ordinary rotifer females that produce, without any aid from male sexual partners, from 20 to 40 generations of female young per year.

Rotifers are a major source of food for other animals in freshwater ecosystems. They also aid in soil decomposition. Rotifer fossils are unknown. They probably evolved from bilateral, ciliated, flatwormlike ancestors, because their pharynx, cilia, and flame cells resemble those of flatworms.

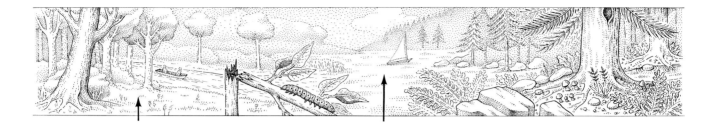

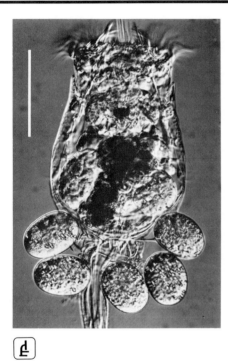

Brachionus calyciflorus, a live freshwater female rotifer. The eggs are attached by thin filaments until they hatch. LM (interference phase contrast), bar = 0.1 mm. [Photograph courtesy of J. J. Gilbert; drawing by L. Meszoly, information from J. J. Gilbert.]

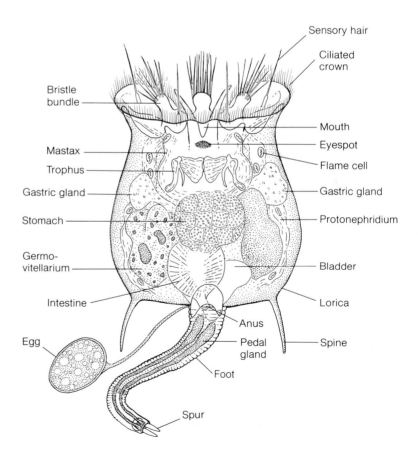

Greek *kinein,* to move; *rynchos,* snout

Campyloderes
Caterla
Centroderes
Echinoderes
Pycnophyes
Semnoderes
Trachydemus

Kinorhynchs are small, free-living, wormlike pseudocoelomates found exclusively in marine habitats. They are named for their mode of locomotion—they pull themselves along by their snouts. Most are less than a few millimeters long. Adults have spiny heads, necks, and eleven or twelve trunk segments called zonites. Their bodies are roughly triangular in cross section. Either the head or both head and neck can retract into the trunk, which is covered with hard cuticle plates. The cuticle between the plates is very flexible; in the retracted position, the plates close over the head and neck. The dorsal plates bear movable median and lateral spines.

Kinorhynchs dwell in silt-covered and muddy bottoms between high and low tidemarks. They have been collected from the Black Sea, the North Sea, the English Channel, the Canary Islands, North America, Zanzibar, Japan, and Antarctica. Kinorhynchs have no direct economic importance to people but, like other marine wormlike animals, they are part of food webs.

About one hundred species of kinorhynchs are known; most are yellow-brown in color. As in gastrotrichs, the fluid of the body cavity serves in respiratory, circulatory, and skeletal functions. The cuticle, which lacks cilia, is not made of cells. Kinorhynchs are unable to swim. They burrow by forcing fluid into their spiny head; the head is extended and anchored. Then, as the head is retracted, it hauls the trunk ahead. Food—diatoms (Phylum Pr-11) or organic matter in mud—is sucked through the anterior mouth by a muscular, barrel-shaped pharynx, from which it passes into the complete digestive tract. Digestion is believed to take place in the foregut. The anus is located at the posterior end of the trunk. Excretion and water balance are regulated by a pair of protonephridia, tubes that contain ciliated cells; these tubes collect dissolved wastes and release them to the exterior through an opening in the eleventh zonite.

The kinorhynch nervous system is composed of several components. A nerve ring circles the pharynx, and a single nerve cord extends down the ventral side of the body. In addition, there are scattered clusters of ganglia, sensory bristles on the trunk, and, in some species, red ocelli, which are eyelike photosensitive organs lying behind the mouth cone.

The sexes of kinorhynchs are separate, but there are no obvious external distinguishing features. Males have paired testes; females have paired ovaries. From these gonads, a duct called the gonoduct opens on the terminal zonite. The ovary is syncytial; it contains both germinal nuclei that undergo meiosis to form eggs and nutritive nuclei that are not directly involved in reproduction. Although copulation has not been observed, reproduction is assumed to be sexual. Fertilized eggs, which develop ex-

ternally, have been seen in shed cuticles. Larvae differ from adults in that they lack spiny heads and complete guts. The larval cuticle undergoes at least five successive molts as the young become adult.

Because no kinorhynch fossils are known, kinorhynch evolution must be inferred by comparing living organisms. Kinorhynchs are similar to rotifers, gastrotrichs, and nematodes in a number of ways—the presence of protonephridia, the proximity of the epidermis and the nervous system, the presence of adhesive tubes (the pedal gland of rotifers and the caudal gland of nematodes) and copulatory spicules. The kinorhynch body is usually

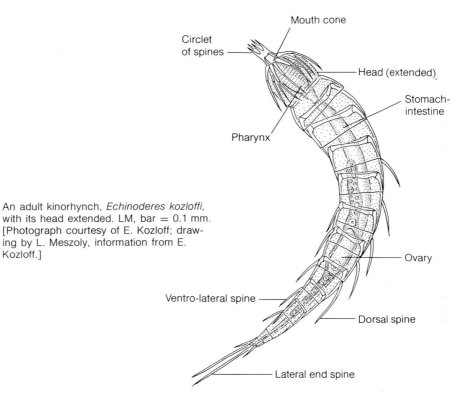

Circlet
of spines

Mouth cone

Head (extended)

Stomach-
intestine

Pharynx

Ovary

Ventro-lateral spine

Dorsal spine

Lateral end spine

An adult kinorhynch, *Echinoderes kozloffi,*
with its head extended. LM, bar = 0.1 mm.
[Photograph courtesy of E. Kozloff; drawing by L. Meszoly, information from E.
Kozloff.]

forked at the hind end; this and its spiny cuticle are reminiscent
to those of gastrotrichs. Because they lack circular muscles,
kinorhynchs resemble nematodes (A-14). The segmentation of
the kinorhynch cuticle superficially resembles the segmentation
of arthropods (Phylum A-27). However, there is no corresponding internal segmentation, which in arthropods includes mesodermal structures. Thus the external similarity of kinorhynchs
and arthropods is due to convergent evolution. All things considered, it is most likely that kinorhynchs, like other pseudocoelomates (Phyla A-9 through A-15), are descended from free-living
primitive flatworms.

A-12 Acanthocephala

Greek *akantha*, thorn; *kephale,* head

Acanthocephalus
Acanthogyrus
Echinorhynchus
Gigantorhynchus
Leptorhynchoides
Macracanthorhynchus
Moniliformis
Neoechinorhynchus
Polymorphus

All acanthocephalans, or spiny-headed worms, are gut parasites of vertebrates (Phylum A-32), usually of carnivores. They are bilaterally symmetrical and either unsegmented or only superficially segmented—their internal organ systems are never segmented. They have a protrusible, spiny proboscis, which is globular or cylindrical. The spines anchor the worms to the gut wall of their hosts. In some of them, the body is also spiny.

More than 600 species of spiny-headed worms have been described. Adult females are usually about 2 cm long; a few are as short as 1 mm, and the largest is longer than 1 meter. Males tend to be smaller than females of the same species. Acanthocephalans may have a wrinkled or a smooth external surface; it may be whitish, yellow, orange, or red.

Acanthocephalans live in soil and in fresh or salt water nearly anywhere in the world—depending on the habitat of their hosts. Because they have no free-living stage, they are seldom seen, but hundreds of adult acanthocephalans may live in a single vertebrate animal. They have been reported from fish, turtles, snakes, amphibians, birds, and mammals such as insectivores, dogs, hyenas, squirrels, shrews, moles, rats, pigs, dolphins, and seals.

The sexes are separate and reproduction is entirely sexual. The female produces eggs in ovarian balls, which are either attached inside posterior to the proboscis or lie free in the body cavity. The males have a pair of testes, from which sperm ducts lead to a penis. Adult worms mature sexually in the gut of their vertebrate host, where they copulate. During mating, seminal fluid from Saefftigen's pouch is injected into the wall of a male structure called the bursa, which thus everts. The bursa holds the female while sperm are released through the penis into her vagina. Secretions of a cement gland are also released through the penis; they cap the posterior end of the female to prevent loss of the sperm. Both fertilization and embryonic development take place inside the female.

After the embryonic stages, cell membranes in the larva break down and most tissues, except for gonads, become syncytial. Through a gonopore, the female releases shelled larvae, called acanthors, into the intestine of her host, from which they are expelled with the host's feces. They are picked up from soil or water by an intermediate host, arthropods such as roaches, grubs, aquatic crustacea (Phylum A-27) or mollusks such as snails (Phylum A-19). The acanthors hatch and metamorphose into a second larval stage, called an acanthella, which grows and develops in the intermediate host. Development stops short of adulthood, and the acanthella enters an encysted resting stage, called a cystacanth. Vertebrates become infected when they ingest an intermediate host harboring a cystacanth.

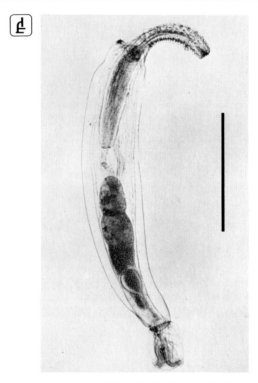

A *Leptorhynchoides thecatus,* a young male parasitic acanthocephalan from the intestine of a black bass, *Micropterus salmoides.* LM (worm fixed and stained), bar = 1 mm. [Courtesy of S. C. Buckner.]

Intermediate hosts are often the prey of other invertebrates. Thus, acanthocephalan larvae often are transferred to two or three different invertebrate hosts, in which they bore through the gut wall and encyst but do not develop further. In these intermediate hosts, called transport hosts, the acanthocephalans cannot develop to adulthood, but they remain infective.

Circulatory, digestive, and respiratory organs are lacking in both adults and larvae. Little is known of acanthocephalan nutrition. They probably absorb digested food from their host through their thin cuticle and circulate nutrients within their body wall

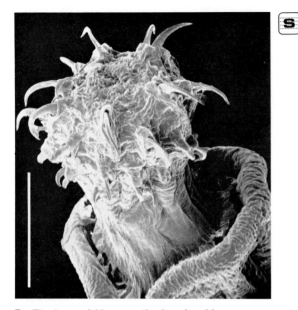

B The larva of *Macracanthorhynchus hirudinaceus* showing its everted proboscis. The hooks are used to penetrate tissues of the worm's beetle hosts. SEM, bar = 0.4 mm. [Courtesy of T. T. Dunagan (*Trans. Am. Micro. Soc.* 90:331, 1969).]

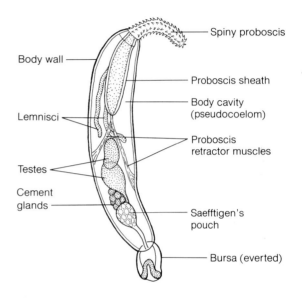

C A young male *Leptorhynchoides thecatus*. [Drawing by L. Meszoly, information from B. Nickol.]

through canals that do not open to either the exterior or their body cavity. The excretory organs consist of ciliated flame bulbs in collecting tubules, from which urine drains into the sperm duct in males, and the uterus in females.

The nervous system is very simple. Nerves extend to the tissues from a ventral cerebral ganglion in the proboscis sheath. The reproductive organs have touch receptors, and there is a sensory pit at the tip of the proboscis.

The innermost body-wall layer contains thin circular and longitudinal muscles. When these muscles contract the body, the proboscis is everted by hydrostatic pressure in the lemnisci, special fluid reservoirs; it is retracted directly by muscles. Larvae move by contracting muscles in the proboscis itself.

The body cavity is a pseudocoel rather than a true coelom, because it develops from the hollow cavity of the blastula itself. The pseudocoel lacks the epithelial lining, called the peritoneum, that is typically present in coelomate animals.

Acanthocephalans show little evidence of relationships to other phyla. However, the ultrastructure of sperm, body wall, and muscles resembles that of rotifers (Phylum A-10), gastrotrichs (Phylum A-9), nematodes (Phylum A-14), kinorhynchs (Phylum A-11), nematomorphs (Phylum A-15), and priapulids (Phylum A-20). They differ from members of these phyla, however; for example, they are unlike nematodes in that they lack a gut and have a proboscis, circular body-wall muscles, ciliated excretory organs, and unique reproductive and developmental features.

A-13 Entoprocta

Greek *entos*, inside; *proktos*, anus

Barentsia
Loxosoma
Loxosomella
Pedicellina
Pedicellinopsis
Urnatella

Entoprocts are an obscure group of small marine animals that do not even have common names. Most are sessile, living in colonies firmly attached by stalks and basal disks to rocks, pilings, shells, or other animals. At the free end is a circle of ciliated tentacles that contract and fold. Depending on the species and the age of the specimen, there are from six to thirty-six tentacles. Their lightly tinted, transparent bodies are called calyxes. Some entoprocts look and creep like hydroid coelenterates (A-3) but, unlike coelenterates, they are unable to withdraw their tentacles into their bodies. The whole animal—tentacles, calyx, and stalk—may be as long as 10 mm.

About 60 species are known. Many form large populations that extend in shallow marine waters to form "animal mats." A few species are solitary, such as *Loxosomella davenportii,* one of the few entoprocts that is mobile as an adult. It somersaults with its tentacles over its basal disk. Entoprocts are widely distributed along the coasts of Africa, South and North America, Asia, Europe, and the Arctic. *Urnatella,* the only freshwater form, oddly enough is found only in two very separate locations—in eastern and midwestern parts of the United States and in India.

Entoprocts feed on particles suspended in water, but they do not filter them out. Diatoms (Phylum Pr-11), desmids (Phylum Pr-14), other plankton and detritus are trapped on the cilia of the tentacles, which move the food into the mouth. Then it passes through a ciliated, U-shaped digestive tract complete with anus. The only muscles in the calyx are sphincters at junctions between parts of the digestive system.

Entoprocts lack hearts and blood vascular systems. Uric acid and guanine, discharged from the stomach surface into the stomach, are excretory products. These are collected in a protonephridium, two ciliated flame bulbs in ducts that lead to the nephroporus, a single excretory pore near the mouth.

The nervous system of entoprocts is simple: a ganglion lying in the loop of the alimentary canal and nerves extending to tentacles, stalk, body, and a creeping horizontal stalk called a stolon. Various specialized cells distributed about the body are thought to be sensitive to chemicals, light, and vibrations.

Entoprocts reproduce both asexually and sexually. They regenerate readily, and colonies commonly arise by budding. Most entoprocts are hermaphrodites. Each individual bears two testes and two ovaries; the reproductive cells exit the body through a common structure, the gonopore. Some species are dioecious, but males and females show no conspicuous external sex differences—both have two gonads opening to a single gonopore, which is located in the vestibule, the area surrounded by the tentacles. Entoproct eggs are probably internally fertilized by

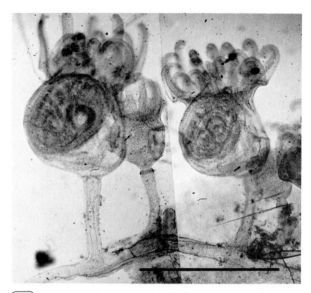

A A living laboratory culture of *Pedicellina australis,* a marine colonial entoproct from European waters. LM, bar = 1 mm. [Courtesy of P. H. Emschermann.]

undulipodiated sperm. The zygotes are incubated in a brood pouch between the gonopore and the anus. They develop into embryos, which are nourished by cells of the parent's body wall. The embryos develop into ciliated, free-swimming larvae resembling the trochophore larvae of annelids (Phylum A-23) and mollusks (Phylum A-19). The larvae settle, attach, and then metamorphose into adults.

Unfortunately, the soft-bodied entoprocts have left no recognizable fossils and lack close living relatives. Some zoologists suggest that, because of resemblances in body and intestine shape, rotifers (Phylum A-10) and entoprocts share a fairly recent common ancestor.

Entoprocts superficially resemble ectoprocts (Phylum A-16), with which they have often been confused. Some people believe that ectoprocts evolved from entoprocts because their larvae and patterns of settling are similar. However, these two phyla differ in many ways. The entoproct mouth and anus both open within the circle of tentacles, whereas the ectoproct anus opens

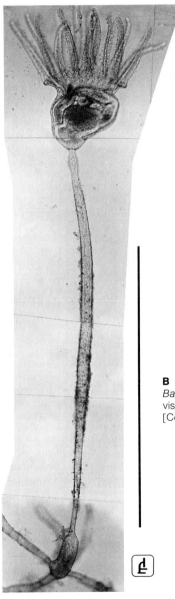

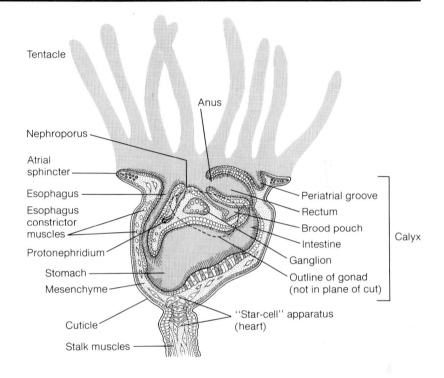

Tentacle

Anus

Nephroporus

Atrial
sphincter

Esophagus

Esophagus
constrictor
muscles

Protonephridium

Stomach

Mesenchyme

Cuticle

Stalk muscles

Periatrial groove

Rectum

Brood pouch

Intestine

Ganglion

Outline of gonad
(not in plane of cut)

Calyx

"Star-cell" apparatus
(heart)

B A single individual of the entoproct
Barentsia matsushimana. Rows of cilia are
visible on the tentacles. LM, bar = 1 mm.
[Courtesy of P. H. Emschermann.]

C A vertical cross section of *Barentsia
matsushimana.* [Drawing by L. Meszoly,
information from P. H. Emschermann.]

outside the circle of tentacles. The body cavity of entoprocts is a
pseudocoelom, whereas ectoprocts have a coelom. Entoproct
tentacles set up a feeding current that enters between the tentacles and leaves at the top of the body, whereas the currents of
ectoprocts enter at the top. Entoprocts trap food in mucus and
sweep it along ciliated tracts into their mouths, whereas ectoprocts use their tentacles to sweep food directly into their
mouths. Entoproct cilia beat in an uncoordinated fashion,
whereas ectoproct cilia beat in regular waves. Entoproct larvae
only superficially resemble the trochophore larvae of ectoprocts.
Entoprocts have protonephridia, whereas ectoprocts lack excretory systems.

A-14 Nematoda

Greek *nema*, thread

Ascaris
Caenorhabditis
Dioctophyme
Diplolaimella
Leptosomatum
Pelodera
Rhabdias
Trichinella
Tricoma
Wuchereria

Nematodes are slender and cylindrical. Because many of them have rounded ends, they are also called roundworms. Although they are covered with a proteinaceous, ornamented cuticle, they are nevertheless typically transparent. They lack both segmentation and cilia (except in sensory organs); this distinguishes them from the annelids (Phylum A-23) and the rotifers (Phylum A-10) respectively. On the other hand, the caudal gland at the posterior end is homologous with the pedal gland of rotifers. Because they lack circular muscles and have abundant longitudinal muscles, nematodes characteristically move by bending or flipping—they never inch along by extending and contracting in the way segmented worms do.

Nematodes have a complete straight digestive tract. They lack a differentiated proboscis, although many have well-developed teeth and are vicious predators, looking like microscopic dragons; others, like the infamous hookworm, have specialized mouthparts that hook them to the intestinal wall of their host. Three germ layers, ectoderm, mesoderm, and endoderm, are formed in embryonic development. They have a body cavity, but it is a pseudocoel rather than a coelom: it is usually not lined and it does not develop by the formation and enlargement of a space between mesodermal layers, but from a space between the embryonic endoderm and ectoderm.

The sexes are separate; in most species, the male is smaller than the female. Reproduction is always sexual. Males have copulatory spicules; some grip the female in mating. In all known cases, fertilization is internal and the gonads, which are single or paired overies and testes, lead out of the worm through a gonopore (in females) or the cloaca (in males). The reproductive capability of nematodes is prodigious. Some females have been known to contain 27 million eggs and extrude them at a rate of 200,000 eggs a day.

There are about 80,000 species of nematodes described in the scientific literature; however, it is estimated that nearly a million living species exist. These worms range from a few millimeters to nearly a meter in length (the female giant nematode *Dioctophyme renale;* the male is only half as long). Many are parasites of man and of domestic plants and animals, which is why some members of the group have been intensively studied.

Free-living roundworms are found in most moist habitats—soils, beach sand, salt flats, the ocean, hot springs, lake water—and they may be present in enormous numbers. As many as a billion individuals per acre have been counted in the top two centimeters of rich soil. Many live in plant tissue; some form root galls, whereas others live in fruits or in bark crevices. Most produce extremely hardy eggs that are well able to wait indefinitely until harsh conditions improve.

There are two classes. The Adenophorea lack phasmids and therefore are also called Aphasmidia. Phasmids are sense organs, possibly chemoreceptors, found in the tail region particularly of parasitic roundworms. Secernentea, or Phasmidia, do have phasmids. Many members of this class—in fact, entire suborders—are parasites of vertebrates or insects. The Strongylina (to which belong hookworms, some gapeworms, hairworms, stomach worms, and lungworms), the Ascaridina (to which belong *Ascaris* and pinworms), and the Spriurina (to which belong the filarial worms that cause the tropical disease elephantiasis) are all phasmidians.

Probably the most famous nematode in the United States is *Trichinella spiralis,* the causative agent of trichinosis—the disease can be acquired by eating infested pork that has not been cooked sufficiently. The minute larvae of these worms are harbored as cysts in the striated muscles of pigs, cats, dogs, rats, and bears. If flesh of an infested animal is eaten by another, the larval cysts are digested, liberating the larvae into the intestine of a new host. About two days after their release, they are sexually mature; male and female mate in the intestine. The females, about 4 mm long, then burrow into the muscles of the intestinal wall and produce hundreds of live larvae. (In this species, the females are ovoviviparous.) The larvae enter the lymph and are carried to the bloodstream. From there they burrow into skeletal muscles, where they coil up and become enclosed in cysts. The cysts may survive, remaining dormant for months, even years, or the host may deposit calcium salts in the parasites' tissues, calcifying the nematodes. Only when the skeletal muscle is eaten are the cysts passed on to another host.

Although inconspicuous to most of us until they cause damage, the activities of nematodes are important in the aeration of soil and the circulation of its mineral and organic components.

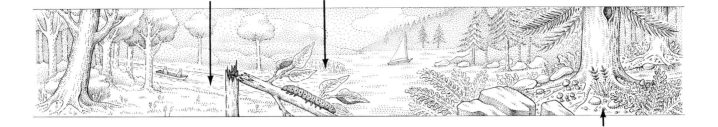

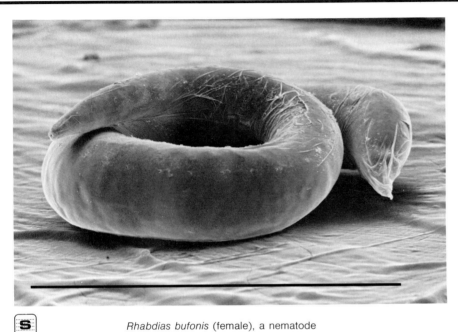

Rhabdias bufonis (female), a nematode
parasitic in the lung of the leopard frog
Rana pipiens. SEM, bar = 1 mm. [Photo-
graph courtesy of R. W. Weise; drawing by
I. Atema, information from R. W. Weise.]

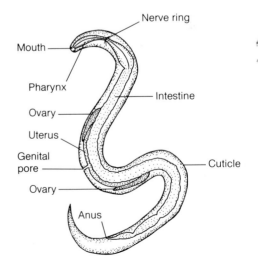

A-15 Nematomorpha

Greek *nema*, thread; *morphe*, form

One of the common names of nematomorphs is *horsehair worm,* stemming from the belief that these slender, cylindrical worms, observed in horse-watering troughs, sprang from horsehairs. Because numbers of adults are commonly seen coiled and entangled with each other, they are also known as *gordian worms,* after Gordius, king of Phrygia. Gordius tied a knot, declaring that whoever untied his intricate knot should rule Asia. Alexander the Great cut the Gordian knot with his sword.

Nematomorphs are wiry and unsegmented brown, black, gray, or yellowish worms. Although 230 species are known they are grouped into a very few genera. They are found all over the world in all kinds of water: oceans, lakes, temperate and tropical rivers, ditches, and even alpine streams. Fewer live in moist soil. In the Gulf of Naples, Vineyard Sound (Massachusetts), Norway, and the East Indies, they can be seen coiling and winding round and round in shallow coastal waters. In marine and fresh waters, they make up only a small fraction of the plankton.

The head end is distinguishable by being lighter than the rest of the body. Depending on species, nematomorphs range from 0.5 mm to 2.5 mm in diameter and from 10 cm to 70 cm in length. Body length depends also on sex: females are longer than males. Round or polygonal thickenings ornament the cuticle, a protective hard covering secreted by the epidermis.

Neither respiratory, circulatory, nor excretory organs are present. Neither adults nor larvae ever eat—their digestive tracts are degenerate, although a part of the digestive tube, the cloaca, is used in reproduction. Instead of ingesting food, they absorb dissolved nutrients through their body wall. In parasitic species, these nutrients are stolen from the host animal. Nematomorphs are very rarely found in the digestive system or urethra of people—they don't seem to cause disease or trouble in any way.

The nematomorph nervous system is similar to that of kinorhynchs (Phylum A-11). They have a circumpharyngeal ring of nerve tissue for a brain, and a single ventral nerve cord. Some adult nematomorphs have eye spots composed of innervated sacs lying beneath transparent cuticle and backed by a pigment ring. Like nematodes (A-14), they have only longitudinal muscles. Liquid in the body cavity, serving as an hydraulic skeleton, permits whiplike swimming and coiling.

Nematomorph sexes are separate. Each individual is either male or female—eggs or sperm made in ovaries or spermaries pass into the cloaca and out through the anus, as in nematodes. Nematomorph adult males are much more active than the females. They crawl or swim with serpentine undulations, especially in the winter months. Females barely move at all. The male wraps around the female and deposits sperm near her cloacal opening. He soon dies. The female then lays millions of eggs in gelatinous strings draped around aquatic plants. From 15 to 80 days later, the eggs hatch into tiny motile larvae that resemble kinorhynchs in having a protrusible, spiny proboscis.

In some way that has not yet been observed, the larvae of freshwater nematomorphs enter the body cavity of some arthropods (Phylum A-27). Larvae may be eaten or drunk by accident, or they may actively bore into animals such as beetles, cockroaches, crickets, grasshoppers, and leeches. Marine nematomorph larvae parasitize both hermit crabs and true crabs. Both marine and freshwater parasitic nematomorphs metamorphose inside their hosts; then the sexually immature worms exit their hosts' bodies in or near water or during rain. How freshwater nematomorphs induce terrestrial hosts to seek water is a mystery. Larvae may pass through one or two hosts, the number of hosts perhaps depending on the species of nematomorph. If worms mature in autumn, they form cysts on waterside grasses and wait until spring to reenter the water. Thus, from egg to adult may take only two months, or as long as fifteen months.

The only marine species found in the United States, *Nectonema agile,* is distinguished from other horsehair worms by a row of slate-colored bristles on each side of its pale whitish or gray-yellow body. This *Nectonema* is most often seen in late summer, from July to October. The chances of finding it are higher on moonless nights when the tide is going out. The details of its geographical distribution are poorly known.

Nematomorphs are pseudocoelomates; like nematodes, rotifers, gastrotrichs, entoprocts, acanthocephalans, and kinorhynchs (Phyla A-9 to A-14), nematomorphs have a pseudocoel, a body cavity not lined with mesoderm. Nematomorphs probably did not evolve directly from any other phylum of this group. Pseudocoelomates are thought to have evolved from acoelomates in several different ways and at different times. Thus, the pseudocoel is not a stage in the development of the true coelom; it evolved independently.

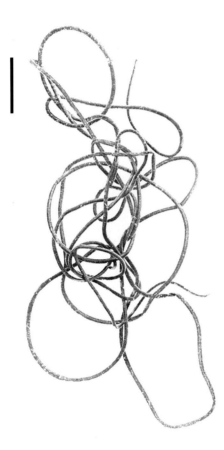

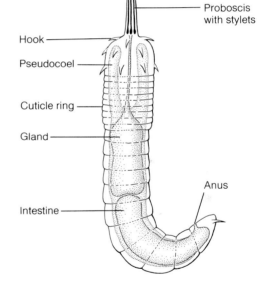

A An adult female *Gordius villoti*, a horsehair worm. Bar = 1 cm. [Courtesy of Trustees of the British Museum (Natural History).]

B Larva of a gordian worm with its proboscis extended. The larva is about 250 μm long. [Drawing by L. Meszoly, information from L. Bush.]

Labels for B:
Hook
Pseudocoel
Cuticle ring
Gland
Intestine
Proboscis with stylets
Anus

A-16 Ectoprocta

Greek *ektos*, outside; *proktos*, anus

Alcyonidium
Bugula
Cristatella
Fredericella
Lichenopora
Membranipora
Pectinatella
Plumatella
Stomatopora
Tubulipora

Ectoprocts are small aquatic animals that have a true coelom and a lophophore, a tentacle-bearing organ that surrounds the mouth. Most are colonial; each individual in a colony is called a zooid. The colonies often form extensive gelatinous or hard mats on algae, rocks, and the bottoms of ships and rafts. Most people are not aware that they are animals because they look so much like seaweeds or moss. In fact, they traditionally have been classified together with the similar entoprocts (Phylum A-13) as Phylum Bryozoa ("moss animals"). Some 5000 species are known. All but 50 are marine.

Ectoproct zooids have U-shaped digestive systems with the mouth and anus both facing up toward the surface. The anus is outside the lophophore. A water hydraulic system is used to extend the ring of tentacles and thus catch phytoplankton, which is their primary source of food. The tentacles of the lophophore are studded with cilia that create currents that bring the food into the mouth. The food passes through an esophagus and a stomach, which leads down to a large cavity, the caecum; an intestine leads back up to the anus.

A retractor muscle anchors these soft parts to the conspicuous nonliving outer covering of the colony. This is composed either of chitin alone or of a chitinous layer with an underlying thick, rigid skeleton made of calcium carbonate. The adults lack respiratory, circulatory, and excretory organs: tissues exchange gases directly with water.

Freshwater ectoprocts all reproduce sexually, and the zooids are hermaphroditic. They are also typically protandric, which means that a given individual can form eggs and sperm alternately, but not at the same time. The colony is usually differentiated so that in the actively reproducing members the lophophore and digestive tract degenerate to provide space for the developing eggs. Most ectoprocts brood large yolky eggs, although some marine species shed small eggs directly into the sea. Some brood them in the coelom, but most brood them externally in a special external chamber called the ovicell. Eggs escape from the coelom either through a simple pore, the coelomopore, or through a pore that is borne on the end of a special organ within the ring of tentacles. When a freshwater ectoproct colony dies in the autumn, it releases statoblasts, armored balls of cells that can survive the winter. In the spring, zooids grow forth from the statoblasts.

Marine ectoprocts can reproduce both sexually and asexually. The eggs cleave radially and form larvae of various kinds; all larvae have a girdle or crown of cilia used for swimming, an anterior tuft of long cilia, and a posterior adhesive sac. At first they are positively phototactic and swim toward light; thus they

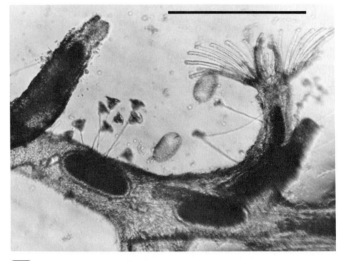

A Two zooids of a colony of *Plumatella casmiana* from a freshwater pond in Colorado, showing statoblasts. These asexually produced balls of cells start new colonies after unfavorable conditions have passed. LM, bar = 0.5 mm. [Courtesy of T. S. Wood.]

escape the brood chamber and disperse. Later they become negatively phototactic; they avoid light and tend to settle in shadows. Many are specifically attracted to certain substrata, such as the surface of the seaweed *Sargassum* (Phylum Pr-12, Phaeophyta). During settling, the adhesive sac everts and secretions are produced that will fasten the animal to the substratum. After attachment, the larval structures retract, tissue is resorbed, and adults develop. Adults can also reproduce by asexual budding, and one can form an entire colony on its own.

There are three classes. The Phylactolaemata are the freshwater ectoprocts. In these, the coelom is continuous among all the individuals in a colony. The zooids are cylindrical. The Stenolaemata comprise the marine species having cylindrical zooids; several whole orders of these became extinct at the end of the Paleozoic Era. The Gymnolaemata include most of the extant marine ectoprocts. They have circular lophophores, which are

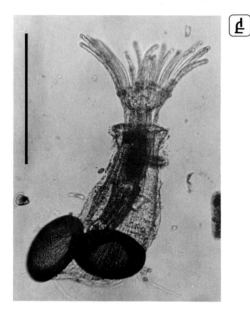

horseshoe-shaped in the other classes. Their colonies contain zooids of different sizes and shapes; that is, they form polymorphic colonies.

Phylum Ectoprocta has a scanty record in the upper Cambrian rocks. From the beginning of the Ordovician, however, thousands of fossil species have been described. *Bugula,* a very common genus now notorious for fouling ship bottoms and pilings, belongs to the order Cheilostomata (in the class Gymnolaemata), which first made its appearance in Cretaceous sediments some 100 million years ago.

Although ectoprocts superficially resemble the entoprocts, the tentacular crown of the entoprocts is not considered to be a lophophore because it surrounds both the anus and the mouth. Furthermore, the body cavity of the entoprocts is a pseudocoel rather than a true coelom. In adult ectoprocts of Class Phylactolaemata, the coelom is divided into three parts. Such a divided coelom is also characteristic of the deuterostome phyla (A-28 through A-32), which are thus suspected of having an ancestor in common with the ectoprocts.

B   A single living zooid of *Plumatella casmiana,* showing the retractile horseshoe-shaped collar, the lophophore, from which tentacles originate. LM, bar = 0.5 mm. [Courtesy of T. S. Wood.]

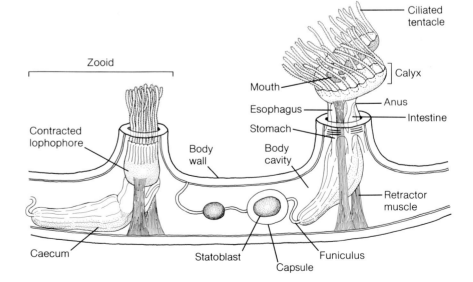

C   Two zooids of *Plumatella casmiana.* [Drawing by P. Brady, information from T. S. Wood.]

201

A-17 Phoronida

Greek *pherein*, to bear; Latin *nidus*, nest

Phoronid worms bear as many as five hundred hollow tentacles on their food-gathering organ, the lophophore. In some species, the lophophore takes the shape of a spirally coiled double ridge. They live permanently in leathery or chitinous blind tubes, which they form from secretions impregnated with calcareous matter or strengthened with sand or shell fragments. All are marine.

Phoronids are from 1 mm to 50 cm in length; their bodies may be pink, orange, or yellow. Some species are solitary, but most live in dense populations. On rocks or pier pilings phoronid tubes can occasionally be observed, interlaced like miniature hard spaghetti in crevices or in empty shells. The sandy or sticklike tubes are also found in tidal flats or between blades of sea grass on shallow bottoms from low-water mark to more than eighteen meters down. The tubes are generally longer than the worms inside, which thus can withdraw their tentacles completely.

Phoronids eat plankton and detritus. A current generated by cilia on the tentacles draws particles toward the mouth, which opens within the double row of tentacles. Food particles enter the mouth, but other bits are rejected—they are carried by cilia to the tips of the tentacles, where they are returned to the water. The mouth opens into a U-shaped, ciliated gut supported in the body cavity by thin membranes called mesenteries. At the other end of the gut is an anus, which opens just outside the spiral of tentacles. A pair of nephridia, tubes lined with cilia, carries wastes and gametes from the coelom and out through nephridiopores located near the anus. A fold on the anal side of the lophophore diverts wastes from the mouth. Digestion probably takes place within the stomach wall rather than in the stomach lumen.

There are no special respiratory organs; gases diffuse through the body surface. However, phoronids do have a closed circulatory system with blood vessels leading to and from the lophophore. These vessels contract, moving the blood through the body. Red blood cells contain the oxygen-carrying protein hemoglobin. No such red blood cells are present in ectoprocts (Phylum A-16). The nervous system, which lies just under the body wall, consists of a ring of nerve tissue around the mouth and nerves that extend to the body and the lophophore. The body surface has sensory cells. A giant nerve fiber coordinates the longitudinal muscles. When the muscles contract, the phoronid rapidly withdraws into its tube.

Most phoronids are hermaphrodites; both testis and ovary hang beside the stomach in the posterior part of the coelom. Gametes pass out from the coelom to the nephridiopores and then, perhaps by a ciliated exterior furrow, to a space enclosed by the tentacles. In this space, external fertilization takes place. Eggs are generally fertilized by sperm that come from another individual. Adults of some species brood the eggs; they hold

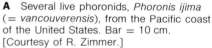

A Several live phoronids, *Phoronis ijima* (= *vancouverensis*), from the Pacific coast of the United States. Bar = 10 cm. [Courtesy of R. Zimmer.]

them among their tentacles probably by means of secretions from the lophophore organ. Brooded or not, eggs develop into actinotrochs, larvae that resemble the trochophore larvae of annelids (Phylum A-23). In brooding species, the actinotrochs spend their early life among the tentacles of the adult; later, they become free-swimming members of the marine plankton. Eventually, actinotrochs settle to the species' preferred type of bottom and metamorphose into adults. Phoronids can also reproduce asexually by budding, and they can regenerate lost lophophores.

Phoronids, ectoprocts, and brachiopods (Phylum A-18) all bear lophophores, and thus probably have ancestors in common. However, some shared adult characters, such as reduced head, secretion of protective coverings, and a U-shaped gut, are obvious adaptations to sessile life and may be at least partly the result of convergent evolution. Annelids (Phylum A-23) and mollusks (Phylum A-19) seem to have ancestors in common with phoronids; the heavily ciliated phoronid larvae have excretory organs (protonephridia) like those in the trochophores (larvae) of the other two phyla.

B A single *Phoronopsis harmeri* extends its tentacles from its sand-encrusted tube; taken from a Pacific coast tidal flat. Bar = 5 mm. [Courtesy of R. Zimmer.]

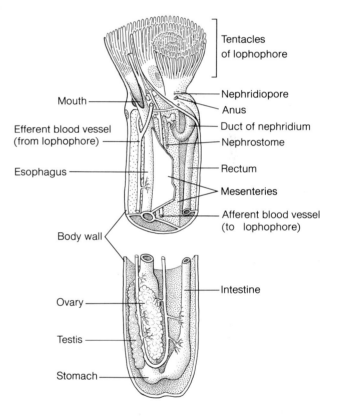

Tentacles of lophophore

Mouth

Nephridiopore

Anus

Efferent blood vessel (from lophophore)

Duct of nephridium

Nephrostome

Esophagus

Rectum

Mesenteries

Afferent blood vessel (to lophophore)

Body wall

Intestine

Ovary

Testis

Stomach

C Cutaway view of a *Phoronis* sp. [Drawing by L. Meszoly, information from R. Zimmer.]

Latin *brachium*, arm; Greek *pous*, foot

Argyrotheca   *Notosaria*
Crania   *Terebratula*
Dallinella   *Terebratulina*
Glottidia
Hemithyris
Lacazella
Lingula
Megathyris

Three living articulate brachiopods, *Terebratulina retusa,* dredged from a depth of about 20 meters in Crinan Loch, Scotland. Bar = 1 cm. [Courtesy of A. Williams; drawing by L. Meszoly, information from A. Williams.]

Often called lamp shells because they resemble Aladdin's lamp, an oil burner on a stalk, these intertidal or subtidal marine animals are relics of a glorious past. They thrived during the Paleozoic Era and lived in many diverse habitats, but members of more recently evolved phyla have come to occupy most of their niches. Because they have calcium carbonate shells, which preserve well, some 30,000 extinct species have been described; only about 260 living species are known. *Lingula,* whose fossils are known from as long as 400,000,000 years ago, may be the oldest genus of animals still extant.

Brachiopods superficially resemble clams (Phylum A-19, Mollusca), but their symmetry is entirely different. Like clams and other bivalves, brachiopods have two apposed hard shells. However, one is ventral (on the belly side) and the other is dorsal (on the back side), whereas the shells of bivalve mollusks are arranged on the left and right sides. Unlike mollusks but like ectoprocts (Phylum A-16) and phoronids (Phylum A-17), brachio-

pods have a lophophore, a food-gathering organ bearing numerous tentacles.

In brachiopods, reproduction is always sexual and the sexes are separate. Eggs or sperm are produced in paired gonads and discharged into the sea water, where fertilization takes place. The fertilized egg develops into a ciliated larva somewhat like the trochophore larva of annelids (Phylum A-23). The free-swimming larva generally develops into a solitary sessile adult attached by a peduncle to the surface of a rock or to another animal. The adult is bilaterally symmetrical and lacks any sort of segmentation.

Brachiopod shells are lined on the interior with soft tissue, called the mantle, whose fine papillae penetrate the shell. A thin loop made of calcium carbonate supports a spiral structure, the lophophore. It attaches to the anterior surface of the body and lies in the space between the mantle lobes. The spiral "trunk" of the lophophore, W-shaped in cross section, is fringed with tenta-

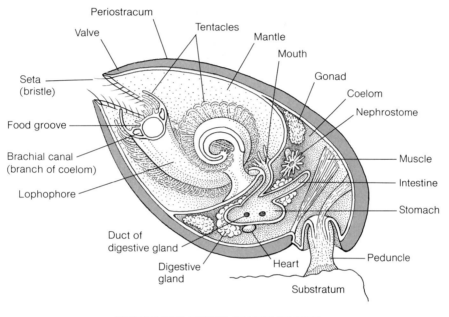

Periostracum
Valve
Tentacles
Mantle
Mouth
Seta
(bristle)
Gonad
Coelom
Nephrostome
Food groove
Brachial canal
(branch of coelom)
Lophophore
Muscle
Intestine
Stomach
Duct of
digestive gland
Digestive
gland
Heart
Peduncle
Substratum

GENERALIZED ARTICULATE BRACHIOPOD

cles; these bear cilia and facilitate respiration by circulating the water in the mantle cavity. The cilia also whip small food organisms into the mouth, which opens near the base of the lophophore into a short gullet leading into a larger stomach and a blind intestine. On the sides of the stomach are one or more pairs of digestive glands. Brachiopods have three pairs of muscles: one serves to open and close the valves of the shell; the other two pairs attach to the peduncle and permit the animal to turn about on it.

Brachiopods are coelomate animals: they develop from three germ layers, and their body cavity is a true coelom formed by enlargement of a mesodermal space. The large fluid-filled coelom harbors the internal organs, which are supported on thin membranes called mesenteries. Branches of the coelom extend into the lophophore and into the lobes of the mantle. Brachiopods have a small contractile heart.

The circulatory system is open—the blood returns to the heart through the tissue spaces rather than through veins. The excretory system consists of a waste-gathering tube, called a nephridium, at each side of the intestine. The end in the coelom has a fringed opening called a nephrostome; the other end drains into the mantle cavity. Although there is a ring of nerves surrounding the gullet, there are no special sense organs such as eyes or ears.

There are two classes of brachiopods. The Inarticulata lack hinges between their valves, which are quite similar to each other. The valves are made of chitinous material containing calcareous spicules. These animals have an anus. The Articulata, on the other hand, lack an anus. Their two valves are very unlike each other, and are attached by a hinge. They are made of calcium phosphate with chitin or scleroprotein. Some species in this class lack a peduncle.

A-19 Mollusca

Latin *molluscus,* soft

Aplysia
Arca
Architeuthis
Argonauta
Busycon
Conus
Crepidula
Cryptochiton
Dentalium
Doris

Haliotis
Helix
Littorina
Loligo
Mercenaria
Murex
Mya
Mytilus
Nautilus
Neomenia

Neopilina
Pecten
Sepia
Teredo
Tridacna
Vema

Mollusks are soft-bodied animals having an internal or an external shell. All mollusks also have a mantle, a fold in the body wall that lines the shell and secretes the calcium carbonate of which the shell is made. Unique to mollusks is the radula, a hard strap made of chitin that is used to gather food by boring or scraping. Mollusks live in aquatic or moist environments. They are well-known inhabitants of mud and sandy flats, and they also dwell in forests, soil, rivers, lakes, and the abyss of the sea.

About 110,000 species of mollusks have been described, making it the second largest phylum after Arthropoda (Phylum A-27). Mollusk shells come in a bewildering, beautiful array of colors and patterns. Some mollusks produce ink; in some cases, it is luminescent. Mollusks come in an enormous range of sizes. The smallest are no bigger than a sand grain; the giant clam *Tridacna* may be 1.3 meters wide; the giant squid *Architeuthis,* the largest invertebrate known, may grow to be more than 20 meters long (including the tentacles).

There are seven classes of mollusks. Members of Class Monoplacophora have one flat shell and live in deep water off the west coast of the Americas, in the Gulf of Aden, and in the South Atlantic. They are better known from the fossil record than from live material. In fact, the first ones to be observed live (*Vema* in our photograph) were taken from an ocean-sediment sampler only in 1977. Aplacophora, the second class, includes the deep-sea, wormlike solenogasters. Class Polyplacophora includes chitons, which have oval bodies covered by eight shelly plates. They live clinging to rocks. Members of Class Pelecypoda, often called bivalves, are the best known mollusks. This group includes clams, mussels, scallops, and oysters. They have two shells, hinged together laterally, and a wedge-shaped foot. They lack tentacles and a head. The class Gastropoda comprises the snails and their relations. The shell is chambered, if there is one; the shell-less gastropods have a stiff internal structure. The class Scaphopoda comprises the tooth shells, which have tusk-shaped shells open at both ends. They burrow into mud or sand head first, leaving the narrow end of the shell exposed above the surface. Class Cephalopoda includes the octopus, squid, and *Nautilus.* They have heads and prehensile arms that encircle beaklike jaws.

In most mollusks, the sexes are separate and fertilization takes place in the water. Land snails, shell-less snail-like marine animals called nudibranchs, and some bivalves are hermaphroditic, although they are usually cross-fertilizing. Some sea slugs and oysters can reverse their sex, from male to female and back again, several times in a season. After courtship, male cephalo-

A *Vema hyalina,* probably the only live monoplacophoran mollusk ever to have been photographed; taken from about 400 meters down, off Catalina Island, California. Although Paleozoic *Vema* and *Neopilina* shells are known, these, the most primitive living mollusks, were first seen live in 1977. Bar = 1 mm. [Courtesy of H. A. Lowenstam and C. Spada.]

pods use a specialized arm to transfer their sperm packs to the female's mantle cavity.

Female octopuses clean, aerate, and defend their eggs. Mollusk eggs are generally deposited in gel clusters, paper cases, bubble rafts, horny capsules, or sand collars. From them hatch free-swimming planktonic trochophore larvae (like those of annelids, Phylum A-23), which metamorphose into veliger larvae (see drawing). Some freshwater clams and chitons brood young, and some periwinkles even "give birth" to live young. The eggs of land snails and cephalopods develop directly into adults without passing through a larval stage.

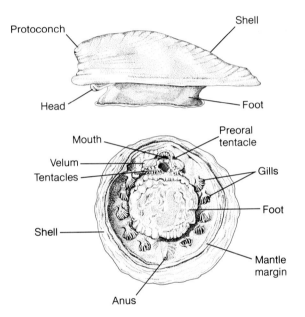

B External morphology of an adult mono-placophoran, *Neopilina* sp. The paired gills manifest segmentation. [Drawing by L. Meszoly, information from J. H. McLean.]

C Veliger larva of a gastropod mollusk; lateral view in swimming position. Generalized, but based on *Crepidula*. [Drawing by L. Meszoly, information from K. E. Hoagland.]

Bivalves and some snails are sedentary filter feeders; their mucus-covered gills entangle food particles and convey them to ciliated flaps, called palps, over the mouth, where they are sorted. Many large mollusks feed on crustaceans, fish, and other mollusks. The ciliated tentacles of scaphopods reach into sand for foraminiferans (Phylum Pr-17). Chitons, snails, and slugs use their toothed radulas to rasp algae and plant cells. The teeth of some chitons are capped with magnetite, an inorganic iron compound. Many mollusks have a structure called a style, which is rotated by cilia to pull food-laden mucus strings into the stomach. The style also releases digestive enzymes. Food is moved through the gut mainly by the action of cilia.

Aquatic mollusks exchange gases through their body surface. The spectacular dorsal tufts of nudibranchs are for respiratory exchange. In some garden and freshwater snails, the mantle functions as an air-breathing lung. Scaphopods lack hearts, gills, and blood vessels. In other mollusks, a dorsal heart pumps blood through vessels and sinuses. Some gastropods and squids have

207

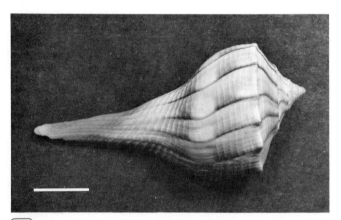

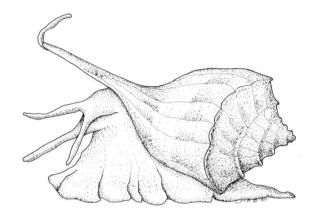

D The shell of *Busycon perversens* Linnaeus, called the lightning whelk or left-handed whelk. Bar = 1 cm. [Courtesy of T. Reeves.]

E *Busycon perversens* in motion. [Drawing by L. M. Reeves.]

accessory hearts as well. Mollusks have well-developed excretory organs. The mollusk nervous system generally consists of a pair of ganglia, nerve cords, and balance organs called statocysts. Cephalopods and scallops have well-developed eyes with cornea, lens, and retina.

Hundreds of mollusk species provide human beings with food. The first recorded aquaculturist, the Roman Sergius Orata, cultured oysters in the first century B.C. Tools, trumpets, sacred and decorative objects, cameos, and buttons are carved from mollusk shells. Freshwater mussels and pearl oysters secrete pearl over irritating particles. The pen (internal shell) of a cephalopod, the cuttlefish *Sepia,* is the cuttlebone on which caged birds condition their beaks. Sepia, a brown pigment used by artists, is prepared from cuttlefish ink. A yellow secretion of *Murex,* a marine gastropod, was the source of the royal, or Tyrian, purple dye used by Phoenicians. Tusk shells served as wampum for American Indians near the west coast. East coast Indians drilled ''roanoke,'' a hard substance used as money, from cockles.

Mercenaria, the famous edible clam or quahog, was the source of purple wampum. Trade routes have been traced from California into the interior by the distribution of abalone shell.

Mollusks are preyed on by many animals, for example, by baleen whales, haddock, walrus, and cod. Some freshwater snails are intermediate hosts to trematode parasites (Phylum A-6), such as those that cause the disease bilharzia (schistosomiasis). Oyster drills and rock- and wood-boring bivalves, such as shipworms, are of economic importance because they attack pilings and wooden ships. Land slugs and snails are sometimes garden pests.

Because their shells are easily preserved, mollusks are well documented in the fossil record. *Nautilus* shells are useful for identifying strata of the lower Cambrian. Bivalves appear by the middle Cambrian; chitons, cephalopods, and gastropods by the upper Cambrian. The oldest tooth-shell fossils come from Devonian sediments. Octopuses appear in the fossil record rather recently—about 65 million years ago, in the Cretaceous. Squids,

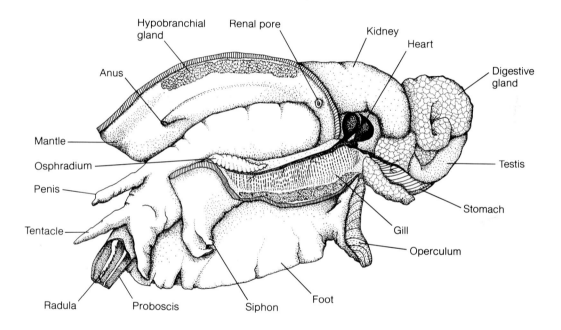

F Internal morphology of *Busycon per-versens,* male. [Drawing by L. M. Reeves.]

the most modern cephalopods, appear later, in the Tertiary.

The relation of mollusks to other phyla is debated. They have a true coelom, which forms a cavity around the heart and gonad. Because they have trochophore larvae and because of their embryology, mollusks are believed by some zoologists to share ancestors with annelids (Phylum A-23) and flatworms (Phylum A-6). Digestion within cells rather than in a stomach cavity, the plan of the nervous system, external cilia, and gliding and creep-ing with ventral waves relate mollusks more strongly to flatworms than to annelids. Known before 1952 only from the lower Cambrian, the monoplacophoran *Vema* and the closely related *Neopilina* are thought by some to closely resemble ancestral mollusks. Others believe that the segmentation of living mono-placophorans derived from secondary replication of body parts, and that ancestral mollusks may have been soft-bodied orga-nisms like trochophore larvae or flatworms.

A-20 Priapulida

Latin *priapulus*, little penis

Halicryptus
Maccabeus
Priapulus
Tubiluchus

Priapulids are short, plump, exclusively marine worms. A small protostomate phylum, not much is known about them. They have a short proboscis called a presoma, which is studded with spiny papillae, and a retractable mouth. The trunk of the body has from thirty to a hundred superficial segments that are covered with spines and warts. *Tubiluchus corallicola* has a long, contractile tail; species of the genus *Priapulus,* after which the phylum is named, have one or two retractable caudal appendages. Other genera, such as *Halicryptus* and *Maccabeus,* lack these caudal appendages. Only eight species of priapulids have been described. The smallest is *Maccabeus tentaculatus,* 2 mm long, and the largest is *Priapulus caudatus,* 8 cm long.

Priapulids burrow or lie partly buried with their mouths lying flush with the bottom. They live in estuaries, and also in water as deep as five hundred meters. *Tubiluchus corallicola* lives in coral sand, silt, and mud in shallow waters of the Caribbean and off Bermuda. Other priapulids are found in the Arctic, off North America north of Massachusetts and California, in the Baltic and North Seas south to Belgium, around Patagonia, in the Antarctic, and in cold deep waters off Costa Rica.

The bodies of priapulids are externally segmented into rings, but there is no internal segmentation. Longitudinal and circular muscles surround the intestine and line the body wall, providing power for forcing the body forward through mud. Priapulids burrow by alternately anchoring their anterior and posterior ends. Unlike other burrowing worms, they do not maintain water currents in their burrows.

Priapulids are carnivorous; they eat polychete annelids (Phylum A-23) and other priapulids. The prey is seized with curved spines that line the mouth and is swallowed whole. During feeding, the proboscis and mouth roll inside and out again, passing food into the muscular, toothed pharynx. Food is digested as it passes down the straight intestine, which leads to the rectum and anus.

Priapulus caudatus has blood; red cells containing the oxygen-carrying pigment hemerythrin circulate in the coelomic fluid. Priapulid respiration is not well understood. The caudal appendage of *Priapulus* may function in gas exchange or chemoreception. However, it can be removed without killing the animal (which regenerates the appendage), so other modes of gas exchange must exist.

Around the pharynx and within the body wall is a collar of nervous tissue called the nerve ring. It connects to a single nerve cord running down the body on the ventral side, which has gan-

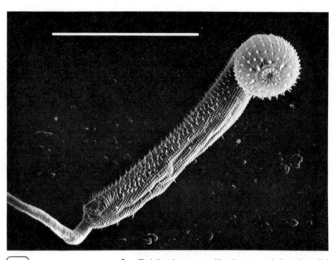

A *Tubiluchus corallicola,* an adult priapulid taken from the surface layer of subtidal algal mats at Castle Harbor, Bermuda. SEM, bar = 0.5 mm. [Courtesy of C. B. Calloway.]

glia from which extend peripheral nerves. Raised bumps, or papillae, emerge from the body surface; they are thought to be sensory organs. The two gonads are tubular. The excretory system consists of a pair of protonephridia, waste-collecting tubes containing ciliated cells. The protonephridia and gonads share ducts that open in nephridiopores, one on each side at the end of the trunk.

In priapulids, the male and female sexes are separate, but they are similar in external appearance. Tube-dwelling *Maccabeus* probably reproduces parthenogenetically, because only females have been found. Along the Scandinavian coast, spawning takes place in winter; fertilization is external—eggs and sperm are shed into the sea. Eggs develop into larvae that are different from adults. They are covered with a lorica of eight plates of cuticle: one is dorsal, one is ventral, three are on each side, and two tiny dorsal and ventral plates lie at the anterior margin of the trunk. They have toes bearing adhesive glands, and they live in mud.

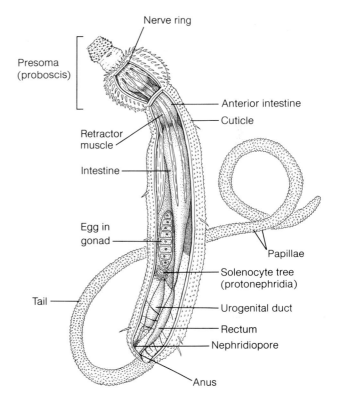

B The presoma of *T. corallicola,* showing the retractile proboscis. SEM, bar = 0.1 mm. [Courtesy of C. B. Calloway (A, B from *Marine Biology* 31:161–174, 1974).]

C Morphology of an adult female *T. corallicola.* [Drawing by L. Meszoly, information from C. B. Calloway.]

After a series of molts, they attain the adult form. Adults continue to shed the chitinous cuticle periodically.

The fossil record of these worms is deficient. The similarities between priapulids and pseudocoelomates (proboscis, superficial segmentation, larval form, and protonephridia containing cilia) are probably superficial. The spacious priapulid body cavity is lined with mesentery, or thin tissue, and is thus considered to be a coelom. The discovery of this fact, in 1961, moved priapulids from pseudocoelomate to coelomate status. However, the evolutionary relationship of priapulids to other coelomate phyla is uncertain.

A-21 Sipuncula

Latin *siphunculus,* little pipe

Aspidosiphon
Dendrostomum
Golfingia
Lithacrosiphon
Onchnesoma
Phascolion
Phascolopsis
Siphonostoma
Themiste

Sipunculans, also called peanut worms, are sea animals. They are bilaterally symmetrical and unsegmented. Many have ciliated bushy tentacles that encircle the mouth. The anterior one-half to one-third of the body consists of an organ called an introvert, which is contractile and studded with minute spines or papillae. Sipunculans can shut down like telescopes; when the introvert is drawn into the plump trunk, they resemble very firm peanut kernels. When burrowing in sand or mud, sipunculans thrust the introvert forward and dilate its tip; when the introvert is retracted, it pulls the body forward.

There are more than 300 species of sipunculans. Some are only a few centimeters long; others are nearly a meter long. Their rubbery, cuticle-covered body walls are of various colors. They may be pearl gray, yellow, or dark brown, and have iridescent tints. They dig mucus-lined burrows that are closed at their lower ends. Some live among the roots of mangroves and eel grass. Some nestle in among reef-forming coral. Others live beneath rocks, in annelid tubes, or in empty mollusk shells. Most sipunculans live between the high- and low-tide marks of warm seas. However, some species are found in polar seas, and others dwell in the abyss as far as 7000 meters down.

Golfingia procera is a parasitic species that sucks out the soft insides of *Aphrodite,* the sea mouse (Phylum A-23, Annelida).

Sipunculans are not very abundant—they probably make up less than 0.5% of the invertebrates in the usual marine benthic communities. The highest density reported, 700 individuals per square meter, was found in intertidal limestone reefs in Hawaii. Sipunculans are active in bioturbation—that is, they disrupt and aerate the sediment in shallow bay bottoms as earthworms do on land.

The spacious body cavity, a true coelom, lacks circulatory and respiratory organs. In some sipunculans, such as *Dendrostomum,* the oxygen carrying respiratory pigment hemerythrin is present in the fluid of the tentacles. (The coelom does not extend into the tentacles.)

Sipunculans feed on diatoms and other protists, larvae, detritus in mud, and algal films gathered from rocks. Some use their mucus-covered tentacles to trap food; cilia on the tentacles waft the food to the mouth. However, dissolved organic matter in sea water may supply as much as 10% of sipunculan nutrition. A ciliated groove moves food through the intestine, which loops forward on itself to open in an anus on the dorsal anterior surface. Metanephridia collect dissolved wastes and release them through nephridiopores near the anus. With this arrangement, sipunculans need not turn in their burrows to defecate. Ammonia is their main nitrogenous waste.

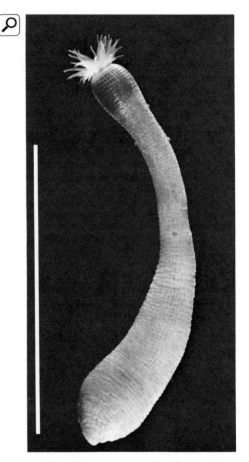

Themiste lageniformis, a peanut worm from the Indian River, Fort Pierce, Florida, with introvert and tentacles extended. Bar = 0.5 cm. The drawing is a cutaway view of a *T. lageniformis* whose introvert is retracted. [Photograph courtesy of W. Davenport; drawing by L. Meszoly, information from M. E. Rice.]

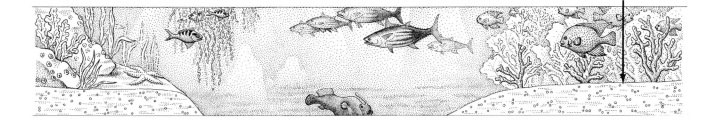

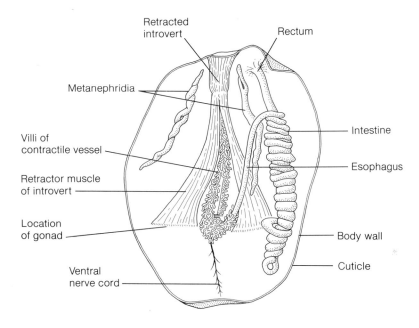

The nervous system of sipunculans is rather simple. A bilobed brain above the esophagus connects to a ventral nerve cord. The body and tentacle surface bears protrusible ciliated parts, presumably sense organs. Sipunculans also have pigmented ocelli, which are ciliated photoreceptor cells.

Sipunculan reproductive tissue lies inconspicuously at the base of the retractor muscles. Although there are male and female peanut worms, they cannot be distinguished externally. Females tend to be more abundant. In females, mature gametes are shed into the coelom where, as oocytes, they accumulate yolk and become ova. Males release sperm, which induce females to discharge their ova through nephridiopores. Fertilization takes place in sea water. Sipunculans have four developmental patterns, depending on species. Some develop directly into adults without passing through a larval stage; others develop indirectly—they form trochophore larvae (like the larvae of annelids, Phylum A-23) that swim free before settling to the bottom. A few species, such as *Aspidosiphon,* reproduce asexually by transverse fission. Still others form distinctive young forms called pelagosphaera larvae, which may remain several months in the plankton.

Sipunculans, echiurans (Phylum A-22), and annelids (Phylum A-23) are all very ancient, and they all have trochophore larvae. Sipunculan fossils from the Cambrian were first described by the famous paleontologist C. D. Walcott. However, these are probably fossils of priapulids. Sipunculans, which are protostomes, may have evolved from annelid ancestors before the annelids developed segmentation.

A-22 Echiura

Greek *echis*, snake; Latin *-ura*, tailed

Echiurans, or spoon worms, are soft, plump, ovoid or cylindrical marine worms. Many have a rough and prickly surface. Most have short, fat trunks bearing a grooved, mobile proboscis. Echiuran bodies are from a few millimeters to about 40 centimeters long, yet they may extend their proboscis enormously, to as long as 1.5 meters. About 130 species have been described. They may be dull brown, gray, red, rose, yellow, or transparent. Green *Bonellia tasmanica* (*Metabonellia tasmanica*) derives its color from its food; the green "bonellin" is a porphyrin pigment derived from the chlorophyll *a* of food algae.

Echiurans dwell in U-shaped tubes or in rock crevices. Many species are found in warm, shallow seas or estuaries, and others in polar seas. Some live at a depth of ten thousand meters in the abyss. Some inhabit abandoned shells or tests belonging to other animals—the brick-red *Thalassema* lives in discarded sand-dollar tests. Few enemies of echiurans are known; one of them, the bat ray (Phylum A-32), uses its flat body like a toilet plunger to pop echiurans out of their burrows. On the shores of the North Sea, echiurans are used for cod bait.

To tunnel, echiurans wedge the anterior of the trunk into the sand or mud; the hind end then is drawn forward and anchored; the anterior wedges forward once more. *Echiurus pallassii* takes some forty minutes to burrow out of sight; other species may work rather more quickly. Spoonworm burrows are distinctive: they descend as far as 50 cm on the diagonal, then run horizontally for 15 cm to about one meter, then vertically to the surface. Mud may wash into the unoccupied part of the burrow, but a pencil-size opening remains through which the spoonworm thrusts its proboscis in search of food when the tide is in.

Food is swept into the mouth along the mucus-coated, ciliated ventral side of the proboscis. Digestion takes place mainly in the cavity of the intestine. Two anal sacs remove waste from coelomic fluid and dump them into the cloaca, an expanded portion of the lower intestine. Echiurans lack a skeleton. Muscular contractions constrict and dilate the body at alternating regions along the body. In this way, they force water through their burrows; this assists the uptake of oxygen and the release of carbon dioxide. Nucleated cells wandering in the coelom carry oxygen bound to hemoglobin. Blood is oxygenated primarily in the proboscis, which has gill-like processes.

Echiurans have neither brains nor sense organs, although nerves loop around the proboscis. The proboscis can be cast off if an animal is disturbed; it is later regenerated.

Echiurans reproduce sexually. The sexes are separate, although in some species males and females look alike. *Bonellia*, on the other hand, holds the record for the most extreme animal

A female echiuran, *Metabonellia tasmanica*, found at low-tide mark in sand under rocks in gullies on the coast of southeastern Australia. Bar = 5 cm. [Photograph courtesy of A. Dartnall, courtesy of Tasmanian Museum and Art Gallery; external view drawn by I. Atema, internal view drawn by P. Brady, information from A. Dartnall.]

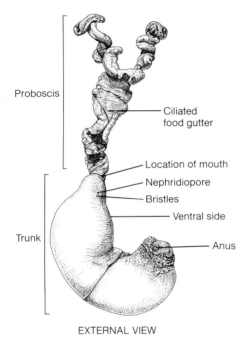

EXTERNAL VIEW

Labels on external view: Proboscis, Ciliated food gutter, Trunk, Location of mouth, Nephridiopore, Bristles, Ventral side, Anus

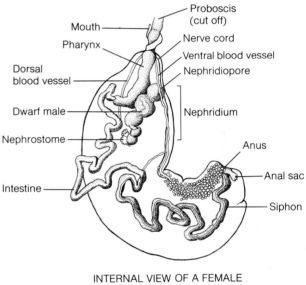

INTERNAL VIEW OF A FEMALE
WITH DWARF MALE

Labels on internal view: Proboscis (cut off), Mouth, Nerve cord, Pharynx, Ventral blood vessel, Dorsal blood vessel, Nephridiopore, Dwarf male, Nephridium, Nephrostome, Intestine, Anus, Anal sac, Siphon

sexual dimorphism. In this genus, the females may be as much as a meter long (proboscis extended), but they lug around a male only a millimeter long. Gametes (eggs and sperm) are formed internally and move through the coelom out into the sea through tubes called nephridia. These nephridia have no excretory functions. Fertilization, usually in the sea, produces trochophore larvae (like those of annelids, Phylum A-23) that metamorphose gradually into the sedentary adults. In the genus *Bonellia,* however, larvae that mature in the sea without coming into contact with adult females of the same species become females; *Bonellia* larvae that do come into contact with adult females are induced to develop into parasitic males, which lodge in nephridia of the females. *Bonellia* eggs are fertilized and undergo early development in these nephridia.

Probably the most famous echiuran is *Urechis caupo,* because it is a year-round source of eggs and sperm that develop easily in the laboratory and are thus useful for studies of invertebrate development. *U. caupo* strains particles as small as 0.04 μm wide from sea water sucked through its thimble-shaped mucus net. This net, from 5 to 21 cm long, is covered with food debris, swallowed, and resecreted as frequently as once every two minutes. *U. caupo* lacks a circulatory system, but pumps sea water through its anus into and out of its thin-walled hindgut. Even if the anus is plugged, oxygen uptake continues, which indicates that respiratory exchange can be maintained across the hemoglobin-containing body wall.

In the Silurian rocks, some 450 million years old, the sandstone imprints of echiuran burrows have been found. The remarkable resemblance of the fossil burrows to those of contemporary echiurans suggests that the phylum was well established by this time. Echiurans probably evolved from polychete annelids (Phylum A-23). Some annelid characteristics shared by many echiurans include setae of chitin, a closed circulatory system (hemoglobin is also found as an oxygen carrier in the coelomic fluid), a complete digestive system open at both ends, nephridia, and the early development pattern via trochophore larvae. The structure of the nervous system and of the body wall and the development of eggs into trochophore larvae relate echiurans not only to annelids but to sipunculans (Phylum A-21) as well.

A-23 Annelida

Latin *anellus*, little ring

Aphrodite   *Nephtys*
Chaetopterus   *Nereis*
Eunice   *Tubifex*
Hirudo
Lumbricus
Macrobdella
Megascolides
Myzostoma

Annelid worms are distinguished by ringlike external segments that coincide with internal partitions containing digestive and reproductive organs repeated in tandem. Annelids are coelomate animals—they have mesoderm-lined body cavities. They have bristles composed of chitin, which may be used for locomotion or, highly modified, for other functions, such as anchoring the worms in their burrows. Many annelids develop from free-swimming ciliated larvae known as trochophores.

Annelids live on land, in the ocean, and in fresh water—they are absent only from Madagascar and Antarctica. They come in many colors; they may be striped or spotted, pink, brown, or purple. Some have colorful gills or cirri. Some are iridescent or luminescent. They range in length from only half a millimeter to three meters (the Australian earthworm *Megascolides*).

Most annelids are active predators or scavengers. Swimming annelids catch fish eggs or larvae. A common habit of marine annelids is to filter-feed from tubes buried in sand or mud. Some trap plankton on structures covered with mucus and cilia, and some pop out of their tubes to seize prey. Some have mucus-covered eversible pharynxes. Some browse on seaweed.

Many annelids burrow incessantly, causing turnover and exposure of detritus and soil and aeration of anaerobic muds and sands. Annelids have extensive muscle systems: circular and longitudinal muscles in the body wall oppose the hydrostatic forces generated by the incompressible coelomic fluid, which functions as a skeleton. Food is pushed through the gut from mouth to anus by peristaltic movements independent of muscle contractions in the body wall.

Annelids have circulatory systems in which the green pigment chlorocruorin or the red pigment hemoglobin carry oxygen in blood vessels and in the coelomic fluid. The blood vessels have valves and move fluid along by contracting. Gas exchange takes place through gills and parapods or through the moist body wall.

The excretory system consists of a pair of ciliated funnels, called nephridia, that remove waste from blood and coelomic fluid from each segment of the body and discharge it through external pores. Annelids have a fairly well-developed nervous system consisting of cerebral ganglia and a ventral nerve cord. Most have eyes, some with retinas and lenses. In addition, many annelids are covered with light-sensitive epidermal cells, and balance organs called statocysts are located near the brain.

There are four classes of annelids: the Polychaeta, mostly marine bristle worms; the Oligochaeta, or terrestrial bristle worms; the Hirudinea, or leeches; and the Myzostomaria, a group of small parasites of echinoderms (Phylum A-29), especially of crinoids. There are about 5300 polychaetes, 3100 oligochetes, 300 hirudinids, and fewer myzostomarians.

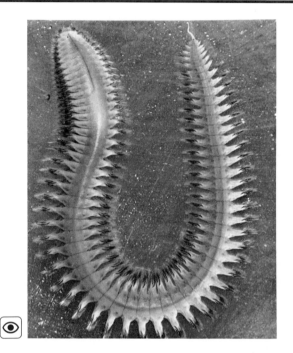

Most polychetes, paddle-footed worms, have a pair of fleshy flaps called parapods on each body segment; each parapod has a bundle of bristles, or setae. Marine polychetes are very common. They include the hairy sea mouse; lugworms, which burrow in sand and mudflats; sabellid and serpulid worms, whose tubes encrust shells, rocks, and algae; peacock worms, which construct tubes of a mosaic of sand or shell; and a few pelagic species. There are also some soil and freshwater species. Oligochetes, also called bristle-footed worms, include the earthworms and a few small freshwater and estuarine forms. The hirudinids with anterior and posterior suckers are popularly called blood suckers. Some are free-living predators; others are parasites of vertebrate and invertebrate animals. Except for the leeches, annelids can regenerate lost parts and can reproduce by budding as well as by sexual means.

Breeding polychetes swarm by millions, their hormones triggered by phases of the moon, the tides, or changes in temperature. The sexes are usually separate, and fertilization is external.

An adult *Nephtys incisa,* a polychete taken from mud under 100 feet of water off Gay Head, Vineyard Sound, Massachusetts. 13 cm long. [Photograph courtesy of G. Moore; external view drawn by I. Atema, internal view of *Nephtys* sp. and trochophore drawn by L. Meszoly, information from M. H. Pettibone.]

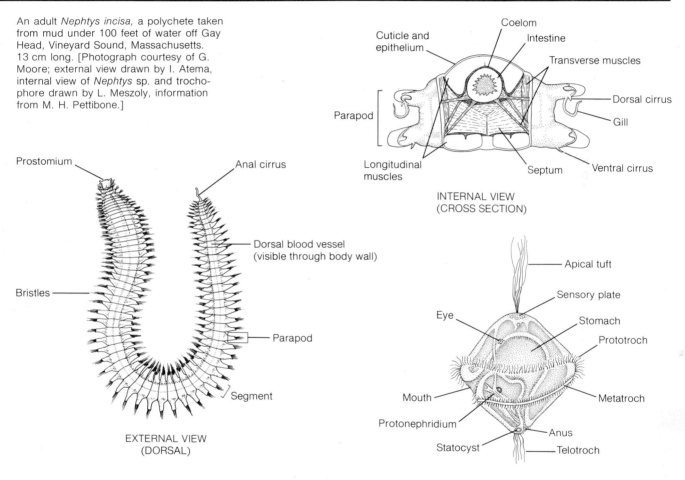

INTERNAL VIEW
(CROSS SECTION)

EXTERNAL VIEW
(DORSAL)

TROCHOPHORE LARVA

Many polychete adults brood their young; in some species, the male protects and aerates the eggs. In many polychete species, individuals are budded off or are transformed into epitokes, special gamete-bearing forms. Epitokes of *Eunice viridis* are eaten in parts of the South Pacific.

Leeches and oligochetes are usually hermaphroditic, but each copulates with another individual. Some leeches attach to their mate with suckers and forcibly drive spermatophores into the partner's body. The eggs are incubated in a cocoon, an adaptation to terrestrial life, and hatch as young adults.

Ancient annelids, being soft bodied, left few fossils. They were probably ancestors of mollusks, sipunculans, and echiurans (Phyla A-19, A-21, and A-22): in all these phyla, eggs hatch into trochophore larvae. Annelids probably are not directly related to other worms. They differ from nemertines, acanthocephalans, nematodes, and nematomorphs (Phyla A-7, A-12, A-14, and A-15) in that they have internally segmented bodies, coeloms, characteristic ventral nervous systems, and complex eyes.

Batillipes
Echiniscoides
Echiniscus
Hypsibius
Macrobiotus
Milnesium

Because of their pawing locomotion, the nineteenth-century English naturalist Thomas Huxley called tardigrades water bears, which they are still called today. Tardigrades move deliberately on four pairs of unjointed stumpy legs, each having four movable claws. They all lack cilia. The head and four body segments of tardigrades are generally covered with a thick, conspicuous, nonchitinous protein cuticle, which is periodically molted. Although tardigrades are coelomate metazoans, the coelom is limited—it lies around the gonad only. The main body cavity is a pseudocoel. From the mouth protrude sharp organs called stylets; new ones are secreted at each molt.

Tardigrades range in length from 50 μm to 1.2 mm. They lack respiratory and circulatory organs; gases and food simply diffuse in their body fluids. Oxygen and carbon dioxide are exchanged directly through the body surface. Tardigrades lack circular body-wall muscles, but have thin, smooth longitudinal muscles attached to the underside of the cuticle. They eat liquid food. Some pierce plant cells with their stylets, ingesting the liquid with a sucking, muscular pharynx. Others prefer to eat the body fluids of rotifers, nematodes, and even other species of tardigrades. A large stomach opens into a short rectum. Dorsal glands, supposedly excretory, also open into the rectum. Excretory granules are left in molted cuticles. The nervous system is rather simple: a dorsal brain surrounds the pharynx, and four ventral ganglia are connected by pairs of longitudinal nerves. Most tardigrades have a pair of eyespots.

Tardigrades are most remarkable for their powers of resistance. They can survive desiccation and temperatures as high as 151°C and as low as −270°C, nearly absolute zero. Species from the arctic, from the tropics, and even from hot springs are known. Tardigrades can turn into dry barrel-shaped, motionless, forms, called tuns because they resemble wine casks. In this state they can survive for as long as a hundred years. They also form thick-walled cysts, different from tuns, in response to hunger and damage. Inside the cyst, the animal contracts and the internal organs degenerate. Cysts (but not tuns) may form in aquatic environments. Future space travel may utilize whatever principle gives tardigrades their great resistance to X-radiation. For human beings, the lethal dose of X-rays is about 500 roentgens; for tardigrades, it is 570,000.

A single ovary or testis lies dorsal to the gut. Although the sexes are separate, the males and females are often difficult to distinguish. Most are females. A few species, such as *Echiniscus,* reproduce parthenogenetically: females lay eggs that grow into females without any male intervention. In species that reproduce sexually, fertilization may be internal or external. In some fresh-

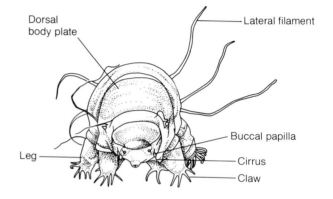

Dorsal body plate

Lateral filament

Leg

Buccal papilla

Cirrus

Claw

water tardigrades, the male injects sperm through the female's genital pore or anus into the space between the new and old cuticle; the fertilized eggs are shed with the old cuticle.

Tardigrade eggs have thin or thick shells. Thick-shelled eggs are produced when conditions for growth are unfavorable. Fertile eggs are often sticky; they can be found attached to moss, algae, bark, and other objects. The eggs develop directly into adults; water bears have no larval stages. Under favorable conditions, after about two weeks of development the young water bears hatch from their egg cases, which they rupture with their stylets. After hatching, tardigrades grow by enlargement of existing cells, rather than by mitotic division to increase the number of cells.

Tardigrades have left few fossils. They are considered to be related to the arthropods (Phylum A-27), particularly the mites, because of their stylets, four pairs of legs, and segmented bodies. Like arthropods, they probably evolved from annelids (Phylum A-23).

Echiniscus blumi, a land-dwelling water bear from Auburn, Placer County, California. Its movable claws enable it to cling to moss and lichens. SEM, bar = 0.5 mm. [Photograph courtesy of R. O. Schuster; drawing by I. Atema, information from R. O. Schuster.]

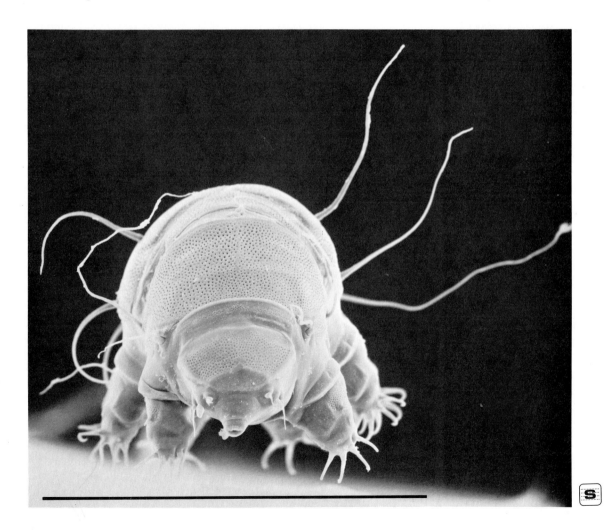

A-25 Pentastoma

Greek *pente*, five; *stoma*, mouth

Armillifera
Cephalobaena
Linguatula
Porocephalus
Raillietiella
Reighardia
Waddycephalus

Members of phylum Pentastoma, called tongue worms because of their shape, have a flat, soft body covered by a chitinous cuticle. Externally, the body appears to be composed of about ninety rings, although it is not segmented internally. All seventy or so species are parasites of vertebrates. They live embedded in the lungs, nostrils, or nasal sinuses of mammals such as dogs, foxes, wolves, goats, and horses, as well as of reptiles such as snakes, crocodiles, lizards, turtles, and even of birds. They are particularly prevalent in tropical and subtropical hosts. Two anterior pairs of leglike hooks with hollow retractile claws hold the worms in place on their hosts.

In the tropics, larval tongue worms occasionally are found in human beings, who acquire them from other vertebrates, generally from domestic animals. Their growth is checked, however, because the nasal tissue responds by forming calcareous capsules around the parasites.

Tongue worms range from a few millimeters long to more than 15 cm long, depending on species. They are colorless, transparent, and glassy except in a pigmented area around their hooks. A projection between the two pairs of hooks bears the mouth. Mucus, epithelial cells, and lymph from the host animal is sucked through the mouth into a swelling foregut and a digestive tract in the form of a straight tube. Excretory, respiratory, and circulatory organs are all absent. The loss of such organs is a common adaptation to parasitic life. The nervous system of tongue worms consists of surface sensory papillae and a ventral nerve cord with pairs of ganglia.

In pentastomes, the sexes are separate and the male is smaller than the female. In the species pictured here, *Linguatula serrata*, the male is 2 cm long and the female is about 13 cm long. Females have a genital pore near the anus, whereas the genital pore of the male is near his mouth. The eggs are minute; they have thick shells and no yolks. Fertilization is internal. The vagina of the female tongue worm is able to hold some half million fertilized eggs. As it fills with eggs, the vagina may stretch until it is a hundred times its former size. The fertilized eggs pass out onto plants with the nasal secretions of the host animal. In order to develop, the eggs must be eaten by herbivorous mammals.

Tongue worms have three larval stages. The first stage develops within the eggs in an intermediate host, a herbivore. Rabbits are the main intermediate hosts for *Linguatula;* fish are intermediate hosts for crocodile pentastomes. In the second larval stage, pentastomes resemble tardigrades (Phylum A-24). These larvae hatch from the eggs in the stomach of the herbivore; they swim and bore with their mouths into the lungs or liver, or into the linings of these organs. In the tissues of the intermediate host, the larvae enter a third stage: they encapsulate to form cysts. It is possible that the cysts are digested when the herbivore host is eaten by a carnivore, and that the pentastome then uses its hooks to move from the carnivore's stomach into its nasal or lung passages. However, there is evidence that the pentastome leaves its cyst while it is still in the carnivore's mouth. In either case, the adults embed themselves in the nasal sinuses or lung tissue, they mate, and the cycle begins again.

The relationship of tongue worms to other animals is difficult to infer because the worms have been so modified as parasites. Some believe that pentastomes descended from early arthropods (Phylum A-27), because members of both these phyla undergo extensive larval development and molt chitinous cuticles at certain stages in their life cycles.

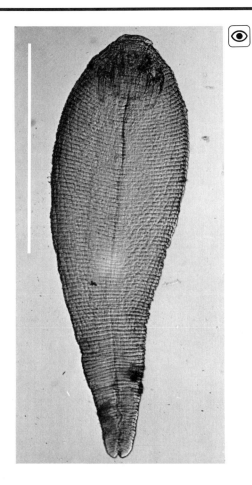

A female living tongue worm, *Linguatula serrata,* that clings to tissues in nostrils and forehead sinuses of dogs. Bar = 5 cm. [Photograph courtesy of J. T. Self; drawing by R. Golder, information from J. T. Self.]

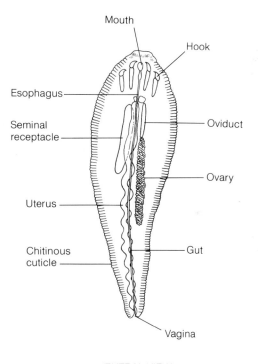

VENTRAL VIEW
OF FEMALE

Mouth
Hook
Esophagus
Seminal
receptacle
Oviduct
Ovary
Uterus
Chitinous
cuticle
Gut
Vagina

A-26 Onychophora

Greek *onyx*, claw; *pherein*, to bear

Mesoperipatus
Opisthopatus
Peripatopsis
Peripatus
Symperipatus

Onychophorans are named for the many pinchers on their legs; they are commonly known as velvet worms. They are small— from 14 mm to 150 mm long—and may be mistaken for arthropods (Phylum A-27) or annelid worms (Phylum A-23). Their bodies are iridescent green, blue-black, orange, or whitish. They have many pairs of unjointed, hollow legs, whose rigidity is maintained by hydrostatic pressure. The body wall has circular, longitudinal, and diagonal muscles like those of annelids. They work against the hydrostatic skeleton (there is no other) to move the animal.

Onychophorans require high humidity. They generally live in forest litter, under logs and rocks, and in tunnels of termites; they venture forth during rain or at night. About ten genera and eighty species have been described. They inhabit tropical India, the Himalayas, Madagascar, the Congo, South and Central America up to Mexico, the West Indies, temperate Australia, South Africa, the Andes, and New Zealand. This distribution corresponds to the modern remnants of the ancient southern continental mass, Gondwanaland, which broke apart some two hundred million years ago. In fact, the fossil record of onychophorans, which traces from the mid-Cambrian, has been used to reconstruct the historical pattern of drifting continents.

The nervous system, which is located on the ventral side, consists of two simple eyes, a brain, and longitudinal cords having transverse connections but lacking the nerve-tissue swellings, called ganglia, found in arthropods and annelids. Onychophorans are carnivorous; they attack and eat arthropods, such as isopods and termites, and mollusks. Disturbed or hunting onychophorans throw out secretions from specialized glands; these secretions congeal into elastic sticky white threads, entangling their prey. The blood is circulated in the body cavity, the haemocoel, by body movements and by the pumping of the muscular tubular heart, which is located on the dorsal side. The coelom itself is vestigial, being reduced to gonoducts and tiny sacs associated with the nephridia (excretory tubes). Oxygen enters all over the body through the skin and is transported from the skin surface by tracheae, canals that permeate the internal organs.

The sexes are separate, and each individual has a pair of testes or ovaries. In some species, fertilization is external; in others, internal. The *Peripatopsis capensis* male deposits spermatophores on the sides or back of the female. Beneath each spermatophore, the skin of the female erodes, permitting sperm to travel to her ovaries. Many species are viviparous; the female nourishes her young by means of a placenta and bears them alive.

Onychophorans interest zoologists especially because they may link two extensive and important phyla, the arthropods and the annelids. Both arthropods and onychophorans have a cuticle that is secreted by the epidermis and is composed of chitin. Their developmental patterns are similar and they both have tracheae, tubes for gas exchange. Like those of arthropods, the jaws of onychophorans are derived from differentiation of appendages. Members of both phyla have tubular hearts that are displaced toward the dorsal side of the body. Onychophorans also resemble annelids. Members of both these phyla molt in patches. They both have paired excretory tubes, called nephridia, that open at the base of each leg. The eyes, when present, are simple, and reproductive tubules are ciliated. Their chief internal organs, suspended in mesentery, are arranged similarly. The mitochondria of the sperm are located between nucleus and axoneme. Onychophorans do have features not seen in either arthropods or annelids. For example, they lack striated muscle tissue, have only a single pair of jaws, and their skin is velvety, being covered with minute papillae. Unlike annelids and arthropods, onychophorans are unsegmented except for their antennae.

Peripatus sp., a blind onychophoran, or velvet worm, taken from a cave in Jamaica. It lacks eyes such as are found at the base of the antennae in species that do not live in caves. Bar = 5 cm. [Photograph courtesy of R. Norton; drawing by L. Meszoly, information from R. H. Arnett.]

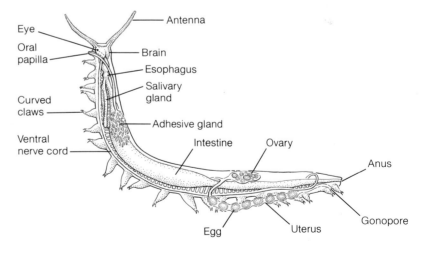

CUTAWAY VIEW OF FEMALE

A-27 Arthropoda

Greek *arthron*, joint; *pous*, foot

Members of the phylum Arthropoda are distinguished by having segmented bodies and segmented appendages. The bodies of most arthropods are made of two or three distinct parts—a cephalum (head), a thorax (chest), and an abdomen. They have a hardened external skeleton, often called the exoskeleton, made of the nitrogen-rich polysaccharide chitin. As the animal outgrows the exoskeleton, it is shed and a new one is regenerated. Many arthropods metamorphose: the egg develops into a larval form that differs considerably from the sexually mature adult. Although arthropods range in size from one hundred micrometers to two decimeters long, most adults are about a millimeter long.

In number of species, the phylum Arthropoda is by far the largest in the animal kingdom. Nearly half a million species of insects alone have been described; some zoologists feel that if the tropical groups were better known there might be as many as ten million living species of insects.

The economic importance of arthropods can hardly be underestimated. Most fruit trees and many crop vegetables rely on insects for pollination. Insects and other arthropods are crucial predators of plant pests and other noxious species. Many cause diseases of plants by transmitting pathogenic fungi and bacteria. Others transmit human pathogens, such as those that cause trypanosomiasis and malaria (Phyla Pr-8 and Pr-19). Arthropods are crucial sources of nutrient for many other animals—and indeed, for some carnivorous plants. Seafood that is not mollusk (Phylum A-19) or fish (Phylum A-32) is generally arthropod.

There are two great groups, subphyla or superclasses, of arthropods, Mandibulata and Chelicerata. In the Chelicerata, the first two body parts are combined into a single cephalothorax; mandibulate arthropods have three distinct body parts.

The mandibulate arthropods include the aquatic gillbreathing crustaceans (Class Crustacea), whose major members are the water fleas (Cladocera), clam shrimps (Conchostraca), ostracods (Ostracoda), fairy shrimps (Anostraca), tadpole shrimp (Notostraca), copepods (Copepoda), and the barnacles (Cirripedia). All of these orders are members of the subclass Entomostraca: they all have a bivalved carapace, or straca. The second subclass of crustaceans, the Malacostraca, are primarily marine organisms, although a few species live in fresh water or on land. In the order Decapoda, the entire cephalothorax is covered by the carapace. Lobster, crayfish, crabs, and shrimp are all decapods. There are two other orders in the Malacostraca: amphipods and isopods, which lack a carapace and have flattened bodies. Most are marine (for example, beach fleas), although the

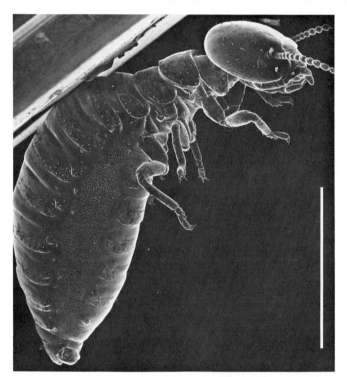

A *Pterotermes occidentis,* the largest and most primitive dry-wood termite in North America. Its colonies are limited to the Sonoran Desert of southern Arizona, southeastern California, and Sonora, Mexico. The swollen abdomen of this pseudergate (worker) covers the large hindgut, which harbors millions of microorganisms responsible for the digestion of wood. SEM, bar = 0.5 mm. [Photograph courtesy of C. Spada; drawings by L. Meszoly, information from W. Nutting.]

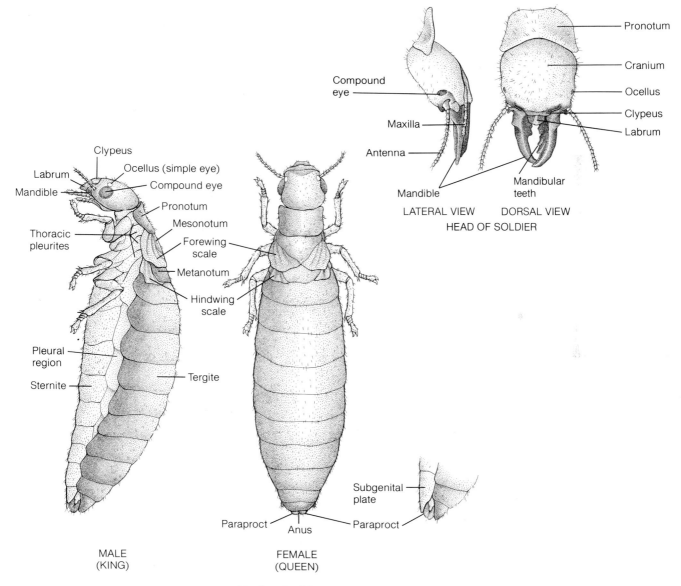

Compound eye

Maxilla

Antenna

Mandible

Pronotum

Cranium

Ocellus

Clypeus

Labrum

Mandibular teeth

LATERAL VIEW   DORSAL VIEW
HEAD OF SOLDIER

Clypeus

Labrum

Mandible

Ocellus (simple eye)

Compound eye

Pronotum

Mesonotum

Forewing scale

Metanotum

Hindwing scale

Thoracic pleurites

Pleural region

Sternite

Tergite

Paraproct   Anus   Paraproct

Subgenital plate

Paraproct

MALE
(KING)

FEMALE
(QUEEN)

ADULT REPRODUCTIVE FORMS

B Soldier of *Pterotermes occidentis,* from south of Tucson, Arizona. Bar = 0.5 mm. [Courtesy of W. Ormerod.]

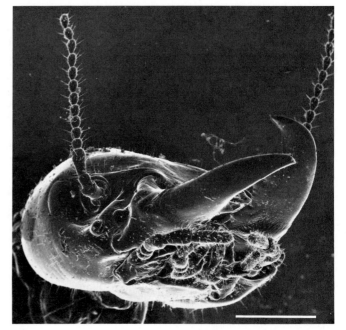

C Head of soldier of *Pterotermes occidentis.* The huge mandibles are used to defend the colony against ants. SEM, bar = 0.5 mm.

common pill bug or sow bug is an isopod that lives on land, in moist soil litter under fallen logs and stones.

There are five other classes in the subphylum Mandibulata: the Diplopoda (millipedes), the Chilopoda (centipedes), the Pauropoda (centipedelike animals having branched antennae and nine or ten pairs of legs), the Symphyla (also similar to centipedes but having from ten to twelve pairs of legs), and the Insecta, by far the largest class of mandibulates.

Insects have three pairs of legs, three body sections, generally one or two pairs of wings, and one pair of antennae. There are some twenty-five orders of insects, and more than six hundred families. Suffice it here to mention some major orders: Isoptera (termites), Collembola (springtails), Ephemeroptera (mayflies), Odonata (dragonflies), Orthoptera (grasshoppers, roaches, and crickets), Hemiptera (true bugs), Homoptera (cicadas and aphids), Lepidoptera (butterflies and moths), Diptera (flies), Siphonaptera (fleas), Hymenoptera (ants, wasps, bees, and chalcids), and Coleoptera (beetles), which contains more than 290,000 described species.

Most members of the second subphylum, the Chelicerata, have six pairs of appendages, of which the first two differ from each other. The first pair, called chelicerae, are grasping and jawlike; the second pair are usually feelerlike or clawlike; the third and other more posterior pairs are usually leglike. Unlike mandibulates, chelicerates lack sensory antennae. They include the classes Pycnogonida (sea spiders), Merostomata (horseshoe crabs), and Arachnida (scorpions, daddy-long-legs or harvestmen, spiders, and the mites and ticks of the order Acarina). Nearly all the arachnids have four pairs of segmented legs. Most are carnivorous; many prey on insects. They play an important role in the balance of nature.

Apparently, the insects arose in the late Paleozoic—by two hundred million years ago, modern cockroaches lived and were fossilized. The phylum Arthropoda itself appeared far earlier; indeed, some fossils of simple joint-footed animals have been seen in the Ediacaran (upper Proterozoic) rocks.

D *Limenitis archippus* mating. These orange and brown viceroy butterflies are found in central Canada and in the United States east of the Rockies. Bar = 7 cm. [Courtesy of P. Krombholz.]

A-28 Pogonophora

Greek *pogon*, beard; *pherein*, to bear

Heptabrachia
Lamellisabella
Oligobrachia
Polybrachia
Riftia
Siboglinum
Spirobrachia
Zenkevitchiana

Pogonophorans, the beard worms, are sessile deep sea worms that produce fixed upright chitin tubes bottom sediments or in wood decaying on the sea floor. They are collected along continental slopes, nearly always in water deeper than 100 meters. They are fairly new to science: the first ones were dredged up off the coast of Indonesia in 1900 and, since then, some 100 species have been reported.

The best samples and greatest diversity of pogonophorans has been found in the western Pacific, as far south as New Zealand and Indonesia, but this is likely to merely reflect the locus of activity of Russian oceanographic vessels. Along the Atlantic coast, beard worms have been brought up from near Nova Scotia down to Florida, the Gulf of Mexico, the Caribbean, and Brazil. They also have been dredged from the eastern Atlantic from Norway to the Bay of Biscay. They are probably distributed around the world in deep, cold ocean waters and in shallow Arctic seas.

Pogonophoran bodies are from 10 cm to about 90 cm long; most are less than 1 mm wide. The tough tubes are usually from 0.1 mm to 2.0 mm wide and open at both ends. The bodies have three sections: a short forepart, a long trunk, and a short rear part. The forepart includes a cephalic lobe, or head, beneath which are long, beardlike tentacles (these give the phylum its name). Depending on species, there may be a single spiral tentacle or as many as 250 of them. The tentacles bear tiny pinnules, extensions of epithelial cells, and are lined with cilia.

It is thought that some pogonophorans feed by extending their tentacles from the mouth of the tube to gather organic detritus and plankton. Suspended particles of food may be trapped on the pinnules, and cilia are thought to drive water from anterior to posterior through the cylinder or funnel made by the tentacles. Gland cells that secrete digestive enzymes have been searched for, but not found. Other pogonophorans may absorb dissolved food through the entire epidermis. The microvillus-covered tentacles also are thought to be important sites of gas exchange. In any case, beard worms have neither mouth, gut, nor anus. Experiments suggest that absorptive feeding by active uptake even within the tube is the rule, at least for small pogonophorans. Very recent work suggest that pogonophorans are nourished symbiotically by the bacteria that cover their bodies.

Posterior to the cephalic lobe, in the forepart, is a glandular region that secretes material for the chitinous tube. The long trunk section bears papillae, raised bumps, and (on most beard worms) short, toothed setae, or bristles. The short hind region, or opisthosoma, is composed of from five to twenty-three short segments bearing setae. The opisthosoma and setae are thought

A *Riftia pachyptila,* pogonophorans of the Class Vestimentifera. Taken at 2500 meters depth off the Galapagos Islands, this is the first photograph of a live colony *in situ.* Bar = 25 cm. [Courtesy of J. Edmond; information from M. Jones.]

to aid in burrowing and in anchoring the open-ended tube. The opisthosoma is easily lost in dredging; in fact, it took a generation after beard worms were discovered to realize that, *in situ,* they all have an opisthosoma.

Pogonophorans have a true coelom, which extends not only into the three body segments but into the tentacles as well. In the opisthosoma, the coelom is segmented just as the body wall is. Beard worms have a closed blood vascular system, and each tentacle contains two blood vessels.

The sexes are separate and externally indistinguishable. In the coelom of the trunk are two cylindrical gonads. Sperm are packaged into spermatophores and probably are released into the sea. At least in some species, embryos are brooded in the dwelling tube of their mother; ciliated embryos have been taken from inside such tubes. Details of fertilization and early development

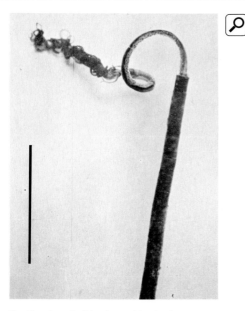

B Front end of body and tentacle crown of *Oligobrachia ivanovi* partly dissected out of its tube. Bar = 1 cm. [Courtesy of A. J. Southward.]

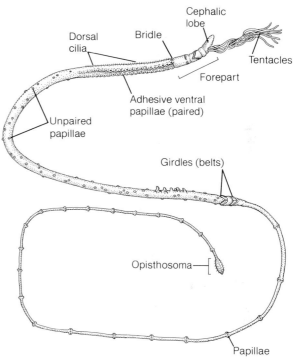

C Diagrammatic and shortened view of pogonophoran removed from tube. [Drawing by L. Meszoly.]

are not well known and no larval forms have been discovered, although gastrulation in *Siboglinum* has been studied.

A fossil *Hyolithellus* from the lower Cambrian rocks in North America, Greenland, and northern Europe has been assigned to Pogonophora. The segmentation of the opisthosoma, the chitinous tube and setae, and the nature of the hemoglobin dissolved in pogonophoran blood suggest that pogonophorans are descended from annelid ancestors. In fact, some workers favor including the pogonophorans in Phylum Annelida; others consider them to be deuterostomes and of independent origin.

The beard worms in our photograph A are much fatter than most pogonophorans and are provisionally assigned to the Vestimentifera, proposed as a class of pogonophora or as a class of Annelida. Some of the characteristics that set Vestimentifera off from other pogonophorans are lateral body-wall folds that meet in the trunk midline to form an exterior tube, absence of setae on the rear part, and thousands of tentacles.

A-29 Echinodermata

Greek *echinos*, sea urchin; *derma*, skin

Arbacia
Asterias
Cucumaria
Echinarachnius
Metacrinus
Ophiura
Pisaster
Solaster
Strongylocentrotus
Thyone

Echinoderms are characterized by the presence of tube feet, bulbed structures belonging to a unique water vascular system that develops from part of the larval coelom. The tube feet serve for locomotion, food handling, and respiration. Water is forced through the vascular system by muscle action.

There are about 6000 living species of echinoderms; they are all marine organisms. Most are intertidal or subtidal, and a few dwell in deep ocean trenches. Thousands of fossil species are known. Familiar members of the group include starfish, sea urchins, sand dollars, sea lilies, and sea cucumbers. Human beings eat some echinoderms; for example, sea urchins (gonads only) and sea cucumbers (also known as bêche-de-mer and trepang).

Adult echinoderms lack heads, brains, and segmentation; most of them are radially symmetrical. The body generally has five symmetrically radiating parts, or arms, reflecting the internal organization of the animal. The body surface is covered by a delicate epidermis, which is stretched over a firm endoskeleton made of movable or fixed calcareous plates. In most cases, the plates are arranged in a genetically determined pattern and bear spines. The plates over certain areas of the arms are perforated; from these areas, called ambulacra, the tube feet project.

The digestive system of most echinoderms is simple and complete: the tract begins with a mouth and ends with an anus, which opens to the exterior. In some species, the anus is lacking.

The circulatory system of echinoderms also radiates in five directions; it circulates a colorless blood that differs little from coelomic fluid. In fact, it is so rudimentary in many species that the coelom itself performs the circulatory and respiratory functions. The coelom is lined by a ciliated membranous tissue called the peritoneum, and the coelomic fluid contains free amoebocytes that engulf foreign particles. In some groups, respiration is by minute gills or papillae that protrude from the coelom into the sea. Some species respire by using their tube feet to exchange fluid with the exterior. The sea cucumbers respire by means of special cloacal structures called respiratory trees. The nervous system consists of nerves in a ring around the mouth and also extending radially out into the body.

Although most echinoderms are able to regenerate lost parts easily, reproduction is not asexual. The sexes are separate, although parents of the opposite sex often look very much alike. The fertilized eggs develop into bilaterally symmetrical, ciliated larvae, which may pass through several distinct stages before they metamorphose into adults.

There are many classes of extinct echinoderms, but only five living classes, which are organized into two subphyla. The first is subphylum Pelmatozoa, which contains only the class Crinoidea,

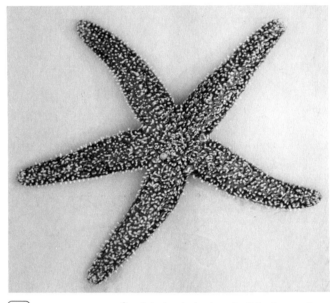

 A *Asterias forbesi*, a starfish. Arm radius of adult = 130 mm.

the sea lilies. The crinoid mouth and anus are both on the upper surface of the body, which is cup shaped over a cup-shaped skeleton. Most are attached to the substrate by a stalk on the aboral surface, that is, the surface away from the mouth. All the other living echinoderm classes belong to the subphylum Eleutherozoa: the sausagelike Holothuroidea (sea cucumbers), the Echinoidea (sea urchins and sand dollars), the Asteroidea (starfish or sea stars), and the Ophiuroidea (brittle stars). These echinoderms lack a stalk and the mouth and anus are on their lower surface, facing the rock or sediment.

Like chordates (Phylum A-32), echinoderms are deuterostomes: the blastopore or opening of the blastula develops into the anus rather than the mouth. The cleavage of the egg, the pattern of blastulation and gastrulation in egg development, the formation of the three germ layers (ectoderm, mesoderm and endoderm), and the existence of a true coelom suggest that echinoderms and chordates have common ancestors.

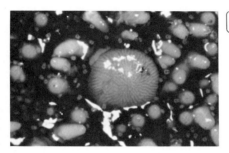
C   A madreporite, the opening through which sea water enters the starfish vascular system.

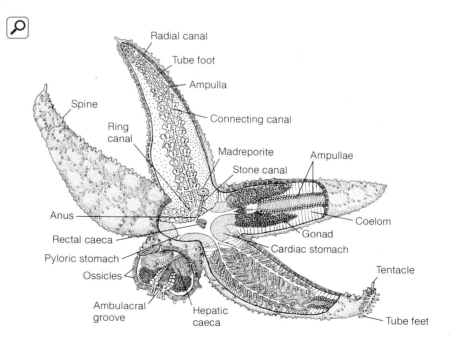

B   Tube feet surrounding the mouth.

D   Morphology of the starfish. *Asterias forbesi.* [Photographs by W. Ormerod; drawing by E. Hoffman, information from H. B. Fell.]

Radial canal
Tube foot
Ampulla
Spine
Connecting canal
Ring canal
Madreporite
Ampullae
Stone canal
Coelom
Anus
Gonad
Rectal caeca
Cardiac stomach
Pyloric stomach
Ossicles
Tentacle
Ambulacral groove
Hepatic caeca
Tube feet

A-30 Chaetognatha

Greek *chaite*, hair; *gnathos*, jaw

Bathyspadella
Eukrohnia
Krohnitta
Pterosagitta
Sagitta
Spadella

Chaetognaths are commonly called arrow worms because of their shape. About fifty species are known, belonging to six genera. The "chaeto" part of their name refers to their remarkable movable hooks, with which they grasp and swallow whole prey—marine animals such as copepods (Phylum—A-27), medusae (Phylum A-3), and tunicates (Phylum A-32) and all kinds of protists. In chaetognaths from upper parts of the ocean, all this food can be seen directly through their transparent intestines. Chaetognaths from deeper waters are brightly colored: red, orange, and pink. The orange bands sometimes seen on *Eukrohnia* may be due to microorganisms that grow on the animal's surface.

Arrow worms are common all over the world in open seas from Spitsbergen, Norway, to the Indo-Pacific Ocean. They are most abundant in warm, shallow seas. Some migrate, living in surface waters in the winter and in deeper waters in the summer. Others migrate daily: up toward the surface waters at night, down to the depths in the day.

Chaetognaths range in length from about 0.5 cm to 15 cm. They are covered with an outer cuticle that, unlike the cuticle of most invertebrate marine animals, lacks chitin. Cilia circulate liquid inside the body, distributing food and wastes. Chaetognaths lack circulatory, respiratory and excretory organs. Oxygen enters directly through the body wall and waste leaves that way.

Rays support one or two pairs of lateral fins and the tail. The worms dart forward or backward by flapping their tails, but not their fins, which serve as stabilizers. Between feedings, a part of the body wall is drawn over the spiny head.

The nervous system contains ganglia, bundles of nerve cell bodies, in the head and on the belly side. The surface of the body has sensitive bumps, called sensory papillae, that make the animal sensitive to touch. The two compound eyes in the head probably are unable to form visual images, although they are sensitive to differences in light intensity.

Each individual worm is hermaphroditic, male and female at the same time. Ovaries, in the trunk, produce eggs; testes, in the tail, produce sperm. The sperm mature before eggs do, and they are released by the rupture of seminal vesicles. Sperm attach themselves to the fins of the worm that produced them or to the fins of a partner. Fertilization takes place inside the chaetognath body. In some species, the eggs are fertilized by sperm from the same individual. Others, such as the benthic arrow worm *Spadella,* cross-fertilize: Two arrow worms approach and lie side by side, facing in opposite directions; each attaches a spermatophore (sperm packet) to the neck of the other, and the sperm stream along each arrow worm's back into seminal receptacles.

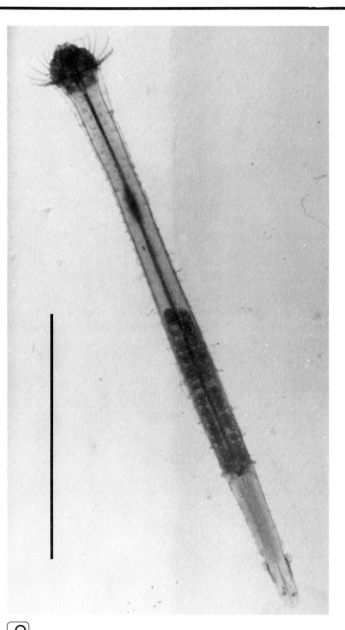

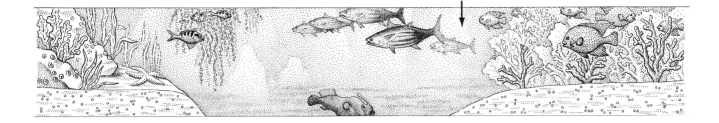

Sagitta bipunctata, a living arrow worm. It probably uses the transparent "fins" as rigid stabilizers and to maintain buoyancy, rather than for swimming. Bar = 5 cm. [Photograph courtesy of G. C. Grant; drawing by I. Atema, information from G. C. Grant.]

Anterior teeth

Posterior teeth

Head

Hooks

Eye

Collarette

Sensory papilla

Ventral nerve ganglion

Intestine

Trunk

Anterior lateral fin

Eggs in ovary

Posterior lateral fin

Anus

Testis

Seminal vesicle

Tail

Tail fin

DORSAL VIEW

In all chaetognaths, embryos develop in the sea into small adult worms without undergoing metamorphosis.

Chaetognaths are an ancient and conservative phylum. Some of their fossils are five hundred million years old, indicating that arrow worms had already evolved by Cambrian times. Their relationships to other animal phyla are not certain. The body wall of adult arrow worms resembles those of gastrotrichs (Phylum A-9), rotifers (Phylum A-10), kinorhynchs (Phylum A-11), and nematomorphs (Phylum A-15), but the resemblance is only superficial.

Chaetognaths are relatives of pogonophorans (Phylum A-28), echinoderms (Phylum A-29), hemichordates (Phylum A-31), and chordates (Phylum A-32). Embryos of these four phyla develop in such a way that the opening in the blastula called the blastopore develops into the anus rather than into the mouth. Only in chaetognaths and chordates does a tail develop posteriorly, behind the anus. Like those of other deuterostomes (Phyla A-28 to A-32), the photoreceptors of chaetognaths develop from cells that have undulipodia containing peripheral double tubules but no central tubules. Other invertebrate animals do not have postanal tails and lack undulipodia in their photoreceptors. Thus, chaetognaths may resemble the invertebrate ancestors of vertebrates.

Chaetognaths are important to marine fisheries. They occasionally prey upon larvae of economically valuable fish, such as herring. However, they also are food for adult herring. Chaetognath species are distributed according to temperature, and they have been used as temperature indicators in tracing the course of ocean currents. The distribution of Sagitta bipunctata, for example, found in waters of the continental shelf off North Carolina, indicates the location of lateral extensions of the Florida Current.

A-31 Hemichordata

Greek *hemi-*, half; Latin *chorda,* cord

Atubaria
Balanoglossus
Cephalodiscus
Glandiceps
Glossobalanus
Ptychodera
Rhabdopleura
Saccoglossus
Spengelia

Hemichordates are small soft-bodied animals that inhabit shallow U-shaped burrows in sandy or muddy sea bottoms; some are found in the open ocean. There are about ninety species. Superficially wormlike, hemichordates are bilaterally symmetrical and lack segmentation. Perhaps their best-known representative is the tongue worm *Saccoglossus.*

There are two basic body plans, corresponding to two classes. The members of Class Enteropneusta have cylindrical bodies. These are the acorn or tongue worms, solitary animals. Most are between 2.5 cm and 250 cm long; *Balanoglossus gigas* is 1.5 meters long. Enteropneust bodies are fleshy and contractile. They have a proboscis, a collar, a trunk, and from 10 to more than 100 pairs of gill slits. A straight digestive tract runs from the mouth, located on the collar, through the conspicuously tripartite coelom. Enteropneusts lack tentacles. There are about 65 species, including *Ptychodera, Balanoglossus,* and *Saccoglossus.*

The members of Class Pterobranchia have vase-shaped bodies. The longest are only about 7 mm long. Most of them are colonial. They have a U-shaped digestive tract, which places the mouth and anus in proximity. The collar, which surrounds the mouth, is extended into pairs of hollow arms bearing ciliated tentacles. Either the pharynx has two gill slits or there are no gill slits at all. Pterobranchs may reproduce either asexually, by budding, or sexually, by biparental sex. There are two orders. In the Rhabdopleurida, each individual has a single gonad. The individuals are connected by a common stalk or stolon, and each animal is enclosed in a secreted tube. Members of the order Cephalodiscida have a pair of gonads. The individuals are free or grouped in a colony covered by a single secreted case.

Hemichordate sexes are separate, but they look alike. The fertile eggs of some species of enteropneusts develop first into a ciliated larvae called tornaria; others develop directly into adults. The pterobranchs have a different sort of larva; it resembles the members of a third class, about which very little information is available, the Planctosphaeroidea. This taxon has been erected to include a transparent spherical pelagic larva about 10 mm wide, having a U-shaped digestive tube, coelomic sacs, and branched bands of cilia. Because of their similarity to the larvae of pterobranchs, planctosphaeroids are classified as hemichordates. However, adult planctosphaeroids are not yet known.

The circulatory system of hemichordates is mostly open. The coelom is divided into three conspicuous chambers—anterior, middle, and posterior. Through the mesosome, the middle part of the coelom, run channels. Circulation is effected by a pulsating heartlike vessel in the protosome, the anterior part of the coelom. It is thought that hemichordates excrete through a pore near the mouth, but there is no experimental proof of this. Nerves spread throughout the epidermis of the middorsal and midventral body and thicken to a dorsal and ventral nerve cord; the two cords are connected by a nerve ring around the gut region; another ring surrounds the proboscis and is connected to the dorsal nerve cord.

The phylum is ancient: fossil pterobranchs have been found in Ordovician rocks some 450 million years old. It may have been the first of the deuterostome phyla—pogonophorans, echinoderms, chaetognaths, hemichordates and chordates (Phyla A-28 through A-32). In all these phyla, the blastopore of the embryo develops into an anus rather than a mouth. Probably all five phyla have a common ancestor.

Like the echinoderms, hemichordates develop from eggs and many become larvae that have ciliated bands. The hemichordates resemble the chordates in that they have gill slits in the throat or pharynx; also, a nerve cord, called the collar cord, that develops from dorsal epidermis of the embryo is sometimes hollow. It was thought for many years that hemichordates have a longitudinal rodlike supporting structure called a notochord; this was enough to classify them in the chordate phylum. Careful study, however, revealed that the so-called notochord of hemichordates is really a buccal pouch, a short anterior projection from the mouth cavity, and is not related to the notochord of chordates. Hemichordates, it is conceded by zoologists, are distinctive enough to deserve a phylum of their own.

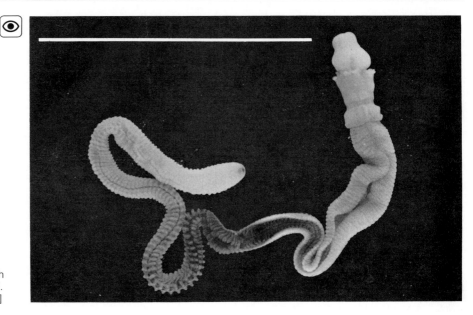

Ptychodera flava, a living acorn worm taken from subtidal sands near Waikiki beach, Hawaii. Bar = 5 cm. [Photograph courtesy of M. G. Hadfield; drawing by I. Atema, information from M. G. Hadfield.]

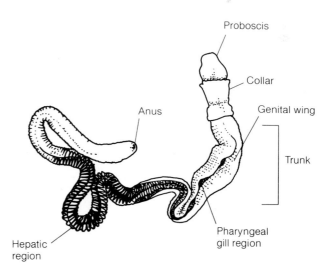

A *Ambystoma tigrinum,* the tiger salamander, belongs to the class Amphibia
and the family Ambystomatidae. It is one
of the most widespread salamanders in
North America. (This one was photographed in Nebraska.) It may grow to be
more than 20 cm long; the adults are
black or dark brown with yellow spots.
Bar = 10 cm. [Courtesy of S. J. Echter-
nacht.]

Members of this phylum, our own, are the best known of all the
animals. In fact, in the minds of many people, chordate is synon-
ymous with animal. All mammals, birds, amphibians, reptiles, and
fish (that is, the vertebrates), along with several obscure groups
of animals (including some invertebrates), belong to this phylum.
There are about 45,000 species, including all the animals of
major economic importance, with perhaps the exception of
some arthropods and mollusks.

Chordates can easily be defined by the presence of three features. One is the single, dorsal nerve cord, which, in mammals,
becomes the brain and spinal cord. A second universal chordate
feature is a cartilaginous rod, the notochord, which forms dorsal
to the primitive gut in the early embryo. This slender bar of cells
contains a gelatinous matrix and is sheathed in fibrous tissue. It
extends the length of the body and persists throughout life in
some of the invertebrate chordates, such as lancelets and lampreys. In the vertebrates, however, in the course of development
it is surrounded and later replaced by the vertebral column. The
third chordate characteristic is the presence, at some stage in
the life cycle, of gill slits in the pharynx or throat. These gill slits
reveal the marine ancestry of the phylum. In the land-dwelling
vertebrates, these slits are present only in the embryo; they close
or transform so that they are absent in the adult animal.

Chordates are bilaterally symmetrical animals that develop
from three embryonic germ layers: endoderm, mesoderm, and
ectoderm. Their bodies are segmented, a fact that is revealed in
the backbone composed of repeated vertebrae. All chordates
have a digestive tract complete with mouth and anus and a well-
developed coelom that develops from the embryonic mesoderm
layer. The internal organs are suspended in this coelom by thin
membranes of tissue called mesentery. All chordates reproduce
sexually; a very few can also reproduce parthenogenetically. In
the vast majority, sexes are separate and large eggs are fertilized
by undulipodiated sperm.

According to most current classifications, there are four subphyla of chordates. Animals of the two acraniate subphyla,
Tunicata and Cephalochordata, lack a brain.

Most tunicates are sessile marine animals; only the larva has a
notochord and a nerve cord, and the adult secretes a tunic, a
tough cellulose sac in which the animal is embedded. There are

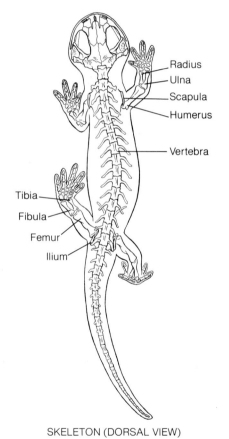

Radius
Ulna
Scapula
Humerus

Vertebra

Tibia
Fibula
Femur
Ilium

SKELETON (DORSAL VIEW)

Esophagus
Heart
Dorsal aorta
Lung

Stomach
Liver

Gall bladder

Testes
Wolffian duct
(transports sperm)
Large
intestine
Kidney
Urinary tubules

Urinary bladder

Cloaca

INTERNAL ORGANS
(VENTRAL VIEW)

B Skeleton of a generalized salamander.
[Drawing by L. Meszoly, information from
R. Estes.]

C Internal view of a generalized salamander. [Drawing by L. Meszoly, information
from R. Estes.]

D *Halocynthia pyriformis,* called the sea peach, is a tunicate belonging to the class Ascidiacea. It is yellow or orange with a tinge of red, and it has a sand-papery surface. It lives in shallow water along the Atlantic coast of North America from Maine northward. Bar = 1 cm. [Courtesy of N. J. Berrill.]

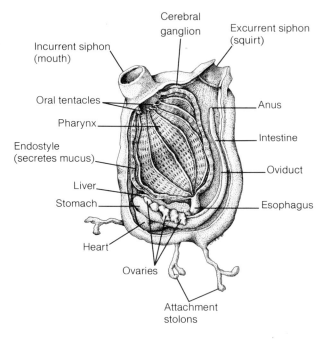

E Cutaway view of a generalized adult noncolonial ascidian tunicate. Although these tunicates are hermaphroditic, the testes are not shown here. [Drawing by L. M. Reeves.]

three classes of tunicates: the Larvacea, which are minute and tadpolelike as adults; the Ascidiacea, which as adults grow a typical tunic; and the Thaliacea, the chain tunicates or salps. The bodies of chain tunicates are barrel shaped and banded by muscles as hoops band a wooden cask. An asexual adult produces a chain of hundreds of buds, which then turn into sexual adults.

The cephalochordates, or lancelets, have a notochord and a nerve cord that persist in the adult and extend the length of the body. They are small, scaleless, fishlike, primitive chordates that belong to only one class, the Leptocardii.

All other chordates are craniates: they have a brain and a skull.

There are two subphyla: the Agnatha, which lack jaws and paired appendages, and the Gnathostomata, which have jaws and usually have paired appendages as well. The ostracoderms, ancient armored fishes with large scales, are a class of agnathids that is entirely extinct. The only living agnathids make up the class Cyclostomata, fishes that lack scales and have a round mouth like a suction cup. Lampreys, hagfishes, and slime eels belong to this group.

Gnathostomes, the jawed chordates, belong to either the superclass Pisces (the fishes) or the superclass Tetrapoda (animals having four limbs). There are two classes of living Pisces.

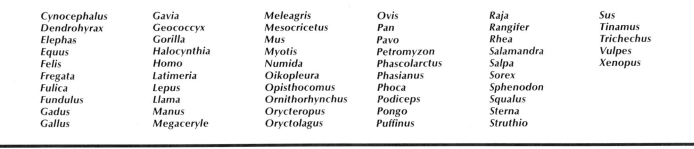

Cynocephalus *Gavia* *Meleagris* *Ovis* *Raja* *Sus*
Dendrohyrax *Geococcyx* *Mesocricetus* *Pan* *Rangifer* *Tinamus*
Elephas *Gorilla* *Mus* *Pavo* *Rhea* *Trichechus*
Equus *Halocynthia* *Myotis* *Petromyzon* *Salamandra* *Vulpes*
Felis *Homo* *Numida* *Phascolarctus* *Salpa* *Xenopus*
Fregata *Latimeria* *Oikopleura* *Phasianus* *Sorex*
Fulica *Lepus* *Opisthocomus* *Phoca* *Sphenodon*
Fundulus *Llama* *Ornithorhynchus* *Podiceps* *Squalus*
Gadus *Manus* *Orycteropus* *Pongo* *Sterna*
Gallus *Megaceryle* *Oryctolagus* *Puffinus* *Struthio*

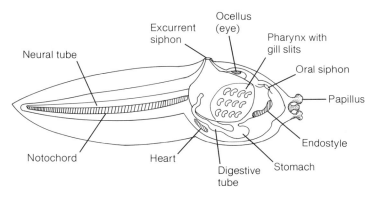

F A tadpolelike tunicate larva.
[Drawing by L. M. Reeves.]

(The placoderms, ancient jawed fishes, are all extinct.) The sharks, skates, and rays belong to the class Chondrichthyes, marine fishes whose scales, called placoid, are each made of a plate of dentine covered by enamel—like teeth. Sharks and their relations lack bones; their skeletons are made instead of a softer, more flexible material, cartilage. All the other fishes belong to the class Osteichthyes, the bony fishes. Their scales, made of bony material, are called cycloid or ctenoid according to whether their outer edge is smooth or spiny. Altogether, there are some 25,000 species of fishes—mainly of the bony kind.

Zoologists recognize four classes of tetrapods. The amphibians (Class Amphibia) lack scales. They respire both through their moist, soft skin and through gills, lungs, or the mouth lining. They lay eggs in water, where they spend at least their early life. There are about 2000 described species, including the frogs, toads, and salamanders.

The featherless reptiles (Class Reptilia) have dry skin covered with scales. They develop from an egg that has internal membranes and is adapted to life on land; it is called the amniote egg. There are about 5000 species, including all the turtles, lizards, snakes, and crocodiles. The famous Mesozoic dinosaurs belonged to this class. Living reptiles (but perhaps not some extinct ones) are poikilothermic, or cold-blooded—they cannot regulate the temperature of their blood very well. Their teeth are generally quite similar to each other. Reptiles breathe through their lungs and are well adapted to life on land.

The feathered reptiles, or birds, constitute Class Aves. There are nearly 9000 living species. They all have land-adapted eggs with calcium carbonate shells; their forelimbs are modified as wings; they have scaly skin with feathers and lack teeth. They are homeothermic—they can regulate their blood temperature internally.

There are about 4500 living species of Class Mammalia, to which we belong. Mammals are homeotherms; some species are better at it than others. They have a four-chambered heart with complete double circulation; that is, the aerated blood of the arteries does not mix with the oxygen-depleted blood of the veins. The skin of most mammals is covered with hair at some stage of life. Mammals nourish their young with milk, secretions produced in mammary glands of the mother. The fertilized egg develops inside the female; in most mammals, a special organ, the placenta, nourishes the developing embryo. Mammals have complex and differentiated teeth.

There are nearly twenty orders of mammals in two great subclasses: the subclass Prototheria, which includes the egg-laying mammals of Australia, and the subclass Theria, which includes all other mammals. The duck-billed platypus and the spiny anteater, both prototherians, have a cloaca (a common channel for digestive, excretory, and reproductive products), a horny beak or bill (no true teeth), shelled and yolky eggs, a pouch, reptilian bones, and poor temperature regulation. Theria consists of two infraclasses—Metatheria (the marsupials) and Eutheria (placental mammals). Most metatherians have an exterior pouch in which the young (born live) suckle for most of their development; they also have a cloaca and a double uterus and vagina. Eutherians have a single vagina; the young undergo considerable development inside the mother before they are born, and are nourished inside her by a special organ called the placenta. Eutherian orders include among others Insectivora (hedgehogs, shrews, and moles), Primates (lemurs, tarsiers, monkeys, apes, and human beings), Chiroptera (bats), Rodentia (squirrels, mice, and porcupines), Carnivora (dogs, cats, and bears), and Pinnipedia (seals and sea lions).

G *Cygnus olor,* the mute swan, the common swan in parks, occasionally established in the wild. Bar = 100 cm. [Courtesy of W. Ormerod.]

Bibliography

General

Barnes, R. D. *Invertebrate zoology,* 3rd ed. W. B. Saunders; Philadelphia; 1974.

Bayer, F. M., and H. B. Owre, *The free-living lower invertebrates.* Macmillan; New York; 1968.

Cloud, P. E., ''Pre-metazoan evolution and the origins of the metazoa.'' In E. T. Drake, ed., *Evolution and environment.* Yale University Press; New Haven, Conn.; 1968.

Grassé, P.-P., ed., *Larousse encyclopedia of the animal world.* Larousse; New York; 1975.

Grassé, P.-P., ed., *Traité de zoologie, anatomie, systématique, biologie,* 17 vols. Masson et cie; Paris; 1948–

Hanson, E. D., *Origin and early evolution of animals.* Wesleyan University Press; Middletown, Connecticut; 1977.

Hyman, L. H., *The invertebrates,* 6 vols. 1: *Protozoa through Ctenophora,* 1940; 2: *Platyhelminthes and Rhynchocoela,* 1951; 3: *Acanthocephala, Aschelminthes, and Entoprocta,* 1951; 4: *Echinodermata,* 1955; 5: *Smaller coelomate groups,* 1959; 6: *Mollusca, 1.,* 1967. McGraw-Hill; New York; 1940–1967.

Nichols, D., J. A. L. Cooke, and D. Whitely, *Oxford book of invertebrates.* Oxford University Press; New York and London; 1971.

Pennak, R. W., *Collegiate dictionary of zoology,* Ronald Press, New York; 1964.

Pennak, R. W. *Freshwater invertebrates of the United States,* 2nd ed. Wiley-Interscience; New York; 1978.

Storer, T. I., R. L. Usinger, R. C. Stebbins, and J. W. Nybakken, *General zoology,* 6th ed. McGraw-Hill; New York; 1979.

A-1 Placozoa

Grell, K. G., and G. Benwitz, ''Die Ultrastruktur von *Trichoplax adhaerens* F. E. Schulze.'' *Cytobiologie* 4:216–240; 1971. (English abstract.)

Miller, R. L., ''Observations on *Trichoplax adhaerens* Schulze 1883.'' *American Zoologist* 11(4):513; 1971.

Miller, R. L., ''*Trichoplax adhaerens* Schulze 1883: Return of an enigma.'' *Biological Bulletin* 141:374; 1971.

Ruthmann, A., ''Cell differentiation, DNA content, and chromosomes of *Trichoplax adhaerens.*'' *Cytobiologie* 15:58–64; 1977.

A-2 Porifera

Bayer, F. M., and H. B. Owre, *The free-living lower invertebrates.* Macmillan; New York; 1968.

de Laubenfels, M. W., *A guide to the sponges of eastern North America.* University of Miami Press; Coral Gables, Florida; 1953.

Newell, N. D., ''The evolution of reefs.'' *Scientific American* 266(6):54–65; 1972.

Rasmont, R., ''Sponges and their world.'' *Natural History* 71(3):62–70; 1962.

A-3 Cnidaria

Bayer, F. M., and H. B. Owre, *The free-living lower invertebrates.* Macmillan; New York; 1968.

Jacobs, W., ''Floaters of the sea.'' *Natural History* 71(7):22–27; 1962.

Kramp, P. L., *Synopsis of the medusae of the world. Marine Biological Association U.K.* 40:1–469. Cambridge University Press; New York; 1961.

Lane, C. E., ''The Portuguese man-of-war.'' *Scientific American* 202(3):158–168; 1960.

Muscatine, L., and H. M. Lenhoff, eds, *Coelenterate biology.* Academic Press; New York; 1974.

Rees, W. J., ed., *The Cnidaria and their evolution.* Academic Press; New York; 1966.

A-4 Ctenophora

Bayer, F. M., and H. B. Owre, *The free-living lower invertebrates.* Macmillan; New York; 1968.

Hardy, A., *Great waters,* Harper & Row; New York; 1967.

Russell-Hunter, W. D., *A biology of lower invertebrates.* Macmillan; New York; 1968.

A-5 Mesozoa

Kozloff, E. N., "Morphology of the orthonectid *Rhopalura ophiocomae.*" *Journal of Parasitology* 55:171–195; 1969.

Lapan, E. A., and H. Morowitz, "The Mesozoa." *Scientific American* 227(6):94–101; 1972.

McConnaughey, B. H., "The Mesozoa," pp. 557–570. In M. Florkin and B. T. Scheer, eds, *Chemical zoology II.* Academic Press; New York; 1968.

Noble, E. R., and G. A. Noble, *Parasitology,* 4th ed. Lea and Febiger; Philadelphia; 1976.

A-6 Platyhelminthes

Croll, N. A., *The ecology of parasites.* Harvard University Press; Cambridge, Mass.; 1966.

Dawes, B., *The Trematoda.* Cambridge University Press, Cambridge, England; 1946.

Erasmus, D. A., *The biology of trematodes.* Crane, Russak; New York; 1972. Arnold; London; 1974.

Smyth, J. D., *The physiology of cestodes.* W. H. Freeman and Company; San Francisco; 1969.

Smyth, J. D., *The physiology of trematodes.* W. H. Freeman and Company; San Francisco; 1966.

Voge, M., "Observations on the habitats of Platyhelminthes, primarily Turbellaria," pp. 455–470. In E. C. Dougherty, ed., *The lower metazoa: Comparative biology and phylogeny.* University of California Press; Berkeley; 1963.

A-7 Nemertina

Bayer, F. M., and H. B. Owre, *The free-living lower invertebrates.* Macmillan; New York; 1968.

Gibson, R., *Nemerteans.* Hutchinson University Library; London; 1972.

Kozloff, E. N., *Seashore life of Puget Sound, the Strait of Georgia, and the San Juan Archipelago.* University of Washington Press; Seattle; 1973.

Roe, P., "The nutrition of *Paranemertes peregrina* (Rhynchocoela: Hoplonemertea). I: Studies on food and feeding behavior." *Biological Bulletin* 139:80–91; 1970.

A-8 Gnathostomulida

Durden, C., J. Rodgers, E. Yochelson, and R. Riedl, "Gnathostomulida: Is there a fossil record?" *Science* 164:855–856; 1969.

Fenchel, T. M., and R. J. Riedl, "The sulfide system: A new biotic community underneath the oxidized layer of marine sand bottoms." *Marine Biology: International Journal on Life in Oceans* 7:255–268; 1969.

Riedl, R. J., "Gnathostomulida from America." *Science* 163:445–452; 1969.

Sterrer, W., "Systematics and evolution within the Gnathostomulida." *Systematic Zoology* 21:151–173; 1972.

A-9 Gastrotricha

Brunson, R. B., "Aspects of the natural history and ecology of the gastrotrichs," pp. 473–478. In E. C. Dougherty, ed., *The lower metazoa: Comparative biology and phylogeny.* University of California Press; Berkeley; 1963.

Brunson, R. B., "Gastrotricha," pp. 406–419. In W. T. Edmondson, H. B. Ward, and G. C. Whipple, eds, *Freshwater biology,* 2nd ed. Wiley; New York; 1959.

D'Hondt, J.-L., "Gastrotricha." *Oceanography and Marine Biology, an Annual Review* 9:141–192; 1971.

Gosner, K. L., *Guide to identification of marine and estuarine invertebrates.* Wiley-Interscience; New York; 1974.

Hummon, W. D., "Biogeography of sand beach Gastrotricha from the northeastern United States." *Biological Bulletin* 141(2):390; 1971.

A-10 Rotifera

Eddy, S., and A. C. Hodson, *Taxonomic keys to the common animals of the north central states.* Burgess; Minneapolis; 1961.

Edmondson, W. T., "Rotifera," pp. 420–494. In W. T. Edmondson, H. B. Ward, and G. C. Whipple, eds, *Freshwater biology,* 2nd ed. Wiley; New York; 1959.

Inglis, W. G., "Aschelminthes," pp. 137–143. In *Encyclopaedia Britannica,* 15th ed. Encyclopaedia Britannica; Chicago; 1974.

Pennak, R. W., "Ecological affinities and origins of free-living acoelomate freshwater invertebrates," pp. 435–451. In E. C. Dougherty, ed., *The lower metazoa: Comparative biology and phylogeny.* University of California Press; Berkeley; 1963.

Pennak, R. W., *Freshwater invertebrates of the United States,* 2nd ed. Wiley-Interscience; New York; 1978.

A-11 Kinorhyncha

Barnes, R. D., *Invertebrate zoology,* 3rd ed. W. B. Saunders; Philadelphia; 1974.

Dougherty, E. C., ed., *The lower metazoa: Comparative biology and phylogeny.* University of California Press; Berkeley; 1963.

Gosner, K. L., *Guide to identification of marine and estuarine invertebrates.* Wiley-Interscience; New York; 1974.

Russell-Hunter, W. D., *A biology of lower invertebrates.* Macmillan; New York; 1968.

A-12 Acanthocephala

Baer, J. G., *Animal parasites.* World University Library; London; 1971. McGraw-Hill; New York; 1971.

Crompton, D. W. T., *An ecological approach to Acanthocephalan physiology* (Cambridge Monograph in Experimental Biology 17). Cambridge University Press; Cambridge, England; 1970.

Nicholas, W. L., "The biology of Acanthocephala." pp. 205–206. In B. Dawes, ed., *Advances in Parasitology 5.* Academic Press; New York; 1967.

Noble, E. R., and G. A. Noble, *Parasitology,* 4th ed. Lea and Febiger; Philadelphia; 1976.

Olsen, O. W., *Animal parasites,* 3rd ed. University Park Press; Baltimore, Md.; 1974.

A-13 Entoprocta

Barnes, R. D., *Invertebrate zoology,* 3rd ed. W. B. Saunders; Philadelphia; 1974.

Gosner, K. L., *Guide to identification of marine and estuarine invertebrates.* Wiley-Interscience; New York; 1974.

Nielsen, C., "Phylogenetic considerations: The protostomian relationships," pp. 519–534. In R. M. Woollacott and R. L. Zimmer, eds., *Biology of bryozoans.* Academic Press; New York; 1977.

A-14 Nematoda

Croll, N. A., and B. E. Matthews, *Biology of nematodes.* Wiley; New York; 1977.

Goodey, J. B., *Soil and freshwater nematodes.* Wiley; New York; 1963.

Lee, D. L., and H. J. Atkinson, *The physiology of the nematodes,* 2nd ed. Columbia University Press; New York; 1977.

Yamaguti, S., *Synopsis of digenetic trematodes of vertebrates,* 2 vols. Keigaku; Tokyo; 1971.

A-15 Nematomorpha

Cheng, T. C., *The biology of animal parasites.* W. B. Saunders; Philadelphia; 1964.

Chitwood, B. G., "Gordiida," pp. 402–405. In W. T. Edmondson, H. B. Ward, and G. C. Whipple, eds, *Freshwater biology,* 2nd ed. Wiley; New York; 1959.

Croll, N. A., *Ecology of parasites.* Harvard University Press; Cambridge, Mass.; 1966.

Noble, E. R., and G. A. Noble, *Parasitology,* 4th ed. Lea and Febiger; Philadelphia; 1976.

A-16 Ectoprocta

Boardman, R. S., A. H. Cheetham, and W. A. Oliver, Jr., eds, *Animal colonies: Development and function through time.* Dowden, Hutchinson & Ross; Stroudsburg, Pa.; 1973.

Larwood, G. P., *Living and fossil Bryozoa.* Academic Press; New York; 1973.

Pennak, R. W., *Freshwater invertebrates of the United States,* 2nd ed. Wiley-Interscience; New York; 1978.

Ryland, J. S., *Bryozoans.* Hutchinson University Library; London; 1970.

Woollacott, R. M., and R. L. Zimmer, eds, *Biology of bryozoans.* Academic Press; New York; 1977.

A-17 Phoronida

Gosner, K. L., *Guide to identification of marine and estuarine invertebrates.* Wiley-Interscience; New York; 1974.

Kozloff, E. N., *Seashore life of Puget Sound, the Strait of Georgia, and the San Juan Archipelago.* University of Washington Press; Seattle; 1973.

MacGinitie, G. E., and N. MacGinitie, *Natural history of marine animals,* 2nd ed. McGraw-Hill; New York; 1968.

A-18 Brachiopoda

Jørgensen, C. B., *The biology of suspension feeding,* Pergamon Press; New York; 1966.

Rudwick, M. J. S., *Living and fossil brachiopods.* Hutchinson University Library; London; 1970.

Russell-Hunter, W. D., *Biology of higher invertebrates.* Macmillan; New York; 1969.

Williams, A., "The calcareous shell of the Brachiopoda and its importance to their classification." *Biological Reviews* (Cambridge Philosophical Society) 31:243–287; 1956.

Williams, A., et al., *Brachiopoda.* Part H (2 vols) of R. C. Moore, ed., *Treatise on invertebrate paleontology.* Geological Society of America; Boulder, Colorado; 1965. University of Kansas Press; Lawrence; 1965.

A-19 Mollusca

Abbott, R. T., *American seashells: The marine Mollusca of the Atlantic and Pacific coasts of North America,* 2nd ed. Van Nostrand Reinhold; New York; 1974.

Lane, F. W., *Kingdom of the octopus.* Jarrolds; London; 1960. Sheridan House; New York; 1960.

Morris, P. A., *A field guide to the shells of the Atlantic and Gulf coasts and the West Indies,* 3rd ed. Houghton Mifflin; Boston; 1973.

Morton, J. E., *Molluscs,* 4th ed. Hutchinson University Library; London; 1967.

Solem, A., *The shell makers.* Wiley-Interscience; New York; 1974.

Yonge, C. M., *Oysters,* 2nd ed. Collins; London; 1966.

A-20   Priapulida

Hammond, R. A., "The burrowing of *Priapulis caudatus.*" *Journal of Zoology* 162:469–480; 1970.

Por, F. D., and H. J. Bromley, "Morphology and anatomy of *Maccabeus tentaculatus.*" *Journal of Zoology* 173:173–197; 1974.

Shapeero, W. L., "Phylogeny of Priapulida." *Science* 133(3456):879–880; 1961.

A-21   Sipuncula

Clark, R. B., "Systematics and phylogeny: Annelida, Echiura, Sipuncula," pp. 1–62. In M. Florkin and B. T. Scheer, eds, *Chemical Zoology 4.* Academic Press; New York; 1969.

MacGinitie, G. E., and N. MacGinitie, *Natural history of marine animals,* 2nd ed. McGraw-Hill; New York; 1968.

Rice, M. E., "Asexual reproduction in a sipunculan worm." *Science* 167:1618–1620; 1970.

Stephen, A. C., and S. J. Edmonds, *The phyla Sipuncula and Echiura.* British Museum (Natural History); London; 1972.

A-22   Echiura

Kohn, A., and M. Rice, "Biology of Sipuncula and Echiura." *BioScience* 21:583–584; 1971.

Risk, M. J., "Silurian echiuroids: Possible feeding traces in the Thorold Sandstone." *Science* 180:1285–1287; 1973.

Stephen, A. C., and S. J. Edmonds, *The phyla Sipuncula and Echiura.* British Museum (Natural History); London; 1972.

MacGinitie, G. E., and N. MacGinitie, Natural history of marine animals. 2nd ed. McGraw-Hill; New York; 1968.

A-23   Annelida

Dales, R. P., *Annelids,* 2nd ed. Hutchinson University Library; London; 1967.

Edwards, C. A., and J. R. Lofty, *Biology of earthworms,* 2nd ed. Chapman & Hall; London; 1972. Halsted Press; New York; 1972.

Laverack, M. S., *The physiology of earthworms.* Macmillan; New York; 1963.

Wells, G. P., "Worm autobiographies." *Scientific American* 200:(6)132–142; 1959.

A-24   Tardigrada

Crowe, J. H., and A. F. Cooper, Jr., "Cryptobiosis." *Scientific American* 225(12):30–36; 1971.

Gosner, K. L., *Guide to identification of marine and estuarine invertebrates.* Wiley; New York; 1974.

Marcus, E., "Tardigrada," pp. 508–521. In W. T. Edmondson, H. B. Ward, and G. C. Whipple, eds, *Freshwater biology,* 2nd ed. Wiley; New York; 1959.

Pennak, R. W., "Ecology of the microscopic Metazoa inhabiting the sandy beaches of some Wisconsin lakes." *Ecological Monographs* 10:537–615; 1940.

Pennak, R. W., *Freshwater invertebrates of the United States,* 2nd ed. Wiley-Interscience; New York; 1978.

A-25   Pentastoma

Nichols, D., J. Cooke, and D. Whiteley, *Oxford book of invertebrates,* Oxford University Press; New York; 1971.

Noble, E. R., and G. A. Noble, *Parasitology,* 4th ed. Lea and Febiger; Philadelphia; 1976.

Self, J. T., "Biological relationships of the Pentastomida." *Experimental Parasitology* 24:63–119; 1969.

A-26   Onychophora

Ross, H. H., *Textbook of entomology,* 3rd ed. Wiley; New York; 1965.

Snodgrass, R. E., "Evolution of the Annelida, Onychophora, and Arthropoda." *Smithsonian Miscellaneous Collection* 97(6): 1–159; 1938.

Snodgrass, R. E., *A textbook of arthropod anatomy.* Cornell University Press; Ithaca, New York; 1952.

A-27   Arthropoda

Borrov, D. J., and R. E. White, *A field guide to the insects.* Houghton Mifflin; Boston; 1970.

Dillon, E. S., and L. S. Dillon, *A manual of common Beetles of eastern North America,* 2 vols. Dover; New York; 1972.

Kaston, B. J., *How to know the spiders,* 3rd ed. Wm. C. Brown; Dubuque, Iowa; 1978.

Pennak, R. W., *Freshwater invertebrates of the United States,* 2nd ed. Wiley-Interscience; New York; 1978.

Zim, H. S., and C. A. Cottam, *Insects: A guide to familiar American species.* Golden Press; New York; 1956.

A-28   Pogonophora

Ivanov, A. V., *Pogonophora.* Consultants Bureau; New York; 1963.

Nørrevang, A., ed., *The phylogeny and systematic position of Pogonophora* (special issue of *Zeitschrift für Zoologische Systematik und Evolutionsforschung*). Verlag Paul Parey; Hamburg and Berlin; 1975.

Southward, E. C., *Pogonophora of the northwest Atlantic: Nova Scotia to Florida* Smithsonian Contributions to Zoology 88:1–29; 1971.

A-29 Echinodermata

Boolootian, R. A., ed., *Physiology of Echinodermata.* Wiley; New York; 1966.

MacGinitie, G. E., and N. MacGinitie, *Natural history of marine animals,* 2nd ed. McGraw-Hill; New York; 1968.

Millott, N., ed., *Echinoderm biology* (Zoological Society of London Symposium 20). Academic Press; New York; 1967.

Nichols, D., *Echinoderms,* 4th ed. Hutchinson University Library; London; 1969.

Nichols, D., *The uniqueness of echinoderms* (Oxford/Carolina Biology Reader). Oxford University Press; New York and London; 1975.

A-30 Chaetognatha

Alvariño, A., "Chaetognaths," pp. 115–194. In H. Barnes, ed., *Oceanography and Marine Biology, Annual Review 3,* George Allen and Unwin; London; 1965.

Eakin, R. M., and J. A. Westfall, "Fine structure of the eye of a chaetognath." *Journal of Cell Biology* 21:115–132; 1964.

Ghirardelli, E., "Some aspects of the biology of the chaetognaths." *Advances in Marine Biology* 6:271–375; 1968.

Grant, G. C., "Investigations of inner continental shelf waters off lower Chesapeake Bay. Part IV: Descriptions of the Chaetognatha and a key to their identification." *Science* 4:107–119; 1963.

A-31 Hemichordata

Barnes, R. D., *Invertebrate zoology,* 3rd ed. W. B. Saunders; Philadelphia; 1974.

Barrington, E. J. W., *The Biology of Hemichordata and Protochordata.* W. H. Freeman and Company; San Francisco; 1965.

Berrill, N. J., *The origin of vertebrates.* Oxford University Press; New York; 1955.

A-32 Chordata

Barrington, E. J. W., *The biology of Hemichordata and Protochordata.* W. H. Freeman and Company; San Francisco; 1965.

Berrill, N. J., *The origin of vertebrates.* Oxford University Press; New York; 1955.

Romer, A. S., *The vertebrate body,* 5th ed. W. B. Saunders; Philadelphia; 1977.

Romer, A. S., *The vertebrate story,* rev. ed. University of Chicago Press; Chicago; 1971.

Young, J. Z., *The life of vertebrates,* 2nd ed. Oxford University Press; New York; 1962.

PLANTAE

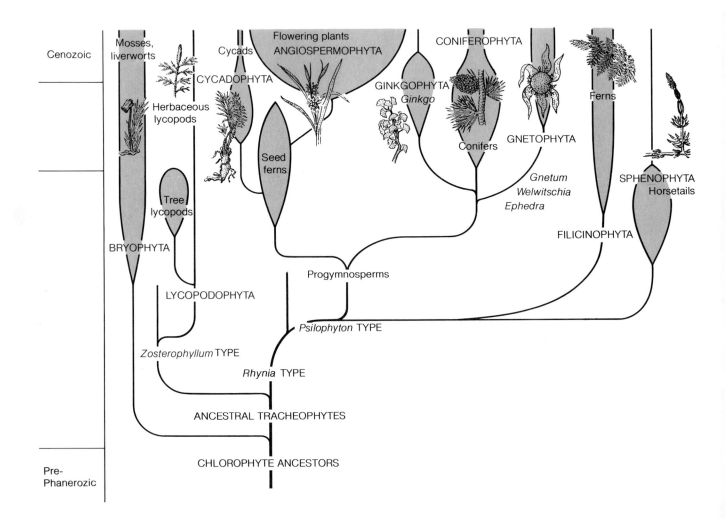

Cenozoic

Pre-
Phanerozoic

Mosses,
liverworts

Herbaceous
lycopods

Cycads

Flowering plants
ANGIOSPERMOPHYTA

CONIFEROPHYTA

CYCADOPHYTA

GINKGOPHYTA

Ginkgo

Ferns

Tree
lycopods

BRYOPHYTA

Seed
ferns

Conifers

GNETOPHYTA

Gnetum
Welwitschia
Ephedra

SPHENOPHYTA
Horsetails

LYCOPODOPHYTA

Progymnosperms

FILICINOPHYTA

Zosterophyllum TYPE

Psilophyton TYPE

Rhynia TYPE

ANCESTRAL TRACHEOPHYTES

CHLOROPHYTE ANCESTORS

PLANTAE

Latin *planta*, plant

Members of the plant kingdom are multicellular, sexually reproducing eukaryotes. Their cells contain green plastids, chloroplasts, which contain such pigments as chlorophylls *a* and *b,* xanthophylls, and other yellow and red carotenoid pigments. Today, they are the major mechanism for transforming solar energy into food, fiber, coal, oil, and other usable forms. Photosynthesis by plants sustains the biosphere, not only converting solar energy into food but producing oxygen as well.

Plants are adapted primarily for life on land, although many live in water during part of their life cycle. Some half a million species are known, and it is believed that there are many still undiscovered plants, especially in the tropics. Furthermore, because many plants resemble each other in form but are chemically different, it is highly likely that this estimate is a low one. The vast majority of plants living today belong to Phylum Angiospermophyta, or flowering plants (Phylum Pl-9).

Within the plant kingdom there are two basic groups: the bryophytes, or nonvascular plants, and the tracheophytes, or vascular plants. The latter are distinguished by conducting tissues called xylem and phloem. Xylem transports water and ions from the roots upward through the plant, and phloem transports photosynthate, sugar, and other products of the leaves throughout the plant. These rigid but metabolically active tissues are absent in the bryophytes.

Unlike the origin of fungi and animals, the origin of plants is generally agreed upon: green land plants descended from green algae (Chlorophyta, Phylum Pr-15). This hypothesis is based on such properties as similar pigmentation (including the presence of chlorophyll *a* and *b* in the chloroplasts), sperm having two undulipodia, and intercellular connections called plasmodesmata. Furthermore, some chlorophytes (such as *Klebsormidium*) have cellulosic walls and patterns of mitotic cell division identical to those of plants. In both, a cell-wall structure called a cell plate or phragmoplast develops perpendicular to the mitotic spindle.

Land plants first appeared in the Devonian Period, as rootless, leafless, and stemless but upright seaweedlike organisms. The earliest for which there is a good fossil record were ancestral tracheophytes of two sorts, represented by the extinct genera *Zosterophyllum* and *Rhynia*. The bryophytes are presumed to have evolved before the appearance and stabilization of vascular tissue—that is, before the appearance of these tracheophytes—although there is no early bryophyte fossil record.

The *Zosterophyllum* types gave rise to lycopods (Phylum Pl-2), a group that speciated extensively at the end of the Paleozoic Era but is now reduced to a few small herbaceous genera. The *Rhynia* types were the ancestors of all the other vascular land plants. Many groups, such as the seed ferns (Cycadofilicales—all extinct) and the horsetails (Sphenophyta, Phylum Pl-3), were far larger and more important in the past than they are now.

Although the flowering plants (Angiospermophyta, Phylum Pl-9) are an enormous group, they are relatively young, having appeared on the scene about 100 million years ago. They are apparently descended from seed ferns. The actual steps that led to the origin of seeds and fruits are not known, but that evolutionary innovation changed the living world by producing an environment in which man and other mammals could survive.

If it seems that there are far fewer plant than animal groups, it is partly because plant taxa are defined by morphological rather than chemical criteria. The differences between many plants are invisible—they produce different chemical compounds called secondary metabolites. Such compounds are not directly required for survival and reproduction, but they play a role in the plant's defenses against fungi, animals, and other plants. They include feeding deterrents, toxins, psychoactive compounds such as the marijuana alkaloids, and nerve poisons such as cyanide. Some plants leak compounds into the soil to prevent plants of other species from growing around them. These poisons and other secondary metabolites, even gaseous compounds, are important in determining the distribution, growth rate, and abundance of plants in natural communities. Thousands of secondary metabolites are known; many are starting materials for the manufacture of drugs.

Plants are distinguished from organisms belonging to the other eukaryotic kingdoms by their life cycles. They develop from embryos, diploid multicellular young organisms supported by sterile or nondeveloping tissue. Unlike animals, most of whose cells are diploid, and fungi, which are mostly haploid or dikaryotic, plants alternate haploid and diploid generations in an orderly fashion. Haploid plants are called gametophytes; diploid plants are called sporophytes. In the bryophytes, the more conspicuous green plant is the gametophyte; the sporophyte is small and brown and different looking. In the tracheophytes, on the other hand, the sporophyte is green and larger and more conspicuous than the gametophyte. The sporophyte generation dominates the life cycle in the most recently evolved phyla: in the flowering plants, the gametophyte is so reduced that, instead of being a separate plant, it is only a small group of cells entirely dependent on the sporophyte.

Pl-1 Bryophyta

Greek *bryon,* moss; *phyton,* plant

Andreaea	*Geothallus*	*Neohodgsonia*	*Scapania*
Anthoceros	*Haplomitrium*	*Pellia*	*Sphaerocarpos*
Bryum	*Hymenophytum*	*Phaeoceros*	*Sphagnum*
Buxbaumia	*Hypnum*	*Physcomitrium*	*Takakia*
Calypogeia	*Lejeunea*	*Plectocolea*	*Targionia*
Carrpos	*Lepidozia*	*Polytrichum*	*Tetraphis*
Conocephalum	*Mannia*	*Riccardia*	*Tortula*
Fossombronia	*Marchantia*	*Riccia*	
Frullania	*Marsupella*	*Ricciocarpus*	
Funaria	*Monoclea*	*Riella*	

The bryophytes are mainly inconspicuous plants growing in moist habitats. They are not fully adapted to life on land because their sperm must swim through water to reach their eggs. Also, because bryophytes lack the fluid-conducting tissues phloem and xylem, they rely on surrounding water to conduct necessary fluids and salts during times of growth. However, most survive periods of desiccation very well. Some bryophytes do have water- and food-conducting tissues, although these are invariably less developed than the xylem and phloem of the tracheophytes. Bryophyte sperm have two forward-directed undulipodia, just as the gametes and vegetative cells of most chlorophyte algae do. The chloroplasts and pigments of the bryophytes also are like those of the chlorophytes: they contain chlorophylls *a* and *b* and carotenoids such as β carotene, and they store starch as food. There are about 24,000 living species.

Unlike tracheophyte plants, the conspicuous and familiar generation of the bryophytes is a green, leafy gametophyte, or haploid organism, on which the diploid sporophyte usually depends for its subsistence. All bryophytes show tissue differentiation: the gametophyte bears multicellular gametangia—either archegonia (which produce eggs) or antheridia (which produce sperm) or both on the same plant. These gametangia are surrounded by sterile layers of tissue. Different bryophytes are distinguished by differences in the extent of dependence of the diploid on the haploid generation, the presence and form of the gametangia, and so forth.

There are three classes of bryophytes. The class Hepaticae, called liverworts, have thallose (leafy) gametophytes and long single-celled rhizoids that grow out from haploid spores. The class is named after the liverlike shape of the thallus, the leafy part of these small plants. As in all members of this phylum, the egg produced by mitotic division in the archegonium is fertilized by sperm. From the resulting zygote, a stalked diploid sporophyte grows. The sporophytes of liverworts are quite simple in construction: they lack stomata, the pores through which, in the leaves of most plants, gases are exchanged with the air. At the tip of the sporophyte, meiotic division takes place, leading to the production of haploid spores. When a spore is released, it falls to the ground, where it germinates into a threadlike protonema, from which an upright gametophyte will grow. This leafy gametophyte, the liverwort itself, differentiates gametangia and the cycle begins again. This haplophase-dominated life cycle is characteristic of all bryophytes. There are about 9000 species of liverworts, living mainly in tropical regions.

The class Anthocerotae, commonly called hornworts, includes

Polytrichum juniperinum, a common ground cover in the mixed coniferous and deciduous forest of New England. Bar = 3 cm. [Photograph courtesy of J. G. Schaadt; drawings by L. Meszoly.]

about 100 living species. Their sporophytes do bear stomata. Anthocerotae differ from the other classes in that the sporophyte keeps growing from a region at its base, called the basal intercalary meristem, as long as conditions are favorable. Other bryophytes stop growing when they reach the height characteristic of each genus.

The largest and best known of the bryophytes belong to the class Musci, the mosses. Our illustration is of *Polytrichum,* one of the most common mosses in the woodlands of the temperate zone. Moss sporophytes have multicellular rhizoids and stomata. The sporophytes release their spores in an often complicated

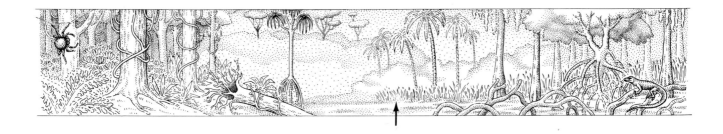

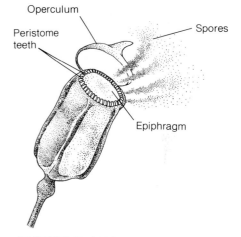

SPOROPHYTE CAPSULE

Operculum

Peristome teeth

Spores

Epiphragm

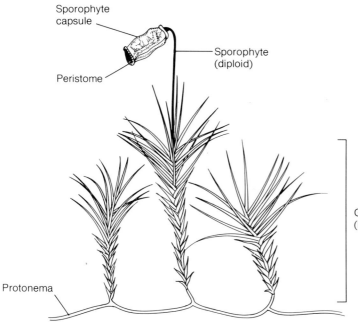

Sporophyte capsule

Sporophyte (diploid)

Peristome

Gametophyte (haploid)

Protonema

sequence of events that include the drying up of the spore capsule. There are about 14,500 species of mosses, in several hundred families. Although most species live in moist tropical environments, the mosses are also by far the most conspicuous bryophytes in temperate North America. The spongy *Sphagnum* that forms in acid boglands is important in the natural development of new soils. It is also used by florists to increase the water-holding capacity of soil, and it is burned as fuel (peat).

Bryophytes have a steady and consistent fossil record dating from the late Paleozoic, but they probably were never the dominant land plant form.

Pl-2   Lycopodophyta

Greek *lykos,* wolf; *pous,* foot; *phyton,* plant

Isoetes
Lycopodium
Phylloglossum
Selaginella

Lycopods, the club mosses and their relatives, are relicts of a glorious past. Only five genera are still living, mainly in the tropics; they comprise perhaps a thousand species altogether. Many more that lived in Devonian times have become extinct. For example, the large, woody lepidodendrids supported forest ecosystems long before the appearance of flowering trees; they dominated the Carboniferous coal forests until they died out some 280 million years ago. Of the two classes found in the fossil record, lepidodendrid trees and herbaceous smaller plants, the latter is the one to which all the living lycopod species belong.

Modern lycopods are not conspicuous and are known mainly to botanists and their students. Most of the tropical species are epiphytes; that is, they live on other plants. There are two well-known genera in the temperate regions. *Lycopodium,* shown in our illustration, has about 200 species and is the most common of the lycopods in the United States. This common club moss remains green all winter and thus is often used in winter holiday decorations as if it were a miniature pine tree. In fact, some call it ground pine, but it is not at all related to pines (Phylum Pl-7). The second well-known genus, *Selaginella,* has more species than any other lycopod genus—about 700. Many grow in moist habitats, but one, the resurrection plant (*Selaginella lepidophylla*), is native to Mexico and the Southwest of the United States. A curious feature of the plant is that it revives upon contact with water even after it has been dried out and apparently dead for months. It can dry out and be rewetted over and over again with no loss of vigor; hence, its name.

Like all plants, lycopods undergo an alternation of haploid and diploid generations. The sporophyte (diploid) plant is the more conspicuous. It consists of branching horizontal rhizoids (underground stems) and an upright part bearing aerial branches and roots. The leaves, or microphylls, of *Lycopodium* are small, and arranged in snug whorls on the aerial branches. Microphylls probably evolved as an outgrowth of the main photosynthetic axis of the plant; they differentiated eventually to form reproductive leaves and other structures. Megaphylls, the leaves characteristic of ferns and more complex plants (Phyla Pl-4 through Pl-9) had probably a different origin (see Phylum Pl-4).

Among the microphylls, some are fertile: they have been modified into sporangia, organs that produce spores. In some species, the fertile and the sterile microphylls are interspersed; both are photosynthetic. In others, such as our *Lycopodium obscurum,* nonphotosynthetic fertile scalelike microphylls are grouped together into cones, called strobili, that form at the tips of the aerial branches.

The club moss *Lycopodium obscurum* is widespread in the central and northeastern United States, in wooded areas under maples, pines, and oaks. Bar = 6 cm. [Photograph courtesy of W. Ormerod; drawings by R. Golder.]

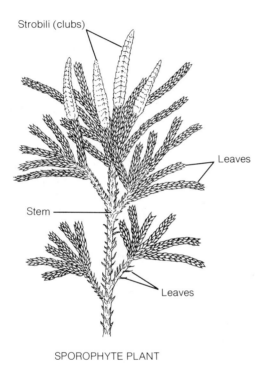

Strobili (clubs)

Leaves

Stem

Leaves

SPOROPHYTE PLANT

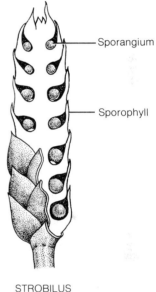

Sporangium

Sporophyll

STROBILUS
(CLUB)

Lycopodium is homosporous, which means that it produces only one kind of spore. Some members of the phylum (*Selaginella*, for example) are heterosporous, forming two kinds of spores, megaspores and microspores. Megaspores germinate into female gametophyte plants, called megagametophytes, inside of which archegonia produce eggs. The microspore germinates into a male gametophyte, or it may simply release sperm. In homosporous lycopods, the spores germinate into gametophytes, which may be small and white little masses of branching tissue or may be green and photosynthetic. These plants which regularly harbor symbiotic fungi in their tissues, may live inconspicuously in the soil for years. The gametophytes produce male organs (antheridia) and female organs (archegonia). Eventually (without meiosis, because the gametophytes are already haploid), sperm develop in the antheridium and eggs form in the archegonium. The sperm, which have two undulipodia, swim down the neck of the archegonium and fertilize the eggs; water is required, if only a thin film of it. The zygote, which may remain attached to the gametophyte, develops into a conspicuous green sporophyte, completing the cycle.

Pl-3 Sphenophyta
(Equisetophyta)

Greek *sphen,* wedge; *phyton,* plant

Sphenophytes, the group of plants to which the common horsetail belongs, are easily recognized by their jointed stems and rough, ribbed texture. The roughness is caused by mineral silica concentrated within the epidermal cells of the green photosynthetic stems; it accounts for the plant's common name, scouring rush. Indeed, many a camper has used horsetails to scour pots.

Sphenophytes, like bryophytes and lycopods (Phyla Pl-1 and Pl-2), are relics of a far more glorious past. In many of the forests of the later Devonian and the Carboniferous period, these plants dominated the vegetation; some of them were huge woody trees reaching several meters in diameter and 15 or so meters in height. There were many different species belonging to many different higher taxa. Today, there are only some five species and all belong to the single genus illustrated here, *Equisetum;* they thrive on salt flats, along the banks of streams, and in moist low parts of wooded areas.

The familiar horsetail represents the diploid, sporophyte generation; it is homosporous, that is, it produces only one kind of spore. At the apex of the stem is a cone, or strobilus, that bears about fifty short branches called *sporangiophores.* Inside each sporangiophore are several sporangia, tissues in which meiosis takes place and spores are eventually produced. The spores of *Equisetum* bear special long coiled flaps called elaters, which are formed by differentiation of the outer spore wall. Elaters uncoil when they dry out, and thus they help the spore to be wind borne. When a spore lands in a place that is sufficiently moist, the elaters coil up and the spore germinates to form a gametophyte, a dot-size green photosynthetic plant.

Gametophytes are either bisexual or male. They have many lobes of tissue emerging from a set of rhizoids that anchors them to the soil. In both types of gametophyte, the upper lobes produce multicellular antheridia, which give rise to sperm. On the sides of the bisexual gametophyte, female multicellular organs, archegonia, are produced and give rise to eggs. The sperm, which bear several undulipodia, fertilize eggs of archegonia either on the same or on other gametophytes. Sperm from male-only gametophytes must swim through water to find a bisexual gametophyte bearing mature eggs; several sperm, even from different plants, can fertilize the eggs on the same small gametophyte. The resulting zygotes then develop into separate diploid sporophytes as the parent gametophyte dies. Adult horsetail plants are often found growing in clusters, reflecting their development from a common gametophyte.

Amerindians, English and Tuscan peasants, and Romans once consumed horsetails, and some modern references list them as edible. However, they are known to poison livestock, especially cattle and horses. The neurotoxin aconitic acid and silica have been suspected, but now it is known that the toxicity is due to the enzyme thiaminase, which breaks down the vitamin thiamine.

A *Equisetum arvense* shoot bearing a strobilus. This horsetail is common in wasteland areas and on silica-rich soils. Bar = 3 cm. [Courtesy of J. G. Schaadt.]

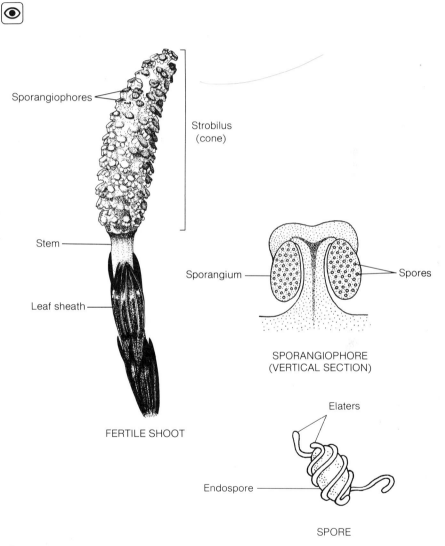

B *Equisetum hiemale,* common even in urban areas. Bar = 15 cm. [Courtesy of W. Ormerod.]

C *Equisetum arvense* strobilus [by I. Atema], sporangiophore [by R. Golder], and spore [by I. Atema].

Pl-4 Filicinophyta

(Pteridophyta, filicinae)

Latin *filix*, fern; Greek *phyton*, plant

Adiantum　　　　*Marattia*
Anemia　　　　　*Marsilea*
Asplenium　　　　*Ophioglossum*
Azolla　　　　　*Osmunda*
Botrychium　　　*Platyzoma*
Cyathea　　　　*Polypodium*
Dennstaedtia　　*Polystichum*
Dicksonia　　　　*Pteridium*
Dryopteris　　　*Pteris*
Hymenophyllum　*Salvinia*

Ferns are familiar vascular land plants that, like bryophytes, lycopods, and sphenophytes (Phyla Pl-1 through Pl-3), reproduce by means of spores, rather than seeds. Unlike those phyla, ferns have leaves of the kind called megaphylls (Greek "large leaf"). Unlike microphylls, which develop directly from the main photosynthetic stem (see Phylum Pl-2), megaphylls are thought to have evolved by the formation and fusion of lateral branches. Megaphylls are developmental and evolutionary precursors not only of fern fronds and the leaves of higher plants (Phyla Pl-5 through Pl-9), but also probably of the specialized reproductive tissues sporangia, cones, and flowers.

Ferns first appeared in the Devonian Period and are abundant in the fossil record from the Carboniferous through the present. They are the most complex, diverse, and abundant of the plant phyla that do not form seeds. Because their fertilization requires the swimming of undulipodiated sperm, ferns are limited to habitats that are at least occasionally moist. About 12,000 living species are known, two-thirds of them found in tropical regions. They are distinguished by the form of their spores and the details of their life cycles. Among the smallest are members of the aquatic genus *Azolla*. These ferns have developed very successful associations with the nitrogen-fixing cyanobacterium *Anabaena* (Phylum M-7), and they are common tropical water weeds. Their fronds are typically 1 cm long, whereas the fronds of large tree ferns may be 500 times as large. Most commonly, fern fronds are compound, divided into leaflets called pinnae; these may be subdivided further into pinnules. The genus *Polypodium*, illustrated here, is represented by nearly 1200 species. Although most are tropical, two, including the common polypody, are extensively distributed in the woods of the northern United States.

The vast majority of ferns are homosporous—they form only one kind of spore. The sporangia grow on various parts of the plant: on the undersurfaces of ordinary fronds, on special stalks, or on special modified fronds (as in *Polypodium*). The sporangia tend to develop in clusters called sori. In many ferns the sori are covered with the indusia, a special tissue that, when dry, shrivels and folds back to release the ripe spores.

Meiosis in the sporangia gives rise to clusters of wind-borne spores. From a spore that has landed in suitable environment grows a green photosynthetic filament, called the protonema. Under the influence of sunlight (the initiation of growth is caused by the absorption of blue light) a heart-shaped gametophyte develops, the prothallus. On its lower surface are numerous rhizoids that form regular and species-specific symbiotic associations with fungi.

A *Osmunda cinnamomea*, the cinnamon fern, very widespread in moist and shady areas, especially along the edges of ponds and streams. Bar = 50 cm.

Fern gametophytes may bear only antheridia (containing sperm), only archegonia (containing eggs), or both. The developmental pattern depends on the species and, in some cases, environmental factors such as crowding. The sperm, which may have from several to thousands of undulipodia, swims to the archegonium; in some cases, it is attracted by chemicals, such as malic acid. It fertilizes the egg and the sporophyte generation begins. The young sporophyte may retain its connection to the gametophyte during its early development, but it soon grows into the large and independent fern plant as the gametophyte tissue dies.

Although heterospory is not common in ferns, it occurs in certain genera such as *Platyzoma*, native to northern Australia. In these plants the smaller male spores form thin gametophytes, and the larger female spores form the more familiar rounded gametophytes.

258

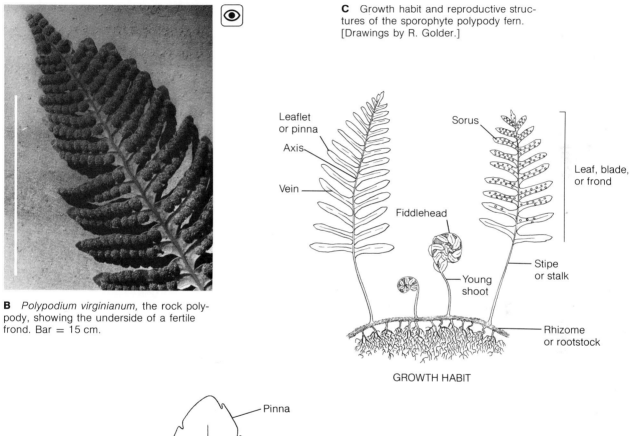

B *Polypodium virginianum,* the rock poly-pody, showing the underside of a fertile frond. Bar = 15 cm.

C Growth habit and reproductive structures of the sporophyte polypody fern. [Drawings by R. Golder.]

Leaflet or pinna

Axis

Vein

Sorus

Leaf, blade, or frond

Fiddlehead

Young shoot

Stipe or stalk

Rhizome or rootstock

GROWTH HABIT

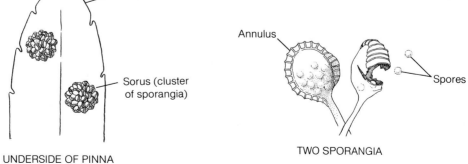

Pinna

Sorus (cluster of sporangia)

UNDERSIDE OF PINNA

Annulus

Spores

TWO SPORANGIA

259

Pl-5 Cycadophyta

Greek *kykos*, a palm; *phyton*, plant

Bowenia
Ceratozamia
Cycas
Dioon
Encephalartos
Macrozamia
Microcycas
Stangeria
Zamia

Some cycads are small shrubby plants superficially resembling little pineapple fruits; others, such as *Cycas* and *Zamia,* are palmlike trees more than three meters high. Their scaly trunks are covered with the bases of shed leaves. The plant's photosynthetic activity takes place in leaves clustered at the apex of the stem. Cycads bear naked seeds; that is, their seeds are not inside ovaries (see Phylum Pl-9, flowering plants, for comparison). Thus, they are gymnosperms (Greek *gymnos,* naked; *sperma,* seed).

Cycads have traditionally been classified with the other gymnosperms, such as conifers (Phylum Pl-7) and ginkgos (Phylum Pl-6). However, cycads differ from other gymnosperms enough to deserve phylum status. Unlike conifers and ginkgos, they have palmlike or fernlike compound leaves, they harbor symbiotic cyanobacteria in special roots, and exhibit sluggish cambial growth.

The cambium is a tissue layer that underlies the outer layer, or bark, of woody vascular plant stems; its cells grow and divide throughout the life of the plant. Cambial growth is a far more active source of new tissue in the conifers than in the cycads. Unlike the conifers and ginkgos, cycads tend to have unbranched stems. They are more closely related to flowering plants through common ancestors among the extinct seed ferns than they are to modern conifers and their relatives.

There are about 100 living species of cycads, grouped into nine genera, found only in tropical and subtropical regions. A diversity of these plants, including *Macrozamia communis,* is found in very dry deserts of Australia. In temperate zones, they are occasionally grown in greenhouses. *Zamia,* the only genus that grows outside in the continental United States, is found in southern Flordia.

Most cycads have large primary roots; some are as long as 12 meters and reach deep into sandy and dry soils for water. Cycads also have secondary roots that grow on or even above the soil. These roots, called coralloid roots, have a layer of greenish tissue just inside the surface epithelium. It harbors cyanobacteria (Phylum M-7) such as *Anabaena,* which lie in such neat rows between the host cells that they look like a tissue layer of the cycad itself. Free-living *Anabaena* is usually a filamentous organism, but it becomes coccoid—the filaments break into single cells—inside the coralloid roots. It is generally believed that the nitrogen-fixing ability of these symbiotic cyanobacteria permits cycads to populate areas whose soils are depleted of nitrates.

Most cycads bear their reproductive structures in conspicuous cones called strobili. All cycads are dioecious, which means that a single plant may bear either male or female cones, but not

A

A, B *Macrozamia communis,* a very young tree from sandy soil near Melbourne, Australia. Bar = 10 cm. [Photograph courtesy of C. P. Nathanielsz and I. A. Staff; drawing by I. Atema.]

Leaves

Stem

Coralloid roots

Root

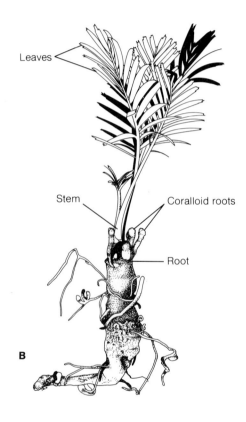

B

C An androstrobilus (male cone) of *Ceratozamia purpusii,* a cycad native to Mexico. Female cones are shorter and plumper. Bar = 50 cm.

both. Female seed cones are called gynostrobili; male cones are called androstrobili. The pollen, produced on the androstrobili, is dispersed by wind. A pollen grain that lands on a gynostrobilus grows a tube toward the female gamete; within the grain, an undulipodiated sperm develops. The sperm swims through the pollen tube to fertilize the egg. The product of fertilization, the zygote, develops into an embryo having two cotyledons, tissues that nourish the embryo. Superficially, this growth pattern is like that of dicotyledonous flowering plants.

Cycads are widely cultivated as ornamental plants. Cooked subterranean stems of *Zamia* once furnished edible starch to Florida's Seminoles. The Bantus of South Africa ate various parts of *Encephalartos*—the interior of the trunk, the center of the ripe female cones (Kaffir bread), and the seed pulp. Sago flour is prepared from the pith of *Cycas circinalis* in India and Sri Lanka, and of *Cycas revoluta* in Japan. Members of the genus *Cycas* are commonly called sago palms. However, true palms, which include other sago-producing trees such as *Oreodoxa oleracea* of the American tropics, belong to Phylum Angiospermophyta (Pl-9) and are monocotyledonous.

Ginkgo biloba is the only living descendant of a group that during the Mesozoic Era was far more extensive. On the northwest coast of the United States, petrified remains of these great Mesozoic trees can still be seen. It is no accident that photographs of ginkgo trees, like ours, are usually encumbered with streets, houses, and telephone wires: there are probably no truly wild ginkgo trees living today. *Ginkgo biloba* has been cultivated on temple and garden grounds in Asia for centuries. It was introduced into the temperate regions of America and Europe from Asia during historical times. It is amazingly resistant to pollution and to insects; this resistance accounts in part for the prevalence of ginkgos in urban settings.

The common name of the living ginkgo is the maidenhair tree, so named, it is said, because of the resemblance of the small bilobed leaves to the fronds of the maidenhair fern (genus *Adiantum*). The deciduous fan-shaped leaves are borne close to the richly branching stems, giving the trees a characteristic silhouette clearly recognizable even from a distance.

They are dioecious; the sexes are on separate plants. On male trees, haploid microspores develop in cones called microstrobili. The microspores are carried by the wind to female trees, where some will land on fleshy ovules, or gynostrobili, the female reproductive structures. Each microspore grows a tube that penetrates the micropyle, the opening into the ovule. Through this tube, a sperm is released; it swims through the watery female tissues to fertilize the egg nucleus—no environmental water is required. The sperm are helically coiled cells that bear hundreds of undulipodia. They develop only after the microspore tubes have entered the micropyles, and only during a few days in the spring. This fact has enabled biologists to conduct elegant ultrastructural studies of the development of the undulipodiated sperm. The embryo develops within a true but naked seed—it is not enclosed in an ovary wall (fruit) as an angiosperm seed is. However, the integument, or outer skin of the ovule, develops after fertilization into a fleshy fruitlike covering.

If ingested, the ginkgo seed causes nausea; however, this seldom happens, because decaying or ''ripening'' seeds release butyric acid and other foul-smelling substances. For this reason, most people prefer to cultivate male ginkgos. In China, however, ginkgo seeds have been used traditionally as a source of food and drugs. The outer fleshy portion is discarded and the inner kernel is roasted. Currently, the major use of the tree is as an urban ornamental plant.

A *Ginkgo biloba* in an urban setting in northeastern United States. Bar = 1 m.

B *Ginkgo biloba* branch showing mature ovules and distinctive leaf shape. Bar = 5 cm. [Courtesy of W. Ormerod.]

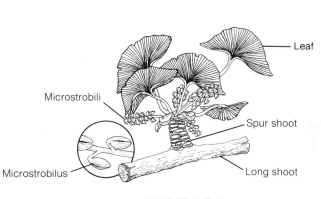

MALE BRANCH

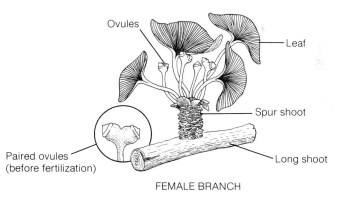

FEMALE BRANCH

C Female and male branches. [Drawings by R. Golder.]

Pl-7　Coniferophyta

Latin *conus,* cone; *ferre,* to bear;
Greek *phyton,* plant

Abies	*Pinus*
Araucaria	*Podocarpus*
Cedrus	*Pseudotsuga*
Cryptomeria	*Sequoia*
Cupressus	*Sequoiadendron*
Juniperus	*Taxodium*
Larix	*Taxus*
Metasequoia	*Thuja*
Picea	*Tsuga*

Most of these cone-bearing plants are trees, although some shrubs and prostrate plants are known. There are about 550 living species of conifers; they are by far the most familiar of the gymnosperms (plants having naked seeds) in the temperate regions. They are grouped into some 50 genera, including *Pinus* (pine), *Taxus* (yew), *Abies* (fir), *Pseudotsuga* (Douglas fir), *Picea* (spruce), and *Larix* (larch). The phylum also includes the largest living plant, *Sequoiadendron gigantea,* the giant sequoia of California. These giants grow to be as much as 100 meters high and 8 meters wide.

Although the conifers dominate many forests of the Northern Hemisphere, they also are quite common in the tropics and in the Southern Hemisphere. Some, like *Araucaria,* the distinctive, curly-branched monkey-puzzle tree, came originally from the Southern Hemisphere but are now cultivated in congenial northern climates, such as that of California.

The leaves of most conifers are needle-shaped, and all are simple (undivided into leaflets). They often have a heavy wax cuticle, and their interior cells form a very compact tissue penetrated by ducts, or resin canals, through which the sap flows. These features are particularly conspicuous in pines that live in arid places. Many species are evergreen and photosynthesize even in winter. Beneath the bark, the growing layer of tissue, or cambium, is very active. In most species, it produces rapid and extensive growth.

Most conifers are monoecious: female and male reproductive structures are borne on the same plant. However, they are heterosporous: two types of sporangia and spores are produced: microsporangia (male) are borne on small cones; megasporangia (female) are borne on larger cones, the familiar pine cones. Pollen, which is generally yellow and dusty, is carried by the wind from the male cone to the female. In the megasporangium, a cell called a megaspore divides several times to produce a small body of tissue called an ovule, which contains the egg. Through the micropyle, the opening into the ovule, a pollen grain grows a tube through which a male gamete nucleus travels to the egg. At no stage are undulipodia present. Fusion of the male and female nuclei may be delayed for as long as a year after pollination. Eventually, fusion does take place and the zygote develops into an embryo.

The seeds of conifers are naked in the sense that the embryos are not covered by an ovary wall, but are embedded in the megasporangium. The embryos can be seen as raised portions on the underside of the female cone scales, which become winged, wind-borne seeds, which usually separate from the cones at maturity. In some pines, however, the seeds do not separate

 A *Pinus rigida,* a pitch pine on a sandy hillside in the northeastern United States. Bar = 1 m.

from the cones; in fact, the cones may cling hard and fast to the parent tree until it is destroyed by fire. Because such pines germinate their seeds only after being subjected to extreme heat, they are important in the repopulation of forests after fires.

The oldest conifers in the fossil record date from the late Carboniferous, about 290 million years ago. It is thought that their remarkable drought-proof leaves evolved in the Permian Period, a time of worldwide aridity.

Conifers are of great economic importance. They are a major source of lumber and their soft wood is the major ingredient in the manufacture of paper. Turpentine, pitch, tar, amber, and resin are all products of conifer metabolism. Old English crossbows were made of the wood of yew trees, and thousands of conifers are cut each year in Europe and the Americas as Christmas trees. Edible seeds of many pines are eaten in Russia and Canada. In the United States *Pinus sabiniana* (Digger pine), *P. coulteri* (Coulter pine), *P. lambertiana* (sugar pine), and several species of pinyon pines are sources of protein-rich pine nuts.

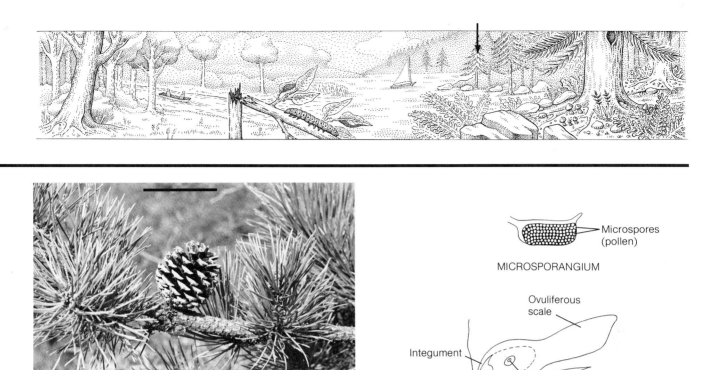

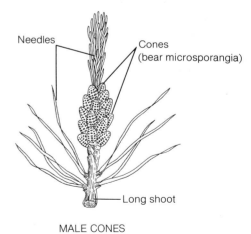

B *Pinus rigida* branch showing a mature female cone. Bar = 10 cm.

MICROSPORANGIUM

Microspores (pollen)

MEGASPORANGIUM

Ovuliferous scale

Integument

Micropyle

Egg

Ovule

Bract

FEMALE CONES

Young cone

Spur shoot

Needle

Ovuliferous scale (bears megasporangium)

Scale leaves

Long shoot

Mature cone

MALE CONES

Needles

Cones (bear microsporangia)

Long shoot

C Reproductive structures of *Pinus rigida*. [Drawings by R. Golder.]

Pl-8 Gnetophyta

Latin *gnetum* from Moluccan Malay *ganemu,*
a gnetophyte species found on the island
of Ternate; Greek *phyton,* plant

Ephedra
Gnetum
Welwitschia

Most people have never seen any of these cone-bearing desert plants. Gnetophytes today number only about seventy species contained in three very different genera—*Welwitschia, Gnetum,* and *Ephedra.* Their cones, unlike those of conifers (Phylum Pl-7), lack resin canals; both female and male cones are compound, having opposite or whorled bracts. The water-conducting xylem cells of gnetophytes are not dead, as those of conifers are. Like other seed-bearing plants, gnetophyte seedlings store their food reserves in leafy structures called cotyledons. Like conifers, the seeds of gnetophytes are naked, that is, not covered with layers of protective coat tissue, and the sperm cells lack undulipodia.

Gnetophytes resemble flowering plants in many ways, and at one time it was thought that they in fact represented a missing link between conifers and angiosperms (Phylum Pl-9). In *Welwitschia,* the male cone contains sterile ovules, female parts, suggesting that, as in many flowering trees, functioning male and female reproductive structures once resided in the same individual. Today they are always separate.

The genus *Welwitschia,* illustrated here, is named for its Austrian discoverer, F. Welwitsch. It contains but one extant species, *W. mirabilis,* which lives only in the deserts of southwestern Africa. It is a bizarre, unmistakable plant, most of which is buried in sandy soil. Lying on the ground are two very long (up to several meters) strap-shaped leaves, which become tattered with age. They continue to grow at their point of attachment during the entire life of the plant, which may live for as long as a century. The leaves are attached to the outer margin of a blunt, woody, cup-shaped stem, which protrudes only a few centimeters above the ground and may be as wide as a meter, depending on the plant's age. Below ground, the stem tapers to a root that extends down to the water table. Both root and stem serve as water storage organs. Several other adaptations of leaf and stem tissues help the plant accommodate to its arid environment.

The genus *Gnetum* comprises about thirty unfamiliar species of woody shrubs, vines, and large-leafed trees found in the tropical deserts and mountains of Asia, Africa, and Central and South America.

Ephedra, the best known gnetophyte genus, comprises about forty Eurasian and North and South American plants, all joint-stemmed shrubs less than 2 meters high, that seasonally lose their scalelike leaves. Superficially, they resemble scrubby

Welwitschia mirabilis (male) sitting on the desert in Southwest Africa. Bar = 25 cm. [Photograph courtesy of E. S. Barghoorn; drawing of plant by I. Atema; drawings of cones by R. Golder.]

cone-bearing desert pines. Known as joint fir or Indian tea, extracts of *Ephedra* are edible. The plants were well known to native Americans as a source of flour and tea, and their medicinal value has been known for some five thousand years. Today, the vasoconstrictor drug ephedrine, extracted from *E. sinica* and *E. equisetina* and related to epinephrine (adrenaline), is widely and effectively used in the treatment of asthma, emphysema, and hay fever.

The three gnetophyte genera are thought to be only remotely related to each other. *Welwitschia* probably evolved more than 300 million years ago from cone-bearing plants ancestral to the modern conifers. This conclusion is based on palynological (fossil pollen) studies of specimens dating back 280 million years.

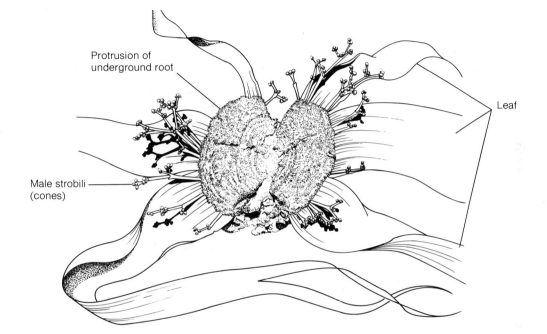

Protrusion of underground root

Leaf

Male strobili (cones)

Bracts

Integument

Bracteole

Egg cell

Integuments

Nodal bract

FEMALE BRACT

GYNOSTROBILUS (FEMALE CONE)

Bracts

Sterile ovule

Androsporangium

Bracteoles

MALE BRACT

ANDROSTROBILUS (MALE CONE)

P1-9 Angiospermophyta

(Anthophyta, magnoliophyta, flowering plants)

Greek *angeion*, little case; *sperma*, seed; *phyton*, plant

Acer	*Cinchona*	*Eucalyptus*
Agave	*Cinnamomum*	*Fagus*
Allium	*Cocos*	*Glycine*
Artocarpus	*Coffea*	*Helianthus*
Avena	*Colchicum*	*Hevea*
Beta	*Datura*	*Hordeum*
Brassica	*Daucus*	*Ipomoea*
Camellia	*Digitalis*	*Lemna*
Chondodendron	*Echinocactus*	*Lilium*
Chrysanthemum	*Elodea*	*Liriodendron*

The angiosperms are the superstars of the plant world. This phylum comprises nearly every familiar tree, shrub, and garden plant that produces flowers and seeds. The common ancestry of the angiosperms is revealed by their common reproductive organ system, the flower. The diversity and abundance of flowering plants today is truly astounding. There are more than 230,000 species grouped into more than 300 families. Some people feel that if we had more botanists the number of angiosperm species would be closer to a million. This vast diversity of plants is virtually unknowable in detail by a single person. How, then, is it organized? Jussieu distinguished plants lacking seed leaves and those bearing one (monocot) or two (dicot) seed leaves.

The structure and the fate of flowers are the best clues to the relationship between flowering plants; fruit morphology is another. Flowers have shown certain evolutionary trends with time. The older and more primitive ones, such as this tulip tree flower, have many parts, indefinite in number. Flowers probably started as a modification of a whorl of megaphylls, or leaves.

The carpel, the female flower part bearing the ovules that, after fertilization, become seeds, is in its most primitive form a folded leaf blade. The ovules, which, in the earliest angiosperms, probably formed in rows along the inner surface of the carpels, have themselves become associated with the carpels in many different ways. With time, the leaf shoots forming the various floral parts shortened and fused, so that today the whorled arrangement is no longer detectable in most species. Probably several carpels fused in the evolution of the more complex flowers. In fact, the radial symmetry of the earliest flowers has been supplanted by bilateral symmetry in some of the more recently evolved plants. It is through these and many other complex morphological changes that the origin and evolution of the various plant families can be best traced.

Angiosperms are heterosporous; they form two kinds of spores. The sporophyte generation has taken over this phylum—the gametophytes have virtually gone into hiding inside the flower. Neither archegonium nor antheridium is present. Microsporogenesis (the formation of male gametophyte) takes place within microsporangia, the pollen sacs borne on the flower anthers, which are sporophyte tissues. The mature male, or microgametophyte, is simply a pollen grain; it consists of only three haploid cells, which are entirely dependent on the sporophyte flower tissues for support.

Megasporogenesis (the formation of the female gametophyte generation) also takes place within the flower. The cells of the developing ovule are all there is of the female gametophyte generation. In the ovule, the megaspore, a single diploid cell, divides

 A *Liriodendron tulipifera*, the tulip tree, in summer in Illinois. Bar = 5 m. [Courtesy of Arnold Arboretum, Harvard University.]

meiotically and mitotically to form several haploid cells, one of which becomes the egg. The others undergo various transformations to form the mature female gametophyte, or embryo sac; the details are complex and yet remarkably uniform in thousands of plants.

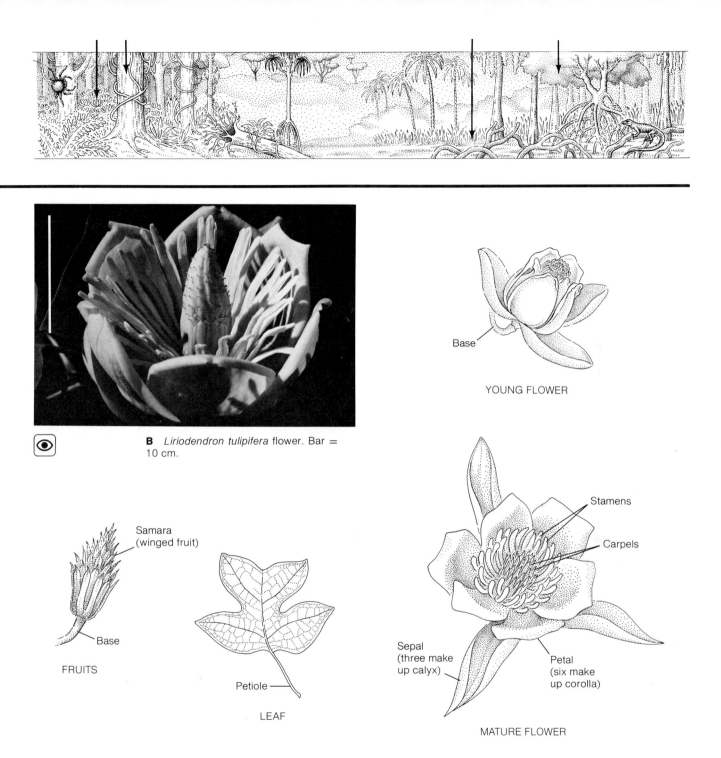

B *Liriodendron tulipifera* flower. Bar = 10 cm.

YOUNG FLOWER

Base

Samara
(winged fruit)

Base

FRUITS

Petiole

LEAF

Stamens

Carpels

Sepal
(three make
up calyx)

Petal
(six make
up corolla)

MATURE FLOWER

C *Liriodendron tulipifera* fruit, flowers, and leaf [drawings by L. Meszoly]; young flower [drawn by R. Golder].

Pl-9 Angiospermophyta

Magnolia
Manihot
Musa
Oryza
Phaseolus
Primula
Prunus
Quercus
Rauwolfia
Rosa

Saccharum
Solanum
Spiradella
Taraxacum
Theobroma
Tradescantia
Trifolium
Triticum
Tulipa
Vitis

Wolffia
Yucca
Zea
Zostera

A male gametophyte, in the form of a pollen tube, grows down into the embryo sac to fertilize the egg nucleus as well as two other nuclei of the embryo sac. This double fertilization leads to the development of the diploid embryo (sporophyte) and its triploid supporting tissue, the endosperm.

All this development activity leads to one of the greatest of evolutionary innovations: the angiosperm seed. This amazingly tough structure not only contains the embryo but provides it with nourishing endosperm; maternal diploid tissue forms a protective seed coat, a complex tissue prepared for many contingencies, depending on the species. Some seed coats require burning or scarring to germinate; some must be chilled or frozen; others must be exposed to certain wavelengths of light.

During the first half of the Cretaceous Period, from an apparently humble beginning, the flowering plants took over the world. Seventy-five million years ago, the great forests of seed ferns, conifers, cycads, and ginkgos gave way to the flowering plants. By the opening of the Cenozoic Era, 70 million years ago, many modern families and even genera of angiosperms had already appeared. The dominance of angiosperms in the plant kingdom is certainly related to the simultaneous evolution of animals, especially of insects (Phylum A-27) and chordates (Phylum A-32).

D *Aster novae-angliae*, the New England aster, blooms purple in meadows and along roadsides. It belongs to the Compositae, the largest of the plant families. Bar = 1.5 cm. [Courtesy of W. Ormerod.]

Bibliography

General

Britton, N. L., and A. Brown, *An illustrated flora of the northern United States and Canada,* 3 vols. Dover; New York; 1970.

Fernald, M. L., *Gray's Manual of Botany,* 8th ed. American Book Company; New York; 1950.

Raven, P. H., R. F. Evert, and H. Curtis, *Biology of plants,* 2nd ed. Worth; New York; 1976.

Scagel, R. F., R. J. Bandoni, G. E. Rouse, W. B. Schofield, J. R. Stein, and T. M. C. Taylor, *Plant diversity: An evolutionary approach.* Wadsworth; Belmont, California; 1969.

Pl-1   Bryophyta

Conard, H. S., *How to know the mosses and liverworts,* W. C. Brown; Dubuque, Iowa; 1956.

Schuster, R. M., *The Hepaticae and Anthocerotae of North America east of the hundredth meridian,* 2 vols. Columbia University Press; New York; 1966.

Smith, G. M., *Cryptogamic botany,* Vol. 2: *Bryophytes and pteridophytes,* McGraw-Hill; New York; 1955.

Pl-2   Lycopodophyta

Cobb, B., *A field guide to the ferns.* Houghton Mifflin; Boston; 1977.

Smith, G. M., *Cryptogamic botany,* Vol. 2: *Bryophytes and pteridophytes.* McGraw-Hill; New York, 1955.

Wherry, E. T., *The fern guide,* Doubleday; Garden City, New York; 1961.

Pl-3   Sphenophyta

Cobb, B., *A field guide to the ferns.* Houghton Mifflin; Boston; 1977.

Smith, G. M., *Cryptogamic botany,* Vol. 2: *Bryophytes and pteridophytes.* McGraw-Hill; New York; 1955.

Wherry, E. T., *The fern guide.* Doubleday; Garden City, New York; 1961.

Pl-4   Filicinophyta

Cobb, B., *A field guide to the ferns.* Houghton Mifflin; Boston; 1977.

Smith, G. M., *Cryptogamic botany,* Vol. 2: *Bryophytes and pteridophytes.* McGraw-Hill; New York; 1955.

Wherry, E. T., *The fern guide,* Doubleday; Garden City, New York; 1961.

Pl-5   Cycadophyta

Bold, H. C., *Morphology of plants,* 3rd ed., Harper & Row; New York; 1973.

Chamberlain, C. J., *Gymnosperms: Structure and evolution.* Dover; New York, 1966.

Dallimore, W., and A. B. Jackson, *A handbook of Coniferae and Ginkgoaceae,* 4th ed., rev. S. G. Harrison. St. Martin's; New York; 1966.

Greguss, P., *Xylotomy of the living cycads.* Akademiai Kiado; Budapest, Hungary; 1968.

Pl-6   Ginkgophyta

Brockman, C. F., *Trees of North America.* Golden Press; New York; 1968.

Chamberlain, C. J., *Gymnosperms: Structure and evolution.* New York; 1966.

Dallimore, W., and A. B. Jackson, *A handbook of Coniferae and Ginkgoaceae,* 4th ed., rev. S. G. Harrison. St. Martin's; New York; 1966.

Sargent, C. S., *Manual of the trees of North America,* Vol. 1 (2 vols.). Dover; New York; 1961. Reprint of 1922 edition; Houghton Mifflin; Boston.

Pl-7   Coniferophyta

Chamberlain, C. J., *Gymnosperms: Structure and evolution.* Dover; New York; 1966.

Dallimore, W., and A. B. Jackson, *A handbook of Coniferae and Ginkgoaceae,* 4th ed. rev. S. G. Harrison. St. Martin's; New York; 1966.

Pl-8 Gnetophyta

Bold, H. C., *Morphology of plants,* 3rd ed. Harper & Row; New York; 1973.

Chamberlain, C. J., *Gymnosperms: Structure and evolution.* Dover; New York; 1966.

Hertrich, W., *Palms and cycads.* C. F. Braun; Alhambra, California; 1951.

Pl-9 Angiospermophyta

Brockman, C. F., *Trees of North America.* Golden Press; New York; 1968.

Gleason, H. A., *The new Britton and Brown illustrated flora of the northeastern United States and adjacent Canada,* 3 vols. Hafner; New York; 1968.

Rickett, H. W., ed., *Wildflowers of America.* Crown; New York; 1963.

Symonds, G., *Shrub identification book.* William Morrow; New York; 1963.

Symonds, G., *Tree identification book.* William Morrow; New York; 1958.

APPENDIX

A List of Genera

This list includes all the genera mentioned in this book, and many others as well. The list names the phylum of each genus and, for many genera, gives the common names by which their members are known.

GENUS	PHYLUM	COMMON NAME
Abies	Coniferophyta	Fir
Acanthamoeba	Rhizopoda	
Acantharia	Actinopoda	
Acanthocephalus	Acanthocephala	
Acanthocystis	Actinopoda	
Acanthogyrus	Acanthocephala	
Acanthometra	Actinopoda	
Acer	Angiospermophyta	Japanese maple, Norwegian maple, sugar maple, mountain maple, red maple, silver maple, black maple
Acetabularia	Chlorophyta	Mermaid's wine glass
Acetobacter	Omnibacteria	
Achlya	Oomycota	
Achnanthes	Bacillariophyta	
Acholeplasma	Aphragmabacteria	
Achromobacter (Alcaligenes)	Omnibacteria	
Acidaminococcus	Fermenting bacteria	
Acrasia	Acrasiomycota	
Actinobacillus	Omnibacteria	
Actinomyces	Actinobacteria	
Actinophrys	Actinopoda	
Actinosphaera	Actinopoda	
Actinosphaerium (Echinosphaerium)	Actinopoda	
Acytostelium	Acrasiomycota	
Adiantum	Filicinophyta	Maidenhair fern
Aerobacter	Omnibacteria	
Aerococcus	Micrococci	
Aeromonas	Omnibacteria	

GENUS	PHYLUM	COMMON NAME
Agardhiella	Rhodophyta	Coulter's seaweed
Agaricus	Basidiomycota	Field mushroom, commercial mushroom
Agrobacterium	Omnibacteria	
Agave	Angiospermophyta	Century plant, sisal, henequen, maguey
Alaria	Phaeophyta	Winged kelp, dapper-locks, honey ware, mirlins
Albertia	Rotifera	
Albugo	Oomycota	
Alcaligenes (Achromobacter)	Omnibacteria	
Alcyonidium	Ectoprocta	
Alginobacter	Omnibacteria	
Alligator	Chordata	American and Chinese alligators
Allium	Angiospermophyta	Chives, leeks, onions, garlic
Allogromia	Foraminifera	
Allomyces	Chytridiomycota	
Alternaria	Deuteromycota	Potato blight
Alysiella	Myxobacteria	
Amanita	Basidiomycota	Caesar's amanita, fly agaric, death cap, destroying angel
Ambystoma	Chordata	Tiger salamander, marbled salamander
Amia	Chordata	Bowfin
Amoeba	Rhizopoda	
Amorphomyces	Ascomycota	
Amphidinium	Dinoflagellata	

274

GENUS	PHYLUM	COMMON NAME	GENUS	PHYLUM	COMMON NAME
Amphipleura	Bacillariophyta		*Ardea*	Chordata	Great blue heron, gray heron, purple heron, goliath heron
Amphiporus	Nemertina				
Anabaena	Cyanobacteria				
Anacystis	Cyanobacteria		*Argonauta*	Mollusca	Paper nautilus, argonaut
Anaplasma	Aphragmabacteria				
Anas	Chordata	Mallard duck, black mallard, gadwall, pintail duck, teal	*Argyrotheca*	Brachiopoda	
			Armillifera	Pentastoma	
			Artemia	Arthropoda	Brine shrimp
Andraea	Bryophyta	Granite moss, black moss	*Arthrobacter*	Actinobacteria	
Anemia	Filicinophyta	Pine fern	*Arthromitus (Coleomitus)*	Fermenting bacteria	
Angiococcus	Myxobacteria		*Artocarpus*	Angiospermophyta	Breadfruit, jackfruit
Anisolpidium	Hyphochytridio-mycota		*Ascaris*	Nematoda	Roundworm of pigs and human beings
Anopheles	Arthropoda	Malaria mosquito			
Anser	Chordata	Greylag goose, snow goose, white-fronted goose	*Ascophyllum*	Phaeophyta	Knotted wrack, yellow tang, sea whistle
Anthoceros	Bryophyta	Horned liverwort	*Aspergillus*	Deuteromycota	Black mold
Antipathes	Cnidaria	Black coral	*Aspidosiphon*	Sipuncula	
Aphanomyces	Oomycota		*Asplenium*	Filicinophyta	Lady fern
Aphrodite	Annelida	Sea mouse	*Astasia*	Euglenophyta	
Apis	Arthropoda	Honey bee	*Asterias*	Echinodermata	Common starfish (of North Atlantic), northern (purple) starfish
Aplysia	Mollusca	Sea hare			
Apodachlya	Oomycota				
Aptenodytes	Chordata	Emperor penguin			
Apteryx	Chordata	Common kiwi, little spotted kiwi, great spotted kiwi	*Asterionella*	Bacillariophyta	
			Asticcacaulis	Omnibacteria	
Arachnoidiscus	Bacillariophyta		*Atolla*	Cnidaria	Crown jellyfish
Araucaria	Coniferophyta	Monkey puzzle tree, bunya pine, Moreton Bay pine, Norfolk Island pine, Paraná pine	*Atubaria*	Hemichordata	
			Austrognatharia	Gnathostomulida	
			Avena	Angiospermophyta	Oat
			Azolla	Filicinophyta	Water fern, pond ferns, mosquito ferns
Arbacia	Echinodermata	Purple sea urchin	*Azomonas*	Nitrogen-fixing aerobic bacteria	
Arca	Mollusca	Ark shells			
Arcella	Rhizopoda	Shelled amoeba	*Azotobacter*	Nitrogen-fixing aerobic bacteria	
Archangium	Myxobacteria		*Babesia*	Apicomplexa	
Archilochus	Chordata	Ruby-throated hummingbird	*Bacillus*	Aeroendospora	
			Bacteroides	Fermenting bacteria	
Architeuthis	Mollusca	Giant squid	*Balaenoptera*	Chordata	Fin, sei, and Bryde's whales, rorqual
Arcyria	Myxomycota				

GENUS	PHYLUM	COMMON NAME	GENUS	PHYLUM	COMMON NAME
Balanoglossus	Hemichordata		*Bradypus*	Chordata	Three-toed sloth, maned sloth
Balantidium	Ciliophora		*Branchioceranthus*	Cnidaria	
Bambusina	Gamophyta		*Branchiostoma*	Chordata	Lancelet, amphioxus
Bangia	Rhodophyta		*Brassica*	Angiospermophyta	Cabbage, broccoli, turnip, black mustard, ruta-baga, kohlrabi
Barbeyella	Myxomycota				
Barentsia	Entoprocta				
Bartonella	Aphragmabacteria				
Bathyspadella	Chaetognatha		*Brucella*	Omnibacteria	
Batillipes	Tardigrada		*Bryopsis*	Chlorophyta	Sea fern
Batrachospermum	Rhodophyta		*Bryum*	Bryophyta	Silver moss
Bdellovibrio	Pseudomonads		*Bubalus*	Chordata	Water buffalo
Beggiatoa	Myxobacteria		*Bubo*	Chordata	Great horned owl
Beijerinckia	Nitrogen-fixing aerobic bacteria		*Bufo*	Chordata	American toad
			Bugula	Ectoprocta	
Beneckea	Omnibacteria		*Busycon*	Mollusca	Knobbed whelk, lightning conch
Beroe	Ctenophora	Thimble jelly, sea mitre			
			Buteo	Chordata	Red-tailed hawk
Beta	Angiospermophyta	Sugar beet, Swiss chard, red beet, mangel, spinach	*Buxbaumia*	Bryophyta	Bug-on-a-stick, elf-cap moss
			Caenorhabditis	Nematoda	
Biddulphia	Bacillariophyta		*Calanus*	Arthropoda	Copepod
Blakeslea	Zygomycota		*Calcidiscus*	Haptophyta	
Blastocladiella	Chytridiomycota		*Calciosolenia*	Haptophyta	
Blastocrithidia	Zoomastigina		*Calicium*	Mycophycophyta	Pear-shaped lichen, stubble lichen
Blatta	Arthropoda	Oriental cockroach			
Blepharisma	Ciliophora		*Callophyllis*	Rhodophyta	Red sea fan
Bodo	Zoomastigina		*Calonympha*	Zoomastigina	
Boletus	Basidiomycota	Edible bolete, lurid bolete, yellow cracked bolete	*Calvatia*	Basidiomycota	Giant puffball
			Calymmatobacte-rium	Omnibacteria	
Bolinopsis	Ctenophora	Lobed comb jelly	*Calypogeia*	Bryophyta	
Bombyx	Arthropoda	Silkworm moth	*Calyptrosphaera*	Haptophyta	
Bonellia	Echiura		*Camellia*	Angiospermophyta	Tea, camellia
Borrelia	Spirochaetae		*Camelus*	Chordata	Arabian (drome-dary) camel, two-humped (bactrian) camel
Bos	Chordata	Cattle, cow, zebu, yak, guar, banteng, kouprey			
Botrychium	Filicinophyta	Moonwort, grape ferns	*Campyloderes*	Kinorhyncha	
			Candida	Deuteromycota	Monilia, thrush
Botrydiopsis	Xanthophyta		*Canis*	Chordata	Dog, gray wolf, coyote, dingo, jackal, red wolf
Botrydium	Xanthophyta				
Botryococcus	Xanthophyta				
Bowenia	Cycadophyta		*Canteriomyces*	Hyphochytridio-mycota	
Brachionus	Rotifera				

GENUS	PHYLUM	COMMON NAME	GENUS	PHYLUM	COMMON NAME
Cantharellus	Basidiomycota	Chanterelle	*Chlamydomonas*	Chlorophyta	
Carchesium	Ciliophora		*Chlamydomyxa*	Labyrinthulamycota	
Cardiobacterium	Omnibacteria		*Chlorarachnion*	Rhizopoda	
Carrpos	Bryophyta		*Chlorella*	Chlorophyta	
Casuarius	Chordata	Cassowary	*Chloridella*	Xanthophyta	
Caterla	Kinorhyncha		*Chlorobium*	Anaerobic photosynthetic bacteria	
Caulerpa	Chlorophyta				
Caulobacter	Omnibacteria				
Cavia	Chordata	Guinea pig	*Chlorococcum*	Chlorophyta	
Cedrus	Coniferophyta	Atlas cedar, Cyprus cedar, deodar cedar of Lebanon	*Chlorodesmis*	Chlorophyta	
			Chloroflexus	Anaerobic photosynthetic bacteria	
Cellulomonas	Actinobacteria		*Chloropseudo-monas*	Anaerobic photosynthetic bacteria	
Centroderes	Kinorhyncha				
Centropyxis	Rhizopoda		*Chondodendron*	Angiospermophyta	
Cepedea	Zoomastigina		*Chondrococcus*	Myxobacteria	
Cephalobaena	Pentastoma		*Chondromyces*	Myxobacteria	
Cephalodiscus	Hemichordata		*Chondrus*	Rhodophyta	Irish moss, carrageen
Cephalothamnion	Zoomastigina		*Chorda*	Phaeophyta	
Cephalothrix	Nemertina		*Chordaria*	Phaeophyta	
Ceratiomyxa	Myxomycota		*Chordeiles*	Chordata	Common nighthawk
Ceratium	Dinoflagellata		*Chordodes*	Nematomorpha	
Ceratomyxa	Cnidosporidia		*Chordodiolus*	Nematomorpha	
Ceratozamia	Cycadophyta	Mexican horncone	*Chromatium*	Anaerobic photosynthetic bacteria	
Cerebratulus	Nemertina				
Cerotocystis	Ascomycota	Dutch elm disease	*Chromobacterium*	Omnibacteria	
Cestum	Ctenophora	Velamen, Venus' girdle	*Chromogaster*	Rotifera	
			Chromulina	Chrysophyta	
Chaetocladium	Zygomycota		*Chroococcus*	Cyanobacteria	
Chaetomium	Ascomycota		*Chroomonas*	Cryptophyta	
Chaetomorpha	Chlorophyta		*Chrysanthemum*	Angiospermophyta	Costmary, ox-eye daisy, pyrethrum
Chaetonotus	Gastrotricha				
Chaetopterus	Annelida	Parchment worm	*Chrysarachnion*	Rhizopoda	
Challengeron	Actinopoda		*Chrysemys*	Chordata	Painted turtle
Chamaesiphon	Cyanobacteria		*Chrysobotrys*	Chrysophyta	
Chantransia	Rhodophyta		*Chrysocapsa*	Chrysophyta	
Chara	Chlorophyta	Stonewort	*Chrysochromulina*	Haptophyta	
Characiopsis	Xanthophyta		*Chytria*	Chytridiomycota	
Chilomonas	Cryptophyta		*Cinchona*	Angiospermophyta	Fever-bark tree
Chironomus	Arthropoda	Midge	*Cinnamomum*	Angiospermophyta	Cinnamon, cassia
Chlamydia	Omnibacteria				

GENUS	PHYLUM	COMMON NAME	GENUS	PHYLUM	COMMON NAME
Ciona	Chordata	Sea squirt	*Collozoum*	Actinopoda	
Citrobacter	Omnibacteria		*Columba*	Chordata	Band-tailed pigeon, some other doves and pigeons
Cladochytrium	Chytridiomycota				
Cladonia	Mycophycophyta	Reindeer lichen, spoon lichen, pixie cup, ladder lichen, British soldiers	*Comatricha*	Myxomycota	
			Conidiobolus	Zygomycota	
			Coniocybe	Mycophycophyta	Mealy stubble lichen
Cladophora	Chlorophyta		*Conocephalum*	Bryophyta	Great scented liver-wort
Clastoderma	Myxomycota				
Clathrulina	Actinopoda		*Conochilus*	Rotifera	
Clavaria	Basidiomycota	Coral mushrooms	*Conocyema*	Mesozoa	
Claviceps	Ascomycota	Ergot fungus	*Conus*	Mollusca	Cone shells
Clevelandina	Spirochaetae		*Corallina*	Rhodophyta	
Closterium	Gamophyta		*Corallium*	Cnidaria	Red coral
Clostridium	Fermenting bacteria	Lock jaw, botulism	*Corvus*	Chordata	Crow, raven, rook, jackdaw
Clypeoseptoria	Deuteromycota		*Corynebacterium*	Actinobacteria	
Coccidia	Apicomplexa		*Coscinodiscus*	Bacillariophyta	
Coccolithus	Haptophyta		*Cosmarium*	Gamophyta	
Coccomyxa	Cnidosporidia		*Cowdria*	Aphragmabacteria	
Cochlonema	Zygomycota		*Coxiella*	Omnibacteria	
Cocos	Angiospermophyta	Coconut	*Crania*	Brachiopoda	
Codium	Chlorophyta	Sea staghorn, oyster thief, spongy cushion	*Craspedacusta*	Cnidaria	
			Crepidula	Mollusca	Slipper shells (boat shells, slipper limpets)
Coelomomyces	Chytridiomycota				
Coeloplana	Ctenophora		*Cristatella*	Ectoprocta	
Coelospora	Apicomplexa		*Cristispira*	Spirochaetae	
Coenonia	Acrasiomycota		*Crithidia*	Zoomastigina	
Coffea	Angiospermophyta	Coffee	*Cronartium*	Basidiomycota	White pine blister rust
Colacium	Euglenophyta				
Colaptes	Chordata	Flicker	*Crotalus*	Chordata	Diamondback rattlesnake, side-winder
Colchicum	Angiospermophyta	Autumn crocus			
Colius	Chordata	Mouse bird	*Crucibulum*	Basidiomycota	Common bird's-nest fungus
Collema	Mycophycophyta	Pulpy, thousand-fruited, tee, blistered jelly, bat's wing, tapioca, crowded pulp, black pepper, dusky jelly, pimpled jelly, granular jelly lichen			
			Cryptochiton	Mollusca	Chiton
			Cryptocercus	Arthropoda	Wood-eating cockroach
			Cryptococcus	Deuteromycota	
			Cryptomeria	Coniferophyta	Japanese cedar
			Cryptomonas	Cryptophyta	

GENUS	PHYLUM	COMMON NAME	GENUS	PHYLUM	COMMON NAME
Cryptosporium	Deuteromycota		*Dermocarpa*	Cyanobacteria	
Ctenoplana	Ctenophora		*Derxia*	Nitrogen-fixing aerobic bacteria	
Cucumaria	Echinodermata	Tar-spot sea cucumber, other sea cucumbers	*Desmarella*	Zoomastigina	
			Desmidium	Gamophyta	
Cupelopagis	Rotifera		*Desulfotomaculum*	Thiopneutes	
Cupressus	Coniferophyta	Cypress	*Desulfovibrio*	Thiopneutes	
Cyanea	Cnidaria	Sea blubber, lion's mane, red winter jellyfish, pink jellyfish	*Desulfuromonas*	Thiopneutes	
			Devescovina	Zoomastigina	
			Diatoma	Bacillariophyta	
Cyanomonas	Cryptophyta		*Dicksonia*	Filicinophyta	Tree fern
Cyathea	Filicinophyta	Tree fern	*Dictydium*	Myxomycota	
Cyathomonas	Cryptophyta		*Dictyostelium*	Acrasiomycota	
Cyathus	Basidiomycota	Striate bird's-nest fungus	*Dictyota*	Phaeophyta	
			Dictyuchus	Oomycota	
Cycas	Cycadophyta	Australian nut palm, sago palm	*Dicyema*	Mesozoa	
			Dicyemennea	Mesozoa	
Cyclotella	Bacillariophyta		*Diderma*	Myxomycota	
Cygnus	Chordata	Mute swan, black swan, whistling swan	*Didinium*	Ciliophora	
			Didymium	Myxomycota	
			Difflugia	Rhizopoda	
Cylindrocapsa	Chlorophyta		*Digitalis*	Angiospermophyta	Foxglove
Cylindrocystis	Gamophyta		*Dileptus*	Ciliophora	
Cymbella	Bacillariophyta		*Dinema*	Euglenophyta	
Cynocephalus	Chordata	Flying lemurs	*Dinobryon*	Chrysophyta	
Cystodinium	Dinoflagellata		*Dinothrix*	Dinoflagellata	
Cytophaga	Myxobacteria		*Dioctophyme*	Nematoda	
Dactylopodola	Gastrotricha		*Dioon*	Cycadophyta	Dioon
Dallinella	Brachiopoda		*Diplocalyx*	Spirochaetae	
Daphnia	Arthropoda	Water flea	*Diplococcus*	Fermenting bacteria	
Dasya	Rhodophyta	Chenille weed	*Diplolaimella*	Nematoda	
Datura	Angiospermophyta	Jimsonweed, thorn-apple	*Diplosoma*	Chordata	
Daucus	Angiospermophyta	Carrot, Queen Anne's lace	*Dipylidium*	Platyhelminthes	Double-pored tapeworm
Deltotrichonympha	Zoomastigina		*Discorbis*	Foraminifera	
Dendrohyrax	Chordata	Tree hyrax, bush hydrax	*Discosphaera*	Haptophyta	
			Distigma	Euglenophyta	
Dendrostomum	Sipuncula		*Doris*	Mollusca	Sea slug (nudibranch)
Dennstaedtia	Filicinophyta	Hay-scented fern			
Dentalium	Mollusca	Tooth shells (tusk shells)	*Dracunculus*	Nematoda	Guinea worm
			Drosophila	Arthropoda	Fruit fly
Derbesia	Chlorophyta		*Dryopteris*	Filicinophyta	New York fern, shield fern

GENUS	PHYLUM	COMMON NAME
Dugesia	Platyhelminthes	Laboratory planarian
Dunaliella	Chlorophyta	
Echinarachnius	Echinodermata	Sand dollar
Echiniscoides	Tardigrada	
Echiniscus	Tardigrada	
Echinocactus	Angiospermophyta	Barrel cactus
Echinochrysis	Chrysophyta	
Echinococcus	Platyhelminthes	Dog tapeworm, hydatid tapeworm
Echinoderes	Kinorhyncha	
Echinorhynchus	Acanthocephala	
Echinosphaerium (Actinosphaerium)	Actinopoda	
Echinostelium	Myxomycota	
Echiuris	Echiura	
Ectocarpus	Phaeophyta	
Edwardsiella	Omnibacteria	
Ehrlichia	Aphragmabacteria	
Eimeria	Apicomplexa	
Elephas	Chordata	Indian elephant
Ellipsoidion	Eustigmatophyta	
Elodea	Angiospermophyta	Waterweed
Eliphidium	Foraminifera	
Elsinoe	Ascomycota	
Embata	Rotifera	
Emiliania	Haptophyta	
Emplectonema	Nemertina	
Encephalartos	Cycadophyta	Kaffir bread (bread "palm")
Encephalitazoan	Cnidosporidia	
Endocochlus	Zygomycota	
Endogone	Zygomycota	Endomycorrhiza fungus
Entamoeba	Rhizopoda	Dysentery amoeba
Enterobacter	Omnibacteria	
Enterobius	Nematoda	Pinworm
Enteromorpha	Chlorophyta	Sea lettuce
Entophysalis	Cyanobacteria	
Ephedra	Gnetophyta	American, Chinese, and longleaf ephedras, joint fir

GENUS	PHYLUM	COMMON NAME
Epipyxis	Chrysophyta	
Equisetum	Sphenophyta	Horsetail
Equus	Chordata	Horse, ass, zebra, donkey, burro, mule
Erwinia	Omnibacteria	
Erythrocladia	Rhodophyta	
Erythrotrichia	Rhodophyta	
Escherichia	Omnibacteria	Colon bacterium
Eucalyptus	Angiospermophyta	Gum tree, ironbark, Australian mountain ash
Eubacterium	Fermenting bacteria	
Euchlanis	Rotifera	
Euchlora	Ctenophora	
Euglena	Euglenophyta	
Eukrohnia	Chaetognatha	
Eunice	Annelida	Palolo worm
Eunotia	Bacillariophyta	
Euphasia	Arthropoda	Krill
Euplectella	Porifera	Venus' flower basket
Euplotes	Ciliophora	
Fagus	Angiospermophyta	Beeches
Fasciola	Platyhelminthes	Sheep liver fluke, cattle liver fluke
Felis	Chordata	Cat, mountain lion, jaguarundi, ocelot, marbled cat, serval, golden cat, pallas' cat, European wild cat, jungle cat, African black footed cat, little spotted cat, African wild cat, margay cat, pampas cat, leopard cat
Ferrobacillus	Chemoautotrophic bacteria	
Flavobacterium	Omnibacteria	
Flexibacter	Myxobacteria	
Flexithrix	Myxobacteria	
Floscularia	Rotifera	

GENUS	PHYLUM	COMMON NAME	GENUS	PHYLUM	COMMON NAME
Fomes	Basidiomycota	Rusty-hoof fomes, a bracket fungus	*Globigerina*	Foraminifera	
Fossombronia	Bryophyta		*Gloeocapsa*	Cyanobacteria	
Fragilaria	Bacillariophyta		*Glossobalanus*	Hemichordata	
Frankia	Actinobacteria	Root-nodule bacterium	*Glottidia*	Brachiopoda	
Fredericella	Ectoprocta		*Gluconobacter*	Omnibacteria	
Fregata	Chordata	Frigate bird	*Glugea*	Cnidosporidia	
Fritschiella	Chlorophyta		*Glycine*	Angiospermophyta	Soybean
Frullania	Bryophyta	Leafy liverwort	*Gnathostomula*	Gnathostomulida	
Fucus	Phaeophyta	Rockweed, wrack	*Gnetum*	Gnetophyta	Gnetum, bawale, longonizia, bulso
Fulica	Chordata	Coot	*Golfingia*	Sipuncula	
Fuligo	Myxomycota		*Gonatozygon*	Gamophyta	
Funaria	Bryophyta	Cord moss	*Goniotrichum*	Rhodophyta	
Fundulus	Chordata	Killifish	*Gonium*	Chlorophyta	
Fusarium	Deuteromycota		*Gonyaulax*	Dinoflagellata	
Fusobacterium	Fermenting Bacteria		*Gonyostomum*	Xanthophyta	
Fusulina	Foraminifera		*Gordionus*	Nematomorpha	Gordian worm
Gadus	Chordata	Atlantic cod	*Gordius*	Nematomorpha	Gordian worm
Gaffkya	Micrococci		*Gorilla*	Chordata	Gorilla
Gallus	Chordata	Chicken and jungle fowl	*Grantia*	Porifera	Purse sponge
			Gregarina	Apicomplexa	Gregarine
Gavia	Chordata	Loons	*Guttulina*	Acrasiomycota	
Gastrodes	Ctenophora		*Guttulinopsis*	Acrasiomycota	
Gastrostyla	Ciliophora		*Guyenotia*	Cnidosporidia	
Geastrum	Basidiomycota	Triplex and crowned earth stars	*Gymnodinium*	Dinoflagellata	
			Haemophilus	Omnibacteria	
			Haemoproteus	Apicomplexa	
Gelidium	Rhodophyta		*Hafnia*	Omnibacteria	
Genicularia	Gamophyta		*Haliclystus*	Cnidaria	
Gephyrocapsa	Haptophyta		*Halicryptus*	Priapulida	
Gelliodes	Porifera		*Halimeda*	Chlorophyta	
Geococcyx	Chordata	Roadrunner	*Haliotis*	Mollusca	Abalone
Geonemertes	Nemertina		*Halisarca*	Porifera	Jelly sponge
Geothallus	Bryophyta		*Halocynthia*	Chordata	Sea peach
Geotrichum	Deuteromycota		*Halobacterium*	Pseudomonads	
Giardia	Zoomastigina		*Halodiscus*	Rhizopoda	
Gigantorhynchus	Acanthocephala		*Halteria*	Ciliophora	
Ginkgo	Ginkgophyta	Ginkgo, maidenhair tree	*Haplognathia*	Gnathostomulida	
			Haplomitrium	Bryophyta	
Glabratella	Foraminifera		*Haplosporidium*	Apicomplexa	
Glandiceps	Hemichordata		*Hartmannella*	Rhizopoda	

GENUS	PHYLUM	COMMON NAME	GENUS	PHYLUM	COMMON NAME
Helianthus	Angiospermophyta	Sunflowers, Jerusa-lem artichoke	*Hypnum*	Bryophyta	Sheet moss, carpet moss
Helicosporidium	Cnidosporidia		*Hypsibius*	Tardigrada	
Heliopora	Cnidaria	Blue coral	*Ichthyosporidium*	Cnidosporidia	
Helix	Mollusca	Edible land snail	*Ikeda*	Echiura	
Hemiselmis	Cryptophyta		*Ipomoea*	Angiospermophyta	Sweetpotato, morning glory
Hemithyris	Brachiopoda				
Hemitrichia	Myxomycota		*Iridia*	Foraminifera	
Heptabrachia	Pogonophora		*Isoachlya*	Oomycota	
Herpetomonas	Zoomastigina		*Isoetes*	Lycopodophyta	Quillwort
Herpetosiphon	Myxobacteria		*Isospora*	Apicomplexa	
Heterodendron	Xanthophyta		*Joenia*	Zoomastigina	
Heteronema	Euglenophyta		*Juniperus*	Coniferophyta	Juniper, red cedar
Hevea	Angiospermophyta	Rubber	*Kalotermes*	Arthropoda	Dry wood termite
Hexamastix	Zoomastina		*Kickxella*	Zygomycota	
Hildebrandia	Rhodophyta		*Klebsiella*	Omnibacteria	
Hillea	Cryptophyta		*Krohnitta*	Chaetognatha	
Hirudo	Annelida	European medicinal leech	*Labyrinthorhiza*	Labyrinthulamycota	
			Labyrinthula	Labyrinthulamycota	
Histomonas	Zoomastigina		*Lacazella*	Brachiopoda	
Histoplasma	Deuteromycota		*Lactobacillus*	Fermenting bacteria	
Hollandina	Spirochaetae				
Homeris	Arthropoda	Lobster	*Lagenidium*	Oomycota	
Homo	Chordata	Human being	*Lamellisabella*	Pogonophora	
Hordeum	Angiospermophyta	Barley	*Laminaria*	Phaeophyta	Oarweed, sea girdle, fingered kelp, tangle, sea belt
Hyalodiscus	Rhizopoda				
Hyalotheca	Gamophyta				
Hydra	Cnidaria	Hydra, green hydra	*Lampea*	Ctenophora	
Hydrogenomonas	Pseudomonads		*Lamprothamnium*	Chlorophyta	Stonewort
Hydrurus	Chrysophyta		*Larix*	Coniferophyta	Tamarack, larch
Hyella	Cyanobacteria		*Latimeria*	Chordata	Lobe-finned fish
Hymenolepis	Platyhelminthes	Dwarf tapeworm of human beings and rodents, tapeworm of chickens	*Latrostium*	Hyphochytri-diomycota	
			Leishmania	Zoomastigina	
			Lejeunea	Bryophyta	
Hymenomonas	haptophyta		*Lemanea*	Rhodophyta	
Hymenophyllum	Filicinophyta	Filmy fern	*Lemna*	Angiospermophyta	Duckweed
Hymenophytum	Bryophyta		*Leocarpus*	Myxomycota	
Hyphochytrium	Hyphochytri-diomycota		*Lepidodermella*	Gastrotricha	
			Lepidozia	Bryophyta	

GENUS	PHYLUM	COMMON NAME	GENUS	PHYLUM	COMMON NAME
Lepiota	Basidiomycota	Smooth lepiota, yellow lepiota, parasol mushroom, shaggy parasol	*Llama*	Chordata	Guanaco, llama, alpaca
			Locusta	Arthropoda	Locust
			Loligo	Mollusca	Common squid
Lepraria	Mycophycophyta		*Loxosoma*	Entoprocta	
Leptomonas	Zoomastigina		*Loxosomella*	Entoprocta	
Leptonema	Spirochaetae		*Lumbricus*	Annelida	Earthworm
Leptorhynchoides	Acanthocephala		*Lycogala*	Myxomycota	Puffball
Leptosomatum	Nematoda		*Lycopodium*	Lycopodophyta	Club mosses
Leptospira	Spirochaetae	Infectious jaundice	*Lyngbya*	Cyanobacteria	
Leptothrix	Omnibacteria		*Maccabeus*	Priapulida	
Leptotrichia	Fermenting bacteria		*Macracanthorhynchus*	Acanthocephala	
Lepus	Chordata	Arctic hare, varying hare, European hare, jack "rabbit"	*Macrobdella*	Annelida	American medicinal leech
			Macrobiotus	Tardigrada	
Leuconostoc	Fermenting bacteria		*Macrocystis*	Phaeophyta	Kelp
			Macrodasys	Gastrotricha	
Leucosolenia	Porifera		*Macromonas*	Chemoautotrophic bacteria	
Licea	Myxomycota				
Lichenopora	Ectoprocta		*Macrozamia*	Cycadophyta	
Lichenothrix	Mycophycophyta		*Magnolia*	Angiospermophyta	Magnolia, cucumber tree, sweet bay
Lichina	Mycophycophyta				
Ligniera	Plasmodiophoromycota				
			Malacobdella	Nemertina	
Lilium	Angiospermophyta	Turk's-cap lily, tiger lily, wood lily, some other lilies	*Mallomonas*	Chrysophyta	
			Manihot	Angiospermophyta	Yuca, cassava, manioc, tapioca
Limenitis	Arthropoda	Viceroy, white admiral, some other butterflies	*Mannia*	Bryophyta	
			Manus	Chordata	Pangolin, scaly anteater
Limnodrilus	Annelida		*Marattia*	Filicinophyta	King fern, tree fern
Limulus	Arthropoda	Horseshoe crab	*Marchantia*	Bryophyta	Common liverwort
Lineus	Nemertina	Bootlace worm	*Marsilea*	Filicinophyta	Water clover
Linguatula	Pentastoma		*Marsupella*	Bryophyta	
Lingula	Brachiopoda		*Mayorella*	Rhizopoda	
Liriodendron	Angiospermophyta	Tulip tree	*Megaceryle*	Chordata	Belted kingfisher
Lissoclinum	Chordata		*Megascolides*	Annelida	Giant earthworm
Lissomyema	Echiura		*Megathura*	Annelida	Polynoid polychaete
Listriolobus	Echiura				
Lithacrosiphon	Sipuncula		*Megathyris*	Brachiopoda	
Lithothamnion	Rhodophyta		*Melanoplus*	Arthropoda	Grasshopper
Littorina	Mollusca	Periwinkle			

GENUS	PHYLUM	COMMON NAME
Meleagris	Chordata	Turkey
Melosira	Bacillariophyta	
Membranipora	Ectoprocta	Sea mat
Membranosorus	Plasmodio-phoromycota	
Mercenaria	Mollusca	Quahog, cherrystone (hard-shell clam)
Mertensia	Ctenophora	Sea walnut
Mesocricetus	Chordata	Golden hamster
Mesoperipatus	Onychophora	
Mesotaenium	Gamophyta	
Metabonellia	Echiura	
Metacrinus	Echinodermata	Sea lily
Metadevescovina	Zoomastigina	
Metasequoia	Coniferophyta	Dawn redwood
Metatrichia	Myxomycota	
Methanobacillus	Methane-ocreatrices	
Methanobacterium	Methane-ocreatrices	
Methanococcus	Methane-ocreatrices	
Methanosarcina	Methane-ocreatrices	
Methylococcus	Chemoautotrophic bacteria	
Methylomonas	Chemoautotrophic bacteria	
Metridium	Cnidaria	Plumose anemone
Micrasterias	Gamophyta	
Microciona	Porifera	Redbeard sponge
Micrococcus	Micrococci	
Microcycas	Cycadophyta	Corcho
Microcyema	Mesozoa	
Microcystis	Cyanobacteria	
Micromonospora	Actinobacteria	
Miliola	Foraminifera	Fire coral
Millepora	Cnidaria	
Milnesium	Tardigrada	
Mischococcus	Xanthophyta	
Mnemiopsis	Ctenophora	Lobed comb jelly
Monas	Chrysophyta	

GENUS	PHYLUM	COMMON NAME
Monilia	Deuteromycota	See *Candida*
Moniliformis	Acanthocephala	
Monoblepharis	Chytridiomycota	
Monocercomonas	Zoomastigina	
Monoclea	Bryophyta	
Monosiga	Zoomastigina	
Morchella	Ascomycota	Morel
Mortierella	Zygomycota	
Mougeotia	Gamophyta	
Mucor	Zygomycota	Bread mold
Murex	Mollusca	Murex or rock shell
Mus	Chordata	House mouse
Musa	Angiospermophyta	Banana, plantain, cooking banana, abaca
Musca	Arthropoda	House fly
Mya	Mollusca	Soft-shell clam, long neck or sand clam, short clam
Mycobacterium	Actinobacteria	
Mycococcus	Actinobacteria	
Mycoplasma	Aphragmabacteria	
Mycosphaella	Ascomycota	
Myotis	Chordata	Little brown bat, some other bats
Mytilus	Mollusca	Common edible mussel, scorched mussel, bent mussel
Myxidium	Cnidosporidia	
Myxobolus	Cnidosporidia	
Myxococcus	Myxobacteria	
Myxostoma	Cnidosporidia	Salmon twist-disease parasite
Myzocytium	Oomycota	
Myzostoma	Annelida	
Naegleria	Zoomastigina	
Nanognathia	Gnathostomulida	
Nautilus	Mollusca	Chambered nautilus
Navicula	Bacillariophyta	
Necator	Nematoda	Hookworm
Nectonema	Nematomorpha	
Nectonemertes	Nemertina	

GENUS	PHYLUM	COMMON NAME	GENUS	PHYLUM	COMMON NAME
Neisseria	Omnibacteria		*Nyctotherus*	Ciliophora	
Nemalion	Rhodophyta	Threadweed	*Obelia*	Cnidaria	
Nematochrysis	Chrysophyta		*Ochetostoma*	Echiura	
Neoechinorhynchus	Acanthocephala		*Ochrolechia*	Mycophycophyta	Cudbear lichen
Neohodgsonia	Bryophyta		*Ochromonas*	Chrysophyta	
Neomenia	Mollusca		*Octomyxa*	Plasmodio- phoromycota	
Neopilina	Mollusca				
Neorickettsia	Omnibacteria		*Oedogonium*	Chlorophyta	
Nephroselmis	Cryptophyta		*Oikomonas*	Zoomastigina	
Nephtys	Annelida	Paddle-footed worm, bristle worm	*Oikopleura*	Chordata	Tunicate
			Oligobrachia	Pogonophora	
			Olpidium	Chytridiomycota	
Nereis	Annelida	Clamworm	*Onchnesoma*	Sipuncula	
Nereocystis	Phaeophyta	Bull or bladder kelp	*Oocystis*	Chlorophyta	
Netrium	Gamophyta		*Ooperipatus*	Onychophora	
Neurospora	Ascomycota	Red bread mold	*Opalina*	Zoomastigina	
Nitella	Chlorophyta	Stonewort	*Ophiocytium*	Xanthophyta	
Nitellopsis	Chlorophyta	Stonewort	*Ophioglossum*	Filicinophyta	Adder's-tongue fern
Nitrobacter	Chemoautotrophic bacteria		*Ophiura*	Echinodermata	Brittle star
Nitrocystis	Chemoautotrophic bacteria		*Opisthocomus*	Chordata	Hoatzin
			Opisthopatus	Onychophora	
Nitrosococcus	Chemoautotrophic bacteria		*Opisthorchis*	Platyhelminthes	Chinese liver fluke
Nitrosogloea	Chemoautotrophic bacteria		*Ornithorhynchus*	Chordata	Duck-billed platypus
			Orycteropus	Chordata	Aardvark
Nitrosolobus	Chemoautotrophic bacteria		*Oryctolagus*	Chordata	Domestic rabbit, Old World gray rabbit
Nitrosomonas	Chemoautotrophic bacteria				
			Oryza	Angiospermophyta	Cultivated rice
Nitrosospira	Chemoautotrophic bacteria		*Oscillatoria*	Cyanobacteria	
Nocardia	Actinobacteria		*Osmunda*	Filicinophyta	Interrupted fern, royal fern, cinna- mon fern
Noctiluca	Dinoflagellata				
Nodosaria	Foraminifera		*Ovis*	Chordata	Domestic sheep, bighorn sheep, moulflons, Dall mountain sheep, Marco Polo sheep, Laristan sheep
Noguchia	Omnibacteria				
Nosema	Cnidosporidia				
Nostoc	Cyanobacteria				
Notila	Zoomastigina				
Notommata	Rotifera		*Oxymonas*	Zoomastigina	
Notosaria	Brachiopoda		*Pan*	Chordata	Chimpanzee
Numida	Chordata	Guinea fowl	*Pandorina*	Chlorophyta	

GENUS	PHYLUM	COMMON NAME	GENUS	PHYLUM	COMMON NAME
Parachordodes	Nematomorpha		*Peronospora*	Oomycota	
Paracoccus	Micrococci		*Petromyzon*	Chordata	Sea lamprey
Paracolobactrum	Omnibacteria		*Phacus*	Euglenophyta	
Paragordius	Nematomorpha		*Phaeoceros*	Bryophyta	Horned liverwort
Paramecium	Ciliophora		*Phaeothamnion*	Chrysophyta	
Paramoeba	Rhizopoda		*Phallus*	Basidiomycota	Common stinkhorn
Paranemertes	Nemertina		*Phascolarctus*	Chordata	Koala
Parmelia	Mycophycophyta	Boulder lichen, crottle, shield lichen	*Phascolion*	Sipuncula	
			Phascolopsis	Sipuncula	
Pasteurella (*Yersinia*)	Omnibacteria	Plague bacterium	*Phaseolus*	Angiospermophyta	String (snap) bean, scarlet runner bean, wax bean, shell bean, mung bean, tepary bean, lima bean
Pavo	Chordata	Peafowl (peacock and peahen)			
Pecten	Mollusca	Scallop	*Phasianus*	Chordata	Ring-necked pheasant
Pectinatella	Ectoprocta		*Philodina*	Rotifera	
Pedicellina	Entoprocta		*Phoca*	Chordata	Harbor seal, hair seal
Pedicellinopsis	Entoprocta				
Pediculus	Arthropoda	Body louse	*Phoronis*	Phoronida	
Pelagophycus	Phaeophyta		*Phoronopsis*	Phoronida	
Pellia	Bryophyta		*Photobacterium*	Omnibacteria	Luminous bacterium
Pelodera	Nematoda		*Phycomyces*	Zygomycota	
Pelodictyon	Anaerobic photosynthetic bacteria		*Phylloglossum*	Lycopodophyta	
			Physarella	Myxomycota	
Pelomyxa	Caryoblastea		*Physarum*	Myxomycota	
Penicillium	Deuteromycota	Blue mold, green mold	*Physcomitrium*	Bryophyta	Urn moss
Penicillus	Chlorophyta	Neptune's shaving brush	*Physalia*	Cnidaria	Portuguese man-of-war
Penium	Gamophyta		*Physoderma*	Chytridiomycota	
Peptococcus	Fermenting bacteria		*Phytophthora*	Oomycota	
Peptostreptococcus	Fermenting bacteria		*Picea*	Coniferophyta	White spruce, black spruce, Norway spruce
Peranema	Euglenophyta		*Pillotina*	Spirochaetae	
Perichaena	Myxomycota		*Pilobolus*	Zygomycota	Dung fungus, cap thrower
Peridinium	Dinoflagellata				
Peripatopsis	Onychophora		*Pinnularia*	Bacillariophyta	
Peripatus	Onychophora	Velvet worm	*Pinus*	Coniferophyta	Pitch pine, yellow pine, long-leaf pine, white pine, jack pine
Periplaneta	Arthropoda	American cock-roach			

GENUS	PHYLUM	COMMON NAME
Pipetta	Actinopoda	
Pisaster	Echinodermata	Ochre seastar, purple seastar, common seastar
Planaria	Platyhelminthes	Freshwater planarian
Planktoniella	Bacillariophyta	
Planococcus	Micrococci	
Plasmodiophora	Plasmodiophoro-mycota	
Plasmodium	Apicomplexa	Malarial parasite
Plasmopara	Oomycota	
Platymonas	Chlorophyta	
Platyzoma	Filicinophyta	
Plectocolea	Bryophyta	
Plesiomonas	Omnibacteria	
Pleurobrachia	Ctenophora	Sea gooseberry, cat's eye
Pleurochloris	Eustigmatophyta	
Plumatella	Ectoprocta	
Pocheina	Acrasiomycota	
Podangium	Myxobacteria	
Podiceps	Chordata	Horned grebe, other grebes
Podocarpus	Coniferophyta	Yellowwood
Polyangium	Myxobacteria	
Polybrachia	Pogonophora	
Polyedriella	Eustigmatophyta	
Polykrikos	Dinoflagellata	
Polymorphus	Acanthocephala	
Polymyxa	Plasmodio-phoromycota	
Polypodium	Filicinophyta	Rock polypody, resurrection fern
Polyporus	Basidiomycota	Sulfur polyporus
Polysiphonia	Rhodophyta	
Polysphondylium	Acrasiomycota	
Polystichum	Filicinophyta	Holly fern, Christmas fern
Polytrichum	Bryophyta	Haircap moss, pigeon wheat

GENUS	PHYLUM	COMMON NAME
Pongo	Chordata	Orangutan
Pontosphaera	Haptophyta	
Porocephalus	Pentastoma	
Porphyra	Rhodophyta	Purple laver, nori
Porphyridium	Rhodophyta	
Postelsia	Phaeophyta	Sea palm
Prasinocladus	Chlorophyta	
Priapulus	Priapulida	
Primula	Angiospermophyta	Primrose
Proales	Rotifera	
Problognathia	Gnathostomulida	
Prochloron	Chloroxybacteria	
Procotyla	Platyhelminthes	Freshwater planarian
Propionibacterium	Actinobacteria	
Prorocentrum	Dinoflagellata	
Prorodon	Ciliophora	
Prostoma	Nemertina	Freshwater nemertine
Proteus	Omnibacteria	
Protococcus	Chlorophyta	
Protoopalina	Zoomastigina	
Protostelium	Myxomycota	
Prunus	Angiospermophyta	Peach, plum, beach plum, pin cherry, black cherry, choke cherry, sweet cherry
Prymnesium	Haptophyta	
Pseudicyema	Mesozoa	
Pseudobryopsis	Chlorophyta	
Pseudocharaciopsis	Eustigmatophyta	
Pseudomonas	Pseudomonads	
Pseudoplasmodium	Labyrinthulamycota amycota	
Pseudotrebouxia	Chlorophyta	Lichen alga
Pseudotsuga	Coniferophyta	Douglas fir
Pteridium	Filicinophyta	Brake, bracken fern
Pteris	Filicinophyta	
Pterosagitta	Chaetognatha	

GENUS	PHYLUM	COMMON NAME	GENUS	PHYLUM	COMMON NAME
Pterotermes	Arthropoda	Sonoran desert termite	*Rhodopseudo-monas*	Anaerobic photosynthetic bacteria	
Ptychodera	Hemichordata		*Rhodospirillum*	Anaerobic photosynthetic bacteria	
Puccinia	Basidiomycota	Wheat rust			
Puffinus	Chordata	Shearwaters	*Rhodymenia*	Rhodophyta	Dulce
Pycnophyes	Kinorhyncha		*Rhopalura*	Mesozoa	
Pyramimonas	Chlorophyta		*Riccardia*	Bryophyta	
Pyrsonympha	Zoomastigina		*Riccia*	Bryophyta	Slender riccia (a liverwort)
Pythium	Oomycota				
Quercus	Angiospermophyta	White oak, black oak, red oak, post oak, live oak, yellow oak, chinquapin	*Ricciocarpus*	Bryophyta	Purple-fringed riccia
			Rickettsia	Omnibacteria	
			Rickettsiella	Omnibacteria	
Raja	Chordata	Skates	*Riella*	Bryophyta	
Raillietiella	pentastoma		*Riftia*	Pogonophora	
Rangifer	Chordata	Reindeer, caribou	*Rosa*	Angiospermophyta	Rose
Rauwolfia	Angiospermophyta	Snakeroot	*Rotaliella*	Foraminifera	
Reighardia	Pentastoma		*Ruminobacter*	Fermenting bacteria	
Renilla	Cnidaria	Sea pansy	*Ruminococcus*	Fermenting bacteria	
Reticulitermes	Arthropoda	Subterranean termite			
Rhabdias	Nematoda	Roundworm of amphibians	*Saccharomyces*	Ascomycota	Bakers' yeast, brewers' yeast
			Saccharum	Angiospermophyta	Sugarcane
Rhabdopleura	Hemichordata		*Saccinobaculus*	Zoomastigina	
Rhabdosphaera	Haptophyta		*Saccoglossus*	Hemichordata	
Rhea	Chordata	Greater (common) rhea	*Sagitta*	Chaetognatha	
			Salamandra	Chordata	European fire salamander
Rhipidium	Oomycota				
Rhizidiomyces	Hyphochytri-diomycota		*Salmonella*	Omnibacteria	
Rhizobium	Nitrogen-fixing aerobic bacteria	Nodule forming bacteria	*Salpa*	Chordata	Salp (tunicate)
			Salvinia	Filicinophyta	Salvinia, water spangles
Rhizochrysis	Chrysophyta				
Rhizoctonia	Deuteromycota		*Sappinia*	Myxomycota	
Rhizomyces	Ascomycota		*Saprolegnia*	Oomycota	
Rhizophydium	Chytridiomycota		*Saprospira*	Myxobacteria	
Rhizopus	Zygomycota	Black bread mold	*Sarcina*	Micrococci	
Rhodomicrobium	Anaerobic photosynthetic bacteria	Budding purple nonsulfur photo-synthesizer	*Sarcinochrysis*	Chrysophyta	
			Sarcocypha	Ascomycota	Some cup fungi
			Sargassum	Phaeophyta	

GENUS	PHYLUM	COMMON NAME	GENUS	PHYLUM	COMMON NAME
Scapania	Bryophyta	Leafy liverwort	*Spadella*	Chaetognatha	
Schistosoma	Platyhelminthes	Blood fluke	*Spartina*	Angiospermophyta	Marsh grass, salt marsh grass
Schizocystis	Apicomplexa				
Schizophyllum	Basidiomycota		*Spengelia*	Hemichordata	
Scolymastra	Porifera		*Sphacelaria*	Phaeophyta	
Scorpio	Arthropoda	Scorpion	*Sphaeractinomyxon*	Cnidosporidia	
Scytosiphon	Phaeophyta	Whip tube	*Sphaerocarpos*	Bryophyta	
Seison	Rotifera		*Sphaerotilus*	Omnibacteria	Sheathed iron bacterium
Selaginella	Lycopodophyta	Spike mosses	*Sphagnum*	Bryophyta	Peat moss, sphagnum
Selenidium	Apicomplexa				
Semaeognathia	Gnathostomulida		*Sphenodon*	Chordata	Tuatara
Semnoderes	Kinorhyncha		*Spirobrachia*	Pogonophora	
Sepia	Mollusca	Common sepia, cuttlefish	*Spirochaeta*	Spirochaetae	Mud spirochete
			Spirodella	Angiospermophyta	Duckweed
Sequoia	Coniferophyta	California redwood	*Spirogyra*	Gamophyta	
Sequoiadendron	Coniferophyta	Giant sequoia	*Spiroplasma*	Aphragmabacteria	Corn stunt disease agent
Serratia	Omnibacteria	Blood of Christ bacterium			
			Spirostomum	Ciliophora	
Shigella	Omnibacteria		*Spirulina*	Cyanobacteria	
Siboglinum	Pogonophora		*Spongia*	Porifera	Bath sponge
Siliqua	Mollusca	Pacific razor clam	*Spongilla*	Porifera	Freshwater sponge
Simonsiella	Myxobacteria		*Spongomorpha*	Chlorophyta	
Siphonostoma	Sipuncula		*Spongospora*	Plasmodiophoromycota	
Solanum	Angiospermophyta	White (''Irish'') potato, Jerusalem cherry, nightshade, eggplant	*Sporocytophaga*	Myxobacteria	
			Sporolactobacillus	Aeroendospora	
			Sporosarcina	Aeroendospora	
Solaster	Echinodermata	Purple sun star, eleven-armed sun star	*Squalus*	Chordata	Spiny dogfish shark
			Stangeria	Cycadophyta	Hottentot's head
Solentia	Cyanobacteria		*Staphylococcus*	Micrococci	
Sordaria	Ascomycota		*Staurastrum*	Gamophyta	
Sorex	Chordata	Masked shrew, long-tailed shrew	*Staurojoenina*	Zoomastigina	
			Stelangium	Myxobacteria	
Sorodiscus	Plasmodiophoromycota		*Stemonitis*	Myxomycota	
Sorogena	Ciliata	Fruiting, ciliate	*Stentor*	Ciliophora	
Sorosphaera	Plasmodiophoromycota		*Sterna*	Chordata	Common tern, roseate tern
			Sticholonche	Actinopoda	

GENUS	PHYLUM	COMMON NAME
Stigeoclonium	Chlorophyta	
Stigmatella	Myxobacteria	
Stoecharthrum	Mesozoa	
Stomatopora	Ectoprocta	
Streptobacillus	Omnibacteria	
Streptococcus	Fermenting bacteria	
Streptomyces	Actinobacteria	
Stromatospongia	Porifera	
Strongylocentrotus	Echinodermata	Western purple sea urchin, giant red sea urchin, green sea urchin
Struthio	Chordata	Ostrich
Stylopage	Zygomycota	
Sulfolobus	Chemoautotrophic bacteria	
Surirella	Bacillariophyta	
Sus	Chordata	Pig, wild boar
Symbiodinium	Dinoflagellata	Zooxanthella
Symperipatus	Onychophora	
Synangium	Myxobacteria	
Synchaeta	Rotifera	
Synchytrium	Chytridiomycota	
Synderella	Zoomastigina	
Synechococcus	Cyanobacteria	
Synechocystis	Cyanobacteria	
Synura	Chrysophyta	
Syracophaera	Haptophyta	
Taenia	Platyhelminthes	Tapeworm
Takakia	Bryophyta	
Taraxacum	Angiospermophyta	Dandelion
Targionia	Bryophyta	
Tatjanellia	Echiura	
Taxodium	Coniferophyta	Bald cypress
Taxus	Coniferophyta	Yew
Telomyxa	Cnidosporidia	
Temnogyra	Gamophyta	
Tenebrio	Arthropoda	Darkling beetle (larva:mealworm)
Teosinte	Angiospermophyta	
Terebratula	Brachiopoda	

GENUS	PHYLUM	COMMON NAME
Terebratulina	Brachiopoda	
Teredo	Mollusca	Shipworm
Tetrahymena	Ciliophora	
Tetramyxa	Plasmodiophoromycota	
Tetranchyroderma	Gastrotricha	
Tetraphis	Bryophyta	Four-tooth moss
Tetraspora	Chlorophyta	
Textularia	Foraminifera	
Thalassema	Echiura	
Thalassicola	Actinopoda	
Thallasiosira	Bacillariophyta	
Thallochrysis	Chrysophyta	
Thecamoeba	Rhizopoda	
Themiste	Sipuncula	
Theobroma	Angiospermophyta	Cacao
Thermoactinomyces	Actinobacteria	
Thermoplasma	Aphragmabacteria	
Thiobacillus	Chemoautotrophic	
Thiobacterium	Chemoautotrophic bacteria	
Thiocapsa	Anaerobic photosynthetic bacteria	
Thiocystis	Anaerobic photosynthetic bacteria	
Thiodictyon	Anaerobic photosynthetic bacteria	
Thiomicrospira	Chemoautotrophic bacteria	
Thiopedia	Anaerobic photosynthetic bacteria	
Thiosarcina	Anaerobic photosynthetic bacteria	
Thiospira	Chemoautotrophic bacteria	
Thiothece	Anaerobic photosynthetic bacteria	
Thiovulum	Chemoautotrophic bacteria	

GENUS	PHYLUM	COMMON NAME	GENUS	PHYLUM	COMMON NAME
Thuja	Coniferophyta	Arbor vitae, white cedar	*Tuber*	Ascomycota	Truffle
Thyone	Echinodermata	Sea cucumber	*Tubifex*	Annelida	"Red worm" sold as aquarium fish food
Tilopteris	Phaeophyta				
Tinamus	Chordata	Gray, solitary, black, great, and white-throated tinamous	*Tubiluchus*	Priapulida	
			Tubipora	Cnidaria	Organ-pipe coral
			Tubulanus	Nemertina	
Tjalfiella	Ctenophora		*Tubularia*	Cnidaria	Oaten-pipes hydroid
Tokophyra	Ciliophora	Suctorian	*Tubulipora*	Ectoprocta	
Tolypella	Chlorophyta	Stonewort	*Tulipa*	Angiospermophyta	Tulip
Tortula	Bryophyta	Wall moss, twisted moss	*Turbanella*	Gastrotricha	
			Udotea	Chlorophyta	
Toxoplasma	Apicomplexa		*Ulothrix*	Chlorophyta	
Trachelomonas	Euglenophyta		*Ulva*	Chlorophyta	Sea lettuce
Trachydemus	Kinorhyncha		*Umbilicaria*	Mycophycophyta	Rock tripe, toad-skin lichen
Tradescantia	Angiospermophyta	Spiderwort			
Trebouxia	Chlorophyta	Lichen alga	*Unicapsula*	Cnidosporidia	
Tremella	Basidiomycota	Orange jelly fungus	*Urechis*	Echiura	Fat innkeeper
Treponema	Spirochaetae	Yaws, syphilis agent	*Urnatella*	Entoprocta	
			Urodasys	Gastrotricha	
Triactinomyxon	Cnidosporidia		*Uroleptus*	Ciliophora	
Tribonema	Xanthophyta		*Urospora*	Chlorophyta	
Trichechus	Chordata	Manatee	*Usnea*	Mycophycophyta	Old-man's beard
Trichinella	Nematoda	Trichina worm	*Ustilago*	Basidiomycota	Corn smut
Trichomitus	Zoomastigina		*Vaucheria*	Xanthophyta	
Trichomonas	Zoomastigina		*Veillonella*	Fermenting bacteria	
Trichonympha	Zoomastigina	Wood-eating hypermastigote	*Velamen*	Ctenophora	
Trichophyton	Deuteromycota	Athlete's foot, cattle ringworm, tinea cruris (jock-strap itch)	*Velella*	Cnidaria	By-the-wind sailor
			Vema	Mollusca	
			Verticillium	Deuteromycota	
			Vibrio	Omnibacteria	
Trichoplax	Placozoa		*Vischeria*	Eustigmatophyta	
Tricoma	Nematoda		*Vitis*	Angiospermophyta	Fox grape, musca-dine grape, frost grape, other grapes
Tridacna	Mollusca	Giant clam			
Trifolium	Angiospermophyta	White clover, red clover, yellow clover			
Triticum	Angiospermophyta	Bread wheat, durum (pasta) wheat	*Volvox*	Chlorophyta	
			Vorticella	Ciliophora	
			Vulpes	Chordata	Kit fox, red fox
Trypanosoma	Zoomastigina		*Waddycephalus*	Pentastoma	
Tsuga	Coniferophyta	Canadian hemlock, Carolina hemlock	*Welwitschia*	Gnetophyta	

GENUS	PHYLUM	COMMON NAME
Wolbachia	Aphragmabacteria	
Wolffia	Angiospermophyta	Watermeal, duck-weed
Woronina	Plasmodiophoro-mycota	
Wuchereria	Nematoda	
Xanthomonas	Pseudomonads	
Xenophyophora	Rhizopoda	
Xenopus	Chordata	African clawed frog
Xenorhabditis	Omnibacteria	Luminescent nematode bacterium
Yersinia (*Pasteurella*)	Omnibacteria	Plague bacterium
Yucca	Angiospermophyta	Beargrass, Spanish bayonet, yucca, Joshua tree
Zamia	Cycadophyta	Coontie, Seminole bread
Zea	Angiospermophyta	Maize, Indian corn
Zelleriella	Zoomastigina	
Zenkevitchiana	Pogonophora	
Zonaria	Phaeophyta	
Zoogloea	Pseudomonads	
Zostera	Angiospermophyta	Eel grass
Zygacanthidium	Actinopoda	
Zygnema	Gamophyta	
Zygogonium	Gamophyta	
Zymomonas	Omnibacteria	Palm wine bacterium

GLOSSARY

Terms generally restricted to certain kingdoms or phyla are indicated by abbreviations in parentheses. However, the terms are not always restricted to those kingdoms and phyla, and do not necessarily apply to all members of the kingdoms and phyla.

aboral Away from the mouth

acervulus Mat of hyphae giving rise to conidiophores closely packed together to form a bedlike mass (F-2)

acoelomate Lacking a coelom (A-1 through A-8)

acritarch Spherical microfossil, possibly the test of a shelled amoeba (Pr-3) or chrysophyte (Pr-4)

actin One of the two major proteins of muscle, in which it makes up the thin filaments (A); also found in filaments that participate in motility of protists (Pr)

actinopod Pseudopod containing filaments and microtubules (Pr-16)

actinospores Spores of actinobacteria (actinomycetes, M-15)

actinotroch larva Free-swimming phonorid larva (A-17)

adhesive organ Organ for attachment, for example, in some platyhelminthes (A-6)

adventitious root Root growing from an abnormal location, as on a stem or leaf (Pl-4 to Pl-9)

aecium Fungal structure consisting of binucleate hyphal cells, which produce spore chains (F-2)

aerobic Requiring gaseous oxygen

aerobiosis Aerobic living

agamete Asexual reproductive cell (Pr-17)

agamogony Series of nuclear or cell divisions giving rise to individuals (agamonts) that are not gametes or are not capable of forming gametes (Pr-17)

agamont Life-cycle stage that does not produce gametes (Pr-17)

alimentary tract Digestive or gastrointestinal tract; a tubular passage that extends from mouth to anus and whose function is to digest and absorb nutrients (A)

alternation of generations Reproductive cycles in which haploid (N) phases alternate with diploid ($2N$) phases (Pr, Pl)

ambulacrum In echinoderms (A-29), a tube-foot lined, ciliated groove leading down the center of each arm (starfish) or over the test (sea urchin); conducts food to the mouth

amniote egg Egg that is isolated from the environment by a waterproof covering, such as a shell, during the period of its development and which is nutritionally self-sufficient, requiring only gas exchange with the outside (A-32)

amoebocyte Cell having amoeboid form of movement; in sponges (A-2), a cell that wanders within the organism (A)

amoeboid Shaped like an amoeba; having a cell form with ever changing cytoplasmic protrusions, or pseudopods (Pr, A)

amphiesmal vesicle Membranous sac underlying test; thought to be responsible for test production in dinoflagellates (Pr-2)

ampulla Internal saclike structures of echinoderm podia (A-29)

anaerobic Requiring the absence of gaseous oxygen

analogous Of structures or behaviors that have evolved convergently; similar in function but different in evolutionary origin

androsporangium Sporangium in which male meiotic products form (Pl)

androspore Meiotically produced spore that grows into a male gametophyte (Pl)

androsporophyll Modified leaf that bears an androsporangium (Pl)

androstrobilus Cone or strobilus bearing microsporangia or pollen sacs (Pl)

aneuploid Deviation from the normal haploid (N) or diploid ($2N$) number of chromosomes (for example, $2N + 1$, $2N - 2$)

anisogamy Formation of gametes that differ in size or morphology

annulus Ring, for example, those on the stem of certain species of mushrooms (F-3)

anterior Toward the front or head end of a cell or organism

anther Pollen-bearing part of the stamen, the male gametophyte of angiosperms (Pl-9)

antheridium Multicellular male sex organ; the sperm-producing gametangium of plants other than seed plants (Pl-1 to Pl-4, Pr-12, Pr-13)

antherozoid Motile male gamete of the Monoblepharidales (Pr-26)

Anthozoa Class of coelenterates (A-3)

antibiotic Substance produced by organisms, typically by bacteria, that injures other organisms or prevents their growth

apical cell Cell at top; for example, the two unciliated cells of the mesozoan infusoriform larva (A-5)

apical pore Opening at the apex of a structure or organism

apical tuft Tuft of cilia on a trochophore larva, for example, of annelids (A-23)

aplanetic Nonmotile (Pr, F)

aplanospore Nonmotile spore (Pr, F)

apothecium Open ascocarp, a base for the spore-bearing structures called asci (F-2)

aragonite A stable form of the mineral calcium carbonate; differs from calcite in the form of its crystals

archegonium Multicellular female sex organ; the egg-producing gametangium of plants other than seed plants (Pl-1 to Pl-4, Pr-12, Pr-13)

ascocarp A fruiting body containing asci (F-2)

ascospore Spores formed by karyogamy and meiosis and contained in an ascus (F-2)

ascus A saclike structure generally containing a definite number of ascospores (F-2)

asexual Of development or reproduction in which the offspring has a single parent

asporogenous Non-spore-forming

astropyle Nipplelike aperture projecting from the central capsule of some actinopods (radiolarians), Phaeodorina (Pr-16)

ATP Adenosine triphosphate; molecule that is the primary energy carrier for cell metabolism and motility

autogamy Union of two nuclei, both derived from a single parent nucleus (Pr)

autotroph An organism that grows and synthesizes organic compounds from inorganic compounds by using energy from sunlight or from oxidation of inorganic compounds

auxospore A diatom cell, that has been released from its rigid test, often zygotic products of fertilization (Pr-11)

axenic Growth in pure culture, that is, in the complete absence of members of other species

axial filament Solid long thin structure that is aligned longitudinally and more or less centrally in a cell or organelle

axial stalk Longitudinally aligned holdfast or other structure that permits standing

axoneme A tubule or shaft of tubules extending the length of an undulipodium or pseudopod

axoplast Granulofibrosal material from which axonemes of axopods emerge and grow (Pr-16)

axopod Permanent pseudopod stiffened by a microtubular axoneme (Pr-16)

axostyle Axial motile structure of members of the Zoomastigina (Pr-8); composed of a patterned array of microtubules and their crossbridges

bacteriophage Virus that parasitizes bacteria

bacteroid Transformed bacterium in a root nodule; site of nitrogen fixation

basal apparatus, basal body (1) Kinetosome (Pr, A, Pl) (2) Thin cylindrical plates found at the base of bacterial flagella (M); see Figure 2 of the Introduction

basidiocarp Fruiting body of a mushroom that bears basidia (F-3)

basidiospore Spore resulting from karyogamy and meiosis and borne on the outside of a basidium (F-3)

basidium Structure bearing on its surface a definite number (species specific) of basidiospores (F-3)

bilateral symmetry Anatomical arrangement in which the right and left halves of an organism or part are approximately mirror images

bilharziasis Schistosomiasis, a severe disease; of human beings; it is caused by infection with blood flukes (trematodes, A-6) and transmitted by snails (A-19)

biogenic Produced by living organisms or their remains

bioluminescence Light generated biochemically and emitted by organisms; for example, *Photobacterium* (M-14), dinoflagellates (Pr-2), fireflies (A-27), and some fish (A-32)

biosphere The part of the Earth's volume that is occupied by living organisms, the Earth's surface

biosynthesis Multienzyme-catalyzed chemical reactions that form organic compounds in organisms

bipinnaria First-stage larva of asteroid echinoderms (starfish, A-29)

blastocoel Cavity of the blastula (A)

blastopore Opening connecting the cavity of the gastrula stage of an embryo with the outside (A); represents the future mouth of some animals (protostomes, A-16 to to A-27) and the anus of others (deuterostomes, A-28 through A-32)

blastula Animal embryo after cleavage and before gastrulation; usually a hollow sphere, the walls of which are composed of a single layer of cells

bothrosome Structure at the labyrinthulid membrane that produces new membrane, sequesters calcium, and filters cytoplasm for the production of the proteinaceous extracellular slime net matrix (Pr-21)

brachiolaria Second larval stage of asteroid echinoderms (A-29); succeeds bipinnaria larva

bract Modified, often colored leaf beneath a flower or flower cluster (Pl-9)

bracteole Small bract (Pl-9)

bridle Pair of oblique ridges of thickened cuticle on the mesosome of pogonophorans (A-28)

budding Asexual reproduction by outgrowth of a bud from a parent cell or body

bud scar Impression left on a twig (Pl) or yeast cell (F-2) after a bud has broken off

bursa Saclike cavity

caecum Cavity, open at one end, usually part of the gastrointestinal tract (A)

calcite A stable form of the mineral calcium carbonate; differs from aragonite in the form of its crystals

calyptra Hood or cap partly or entirely covering the capsule (spore container) of a moss (Pl-1)

calyx flower part; the sepals, the cup-shaped outer of two series of floral leaves (Pl-9)

cambium Cylindrical sheath of dividing cells that produce phloem and xylem (Pl-2 to Pl-9)

capillitium System of sterile threads in the fruiting body of true slime molds (Pr-23)

capitulum Dense tuft of branches at the apex of a shoot in some mosses (Pl-1)

carapace Hard covering on the back of turtles and crustaceans (A)

carboxysome Organelle inside plastids; thought to harbor the CO_2-fixing enzyme ribulobiphosphocarboxylase

cardiac stomach The stomach closer to the mouth, for example, in echinoderms (A-29)

carnivore An organism that obtains food by eating live animals

carotenoid Member of a group of red, orange, and yellow hydrocarbon pigments found in plastids (Pr, Pl)

carpel Flower part; structure enclosing an ovule of an angiosperm; typically divided into ovary, style, and stigma (Pl-9)

carpospores Meiotically produced spores of rhodophytes (Pr-12)

caryoid granules Small nucleuslike intracellular spheres (Pr)

catabolism Metabolic breakdown of organic compounds to release energy and endproducts

catalase Enzyme that catalyzes the breakdown of hydrogen peroxide to water and oxygen

cell plate Phragmoplast; membranous structure that forms at the equator of the spindle during cell division and represents the site of formation of the new cell wall between the daughter cells (Pr, Pl)

cell wall Structure external to the plasma membrane produced by cells; generally rigid and composed primarily of cellulose and lignin in plants (Pl), chitin in fungi (F), and peptidoglycans (diaminopimelic acid, muramic acid, and peptides) in bacteria (M); it is absent (A, Pr) or of various composition in protoctists (Pr)

cellulose Polysaccharide made of glucose units; chief constituent of the cell wall in all plants (Pl) and chlorophytes (Pr-15)

cement gland In acanthocephalans (A-12), a gland that secretes a cement into the vagina to close it off; in gastrotrichs (A-9) and rotifers (A-10), it secretes a substance by which the animals temporarily attach themselves to objects

central capsule Spherical structure that encloses the central part of cells of actinopods (Pr-16)

centriole Small barrel-shaped organelle seen at each pole of the spindle, formed during cell division (A, some Pr), homologous to kinetosome; see Figure 2 of the Introduction

centromere Kinetochore; a proteinaceous structure at a constricted region of a chromosome; holds sister chromatids together and is the site of attachment of the microtubules forming the spindle fibers during cell division (Pr, A, Pl)

cephalic lobe Anterior part of the protosome region of Pogonophora (A-28)

cephalothorax The fused head and thorax of some arthropods (A-27)

cephalum Head (A)

cerebral organ Sense organ of unknown function in nemertines (A-7)

chelicera In some arthropods, the first pair of appendages; for example, those used by arachnids for seizing and crushing prey (A-27)

chemoorganotroph Organism requiring organic compounds as a source of both energy and carbon

chemosynthesis Production of organic compounds by using inorganic ones as the source of both carbon and energy

chitin Tough, resistant, nitrogen-containing complex polysaccharide, a polymer of glucosamine; component of exoskeletons (A) and cell walls (Pr, F)

chloroplast Green plastid, a membrane-bounded photosynthetic organelle containing chlorophylls a and b (Pr-5, Pr-6, Pr-14, Pr-15, Pl)

choanocyte Collared cell having undulipodia (Pr-8, A-2)

chromatid Longitudinal half of a chromosome; seen when chromosomes appear during prophase of mitosis in eukaryotes

chromatin Material of which chromosomes are composed; made of nucleic acids and protein, it stains deep red with Feulgen reagent

chromatophore Pigment containing cell or organelle

chromoneme DNA strands (M, Pr-2)

chromosome Intranuclear organelle made of chromatin; visible during cell division; chromosomes contain most of a cell's genetic material (Pr, F, A, Pl)

chrysolaminarin vesicle Membranous intracellular sac filled with oil (Pr-4, Pr-9, Pr-11, Pr-12)

chrysoplast Yellow plastid, the membrane-bounded photosynthetic organelle of chrysophytes (Pr-4)

cilium Short undulipodium, intracellular but protruding organelle of motility (Pr, A, Pl); see *flagellum* and Figure 2 of the Introduction

cingulum Plate of the equatorial groove of a dinoflagellate test (Pr-2)

cirrus Tuft-shaped organelle formed from fused cilia (Pr-18)

cleavage Successive cell divisions of a zygote that form the blastula (A)

cloaca Exit chamber from gastrointestinal tract; it also serves as the exit for the reproductive and urinary systems (A)

cloaste Ribbed lorica (Pr-8)

clypeus In some insects, a median plate that connects the labrum (upper lip) to the head (A-27)

cnid Polar filament or thread (Pr-20)

coccolithophorid Haptophyte (Pr-5) bearing coccoliths, small, numerous surface plates made of calcium carbonate; often abundant fossils in chalk

coccus Spherical organism (M, Pr)

coelom Body cavity lined on all surfaces by mesoderm (A-16 to A-32)

coenocytic Of a mass of cytoplasm containing nuclei but lacking cell membranes or walls (Pr, F, Pl) see *syncytium*

colony Cells or organisms of the same species living together in permanent but loose association

comb plate Ciliated body plate used by ctenophores for swimming (A-4)

companion cell Small, specialized parenchyma cells found next to conducting sieve tubes in the phloem of vascular plants (Pl-2 to Pl-9)

conceptacle Small saclike structure composed of gametangia, gamete-producing organs (Pr-12)

conidium Vegetative spore borne on a special branch of a hypha (F)

conidiophore Specialized hypha that bears conidia (F)

conidiospore Spore formed asexually, usually at the tip or side of a hypha (F)

conjugation The transmission of genetic material from a donor to a recipient cell (M); fusion of nonundulipodiated gametes or gamete nuclei (Pr-13, Pr-14, F)

connecting canal Canal leading from the radial canal to the tube feet (A-29)

conodont Paleozoic fossil interpretable as extinct cyclostome (for example, lamprey) tooth (A-32) or invertebrate remains (A-8)

contractile sheath Outer covering capable of contraction; for example, the covering of the tail of ascidians (A-32)

convergence, convergent evolution Independent evolution of similar structures for similar functions in taxa that are not directly related

copulatory spicule Secreted cuticular rod used in mating (A)

coralloid root In cycads (Pl-5), specialized root harboring green tissue layers composed of cyanobacteria (M-7)

corolla Flower part; the petals, the inner of two series of floral leaves (Pl-9)

corona Crown or crown-shaped structure

cortex Outer layer of an organism or organ; usually composed of cells (A, Pl) or protein complexes (Pr-18)

cortical layer Cortex or outer layer

costa Highly motile intracellular rod, nonmicrotubular (Pr-8)

cotyledon Leaflike structure of seeds; provides food for embryo (Pl-9)

crustose of low-lying and crusty lichens (F-5)

crystalline inclusions Regularly arrayed or crystal-shaped structures inside cells

crystalline style Enzyme-releasing organ of the digestive system of bivalve mollusks (A-19); rotated by cilia, it grinds algal food

cuticle Outer layer or covering, usually composed of metabolic products rather than of cells (A, Pl)

cyanophage Virus that multiplies and grows in cyanobacteria (M-7)

cyst Encapsulated form (often a dormant stage) of one or several organisms; formed in response to extreme environmental conditions (Pr, A)

cystocarp Carposporophyte, the asexual generation of Floridean rhodophytes (Pr-13); it grows parasitically on the female gametophyte

cytochrome Small protein that contains iron heme and acts as electron carrier in respiration and photosynthesis

cytokinesis Division of cell cytoplasm; distinguished from karyokinesis, division of a cell nucleus (Pr, F, A, Pl)

cytoplasm In a cell, the fluid, ribosome-filled portion exterior to the nucleus or nucleoid

cytostome Ingestive opening or mouth of a protist (Pr)

daughter cells Two genetically and morphologically similar products of equal cell division (M) or mitosis (Pr, F, A, Pl)

deciduous Shed or sloughed off with season (for example, leaves of a temperate-zone tree; antlers)

desmosome Intercellular membranous junction fastening cells together in tissues (A)

detritus Loose natural material, such as rock fragments or organic particles, that results directly from disintegration of rocks or organisms

Deuterostoma Series of Grade Coelomata; includes all animals in which the blastopore becomes the anus (A-28 to A-32)

dichotomous venation Branching or forking veins, as on wings, leaves, or fronds

dicotyledon Member of a subphylum of angiosperms (Pl-9); a plant having two cotyledons in the seed

dictyosome Golgi apparatus or Golgi body (A); parabasal body (Pr-8); a layered, cuplike organelle composed of modified endoplasmic reticulum; plays a part in the storage or secretion of metabolic products

dikaryon Organism composed of mycelium containing pairs of closely associated nuclei, each derived typically from a different parent (F-2, F-3)

dimorphism Morphological or genetic difference between two individual members of the same species; for example, between males and females or between winter and summer forms

dioecious Having male and female structures on different individuals of the same species (Pr, Pl)

diplanetic Of species that produce two types of zoospores and in which there are two swarming periods (Pr-27)

diploid Of cells in which the nucleus contains two sets of chromosomes; 2N (Pr, F, A, Pl)

diplophase That part of the life cycle in which the individuals are diploid (have a double set of chromosomes)

divergence Evolutionary term; accumulation of dissimilar characters in organisms related by descent but exposed to unlike environments for many generations

DNA Deoxyribonucleic acid; a long molecule composed of nucleotides in a linear order that constitutes the genetic information of cells; capable of replicating itself and of synthesizing RNA

dorsal Toward the back side (Pr, A)

ectoderm Outermost layer of body tissue (A, Pl); in particular, the outermost layer of tissue formed in gastrulation (A)

ectoplasm Outer, more or less rigid, and granule-free layer of cytoplasm (Pr)

elater Clubbed hygroscopic cell or band attached to the spore of a horsetail (Pl-3) or the capsule of a moss (Pl-1); aids in dispersing spores

embryo Early developmental stage of multicellular organisms produced from a zygote, or fertilized egg (A, Pl)

embryo sac Female gametophyte consisting, at maturity, of an egg nucleus and accessory nuclei (Pl-9)

endoderm Innermost layer of body tissue formed in the gastrulation of an embryo (A)

endoplasm Inner, relatively fluid central portion of cytoplasm (Pr)

endoplasmic reticulum Extensive system of membranes inside eukaryotic cells; called rough if coated with ribosomes, called smooth if not (Pr, A, Pl)

endosome Intranuclear body, also called nucleolus or karyosome, composed of precursors of ribosomes (Pr, A)

endosperm Tissue surrounding plant embryo; contains stored food; develops from the union of haploid (N) male nucleus and two haploid (2N) polar bodies of female and is therefore triploid (3N) (Pl-9)

endospore Desiccation- and heat-resistant spore produced inside bacteria (M-2, M-4, M-11)

endosymbiont Organism living inside an organism of a different species; may be intracellular or intercellular

epibiotic Living on the surface of another organism

epicone Upper surface (hemisphere) of a dinoflagellate test (Pr-2)

epiphragm Membrane closing the mouth of a moss spore capsule (Pl-1)

epiphyte Epibiotic plant that depends on host plant of another species for support but not for nutrition

epipod Pseudopod formed at the posterior region of an amoeboid cell (Pr, A)

epithelial tissue Tissue that covers the inner or outer surface of a body or structure (A, Pl)

epivalve Upper test or shell (Pr)

esophagus The part of the digestive tract between the pharynx and the stomach (A)

estuary Passage, as at the mouth of a lake or a river, where ocean tide meets a freshwater current

eucarpic Of the mode of development in which reproductive structures form on certain portions of the thallus while the thallus itself continues to perform its somatic functions (Pr, F)

eukaryote A cell having a membrane-bounded nucleus, organelles such as mitochondria and plastids, and several chromosomes in which the DNA is coated with histone proteins (Pr, F, A, Pl)

exospore Externally borne reproductive structure (cell) not necessarily heat- or desiccation-resistant (M-7)

facultative autotroph Organism that, depending on conditions, can grow either by autotrophy (photo- or chemosynthesis) or by heterotrophy

false branching Growth habit of a filamentous structure in which apparent branching is caused by slippage between two rows of cells; only two growth points are simultaneously active on any single cell (M, Pr)

falyx Structure of opalinids (Pr-8); composed of rows of closely set kinetosomes and their undulipodia

fauna Animal life

fermentation Anaerobic respiration; degradation of organic compounds in the absence of oxygen, yielding energy and organic end products (the ferment); organic compounds are the terminal electron acceptors in all fermentations

fertilization Fusion of two haploid cells, gametes, or gamete nuclei to form a diploid zygote (Pr, F, A, Pl)

fiddlehead Emergent young fern frond (Pl-4)

filopod Pseudopod composed entirely of ectoplasm; typically thin and pointed, with microfibrillar ultrastructure (Pr-3, Pr-17)

fission Asexual reproduction by division of cells or organisms into two or more parts of equal or nearly equal size

flagellar swelling Hypertrophied sac region at the base of an undulipodium; caused by the local swelling of plasma membrane (Pr)

flagellate Prokaryote motile via flagella. See *mastigote*

flagellum (1) Long thin solid extracellular organelle of bacterial motility; composed of a protein, flagellin (2) Undulipodium; long, fine, intrinsically motile intracellular structure used for locomotion or feeding; covered by plasma membrane and underlain by regular array of nine doublet microtubules and two central microtubules composed of tubulin, dynein, and other proteins, not flagellin (Pr, A, Pl); flagella are longer than cilia but have the same internal structure; the term *undulipodium* refers to both the flagella and the cilia of eukaryotes

flame cell Ciliated cell at the inner end of a protonephridial tube; the flickering tuft of cilia may resemble a flame (A)

flora Plant life

foliose Of lichens that produce leafy structures (F-5)

frond Leafy part, as of a fern (Pl-4) or seaweed (Pr-12, Pr-13)

fruticose Of lichens that bear upright fruiting bodies (F-5)

fruiting body Structure that contains or bears cysts, spores, or other generative structures (M, Pr, F, Pl)

frustule Asexual planulalike bud that eventually develops into a polyp (A-3); half of a diatom test (Pr-11)

funiculus Body structure resembling a cord (A)

fusion Union of two cells (Pr, A) or hyphae (F)

fusule Aperture through which an axopod extends (Pr-16)

gametangium Organ or cell in which gametes are formed (Pr, F, Pl)

gamete Mature haploid reproductive cell whose nucleus fuses with that of another gamete of an opposite sex to form a zygote (Pr, A, Pl)

gametic meiosis Meiosis that occurs directly prior to gametogenesis

gametogenesis Production of gametes (Pr, A, Pl)

gametophyte Haploid ($1N$) gamete-producing generation, as in plants having alternation of generations (Pl)

gamogony Series of cell or nuclear divisions leading eventually to gamonts, individuals that produce gamete nuclei or gametes capable of fertilization (Pr-17)

gamont Gamete-producing form of an organism; sexually differentiated individual (Pr-17)

ganglion Aggregate of bodies of nerve cells (A)

gap junction Intercellular, membranous, discontinuous junction fastening cells together in tissues; thought to regulate the flow of ions between cells (A)

gastric gland Multicellular epithelial organ that secretes enzymes for extra-cellular digestion (A)

gastrovascular cavity Cavity in which both digestion and circulation take place (A)

gastrula Embryo in the process of gastrulation, during which the blastula with its single layer of cells becomes a three-layered embryo (A)

GC ratio Proportion of guanine and cytosine base pairs in the total number of base pairs in any DNA; a wide difference in GC ratio is taken to preclude a phylogenetic relationship between organisms (M, F)

gemma Asexual reproductive structure, a small mass of vegetative tissue that can develop into a new individual (Pr-27, Pl-1)

gemmule Asexual reproductive structure, a small mass of vegetative tissue that can be released and develop into a new individual (A-2)

generative nuclei Nuclei capable of further division or of fertilization followed by further division (Pr, Pl)

genital pore Exit aperture for fertilized eggs (A)

genome Complete set of genetic material of an organism

germ cells Ova (eggs), spermatozoa (sperm); gametes; cells requiring fertilization; sex cells

germ layer Any of the three primary cell layers (ectoderm, mesoderm, endoderm) formed from a blastula during gastrulation (A)

germinal cells Precursors to gametes or to asexual reproductive cells (Pr, A, Pl)

germinating cyst Cyst in the process of releasing its spores or converting to a growing structure

germs microorganisms that cause disease

gill Respiratory organ used for uptake of oxygen and release of CO_2 by aquatic organisms (A)

girdle Equatorial groove or other structure (Pr, A); cingulum of a dinoflagellate (Pr-2); one of a pair of ridges dividing the trunk of a pogonophoran (A-28)

girdle lamella Centrally or equatorially located flat intracellular structure (Pr)

glycogen A long-chain carbohydrate composed of glucose units (Pr, A)

glycolysis Metabolic pathway in which glucose is broken down into organic acids and CO_2, releasing energy

Golgi body See *dictyosome*

gonad Animal organ composed of tissues that produce gametes; ovary (produces eggs) or testis (produces sperm)

Gram negative Failing to retain the purple stain when subjected to the Gram staining method; indicates the presence of certain component layers in the cell wall (M)

Gram positive Retaining the purple stain (crystal violet) when subjected to the Gram staining method; indicates absence of a certain component layer in the cell wall (M)

granum Inclusion in a plastid; seen as a minute grain under the light microscope, but known to consist of stacked thylakoids (Pr, Pl)

gynosporophyll Leaflike appendage bearing female sporangia in seed plants (Pl-5 to Pl-9)

gynostrobilus Female cone bearing megasporangia, ovules, or seeds (Pl-5 to Pl-9)

haploid Of cells in which the nucleus contains only one set of chromosomes; $1N$ (Pr, F, A, Pl)

haplophase That part of the life cycle in which the individuals are haploid (have a single set of chromosomes)

haptonema Coiled, threadlike microtubular structure often used as a holdfast (Pr-5)

haustorium Projection of a hypha, generally into plant vascular tissue; it penetrates between and into the cells of the invaded tissue to absorb nutrients (F)

hemerythrin Iron-containing respiratory pigment used for storage of oxygen in the blood of various invertebrates (A)

hemocoel Fluid-filled body cavity that functions as part of the circulatory system (A)

hemoglobin Iron-containing protein used for the transport or storage of oxygen; found in the blood of animals (A), the root nodules of some legumes (Pl) that have symbiotic bacteria (M), and in some strains of *Paramecium* and *Tetrahymena* (Pr)

herbivore Organism that obtains nutrition and energy by eating plants or algae

hermaphroditic Organism that simultaneously possesses both male and female reproductive organs (Pr, A); dioecious (Pl)

heterokaryon Organism composed of mycelium containing pairs of closely associated nuclei, the members of each pair known to be derived from different parents (differing in genotype) (F)

heterokont Biundulipodiated cell in which the two undulipodia are unequal in length (Pr)

heterophasic Of a life cycle that includes different stages, such as morphologically distinct haplo- and diplophases (Pr, F, A, Pl)

heterosporous Individual that simultaneously forms spores of two kinds; typically, microspores and megaspores (Pl)

heterotrichous Hairy or filamentous with hairs or filaments of more than one type (Pr-18, Pl)

heterotroph Organism that obtains carbon and energy from organic compounds ultimately produced by autotrophs; saprobes, parasites, carnivores, and others (M, Pr, F, A)

histones Class of positively charged chromosomal proteins that bind to DNA, tend to be lysine and arginine-rich, and stain with fast green

histology Study of tissue (F, Pl, A)

holdfast Organ or organellar structure for attachment (Pr, F, A)

holocarpic Of the mode of development in which the thallus is entirely converted into one or more reproductive structures (Pr, F)

homeotherm Warm-blooded organism capable of regulating its body temperature around a set point fairly independently of external temperatures (A-32)

homologous Of structures or behaviors that have evolved from common ancestors, even if the structures or behaviors have diverged in form and function

homosporous Individual that forms only one kind of spore, either megaspores or microspores; monoecious (Pl-4 to Pl-9)

homofermentation Fermentation of glucose or another simple sugar to produce only a single kind of acid (M)

host Organism that provides nutrition or lodging for symbionts or parasites

hydra Common name of a large genus of abundant freshwater coelenterates (A-3)

hydrogenase Enzyme that catalyzes the breakdown of organic compounds and the simultaneous release of hydrogen (H_2) (M)

hydrostatic pressure Pressure exerted by a liquid on the walls of a container

Hydrozoa Class of coelenterates (A-3)

hymenium Fertile layer of tissue consisting of asci or basidia (F-2, F-3)

hypermastigote Motile heterotroph with up to many thousands of undulipodia (Pr-8)

hypersaline Saltier than sea water ($> 3.4\%$ NaCl)

hypertrophy Overgrowth; unusual increase in size or number

hypha Threadlike tubular filament, a component of a mycelium (M-15, Pr, F)

hypocone Lower surface (hemisphere) of a dinoflagellate test (Pr-2)

hypothecium Thin layer of interwoven hyphae between the hymenium and the apothecium (F-2)

hypovalve Lower test or shell (Pr)

indusium Reproductive tissue on a fern frond; outgrowth in which sori develop (Pl-4)

infusoriform larva Ciliated immature mesozoan (A-5) produced by the rhombogen germ cell; a swarm larva

inner envelope Membrane on the inside of a cell wall or other structure; plasma membrane

integument Skin or enveloping layer of an organism (Pr, F, A, Pl)

intestinal diverticula Blind, often branched sacs developed from intestine (A-6, A-7, A-19)

introvert Slender anterior body part that can be turned inside out and completely withdrawn into the trunk of the body (A-21)

iridescence White light that is reflected to produce a rainbow effect

isogamy The formation of gametes of opposite mating types that are alike in size and morphology (Pr)

isoprenoid The C_5H_{10} unit of a class of organic compounds. They are synthesized from multiples of isopentenyl pyrophosphate; phytol, carotenoids, and steroids are isoprenoids

isokont Undulipodiated cell in which the undulipodia are of equal length (Pr)

karyogamy Fusion of nuclei, often in zygote formation (Pr, F, A, Pl)

karyokinesis Nuclear division (Pr, F, A, Pl)

keratin One of a class of tough, fibrous proteins, components of hair, skin, claws, and horn (A)

kinetochore See *centromere*

kinetoplast Intracellular DNA-containing structure, a modified mitochondrion with an associated kinetosome characteristic of the class Kinetoplastida (Pr-8; for example, *Trypanosoma*)

kinetosome Organelle at the base of all undulipodia and responsible for their formation; like a centriole, its cross section shows a characteristic circle of nine triplets of microtubules (see Figure 2 of the Introduction); centriole with an associated axoneme (Pr, A, Pl)

labrum Animal mouthpart; upper lip of insects (A-27)

lamella Flattened saclike structure (Pr, F, A)

larva Immature form of an animal; distinguishable morphologically from the adult (A-27, A-32)

leaf primordium Lateral outgrowth that will become a leaf (Pl)

leaf sheath Tissue surrounding the base of a leaf (Pl-5 to Pl-9)

lemniscus Fluid-filled reservoir used by an acanthocephalan worm to evert the proboscis (A-12)

leucosin Storage oil usually found in the membranous vesicles of chrysophytes (Pr-4)

lipid One of a class of organic compounds soluble in organic and not aqueous solvents; includes fats, waxes, steroids, phospholipids, carotenoids, and xanthophylls

lipid vesicle Intracellular sac containing lipids

lophophore Food-trapping ridge surrounding the mouth; bears hollow, ciliated tentacles (A-16 to A-18)

lophotrichous Bearing polar bundles of flagella (M), or undulipodia (Pr-8)

lorica Secreted protective covering, test, shell, valve, or sheath (Pr, A)

luminescence See *bioluminescence*

macrogametes Large gametes; in most cases, female (Pr, Pl)

macronucleus Larger of the two types of nuclei in ciliates (Pr-18); site of mRNA synthesis; contains many copies of each gene; required for asexual growth and division

madreporite External aboral terminus of the echinoderm water vascular system; connects to the stone canal (A-29)

mandible Jaw (A)

mantle Covering or coat; body wall that secretes a shell (A-19)

manubrium Protrusion that bears the mouth (for example, in hydrozoans, A-3)

mastax Horny, toothed chewing apparatus in the pharynx of rotifers (A-10)

mastigoneme Lateral appendage of undulipodia; also called flimmer, tinsel, and scale

mastigote Eukaryotic microorganism motile via undulipodia, eukaryotic "flagellate"

mating type Designation of an individual organism to distinguish it from others capable of mating or fusing with it; individuals of the same mating type cannot mate with each other; in many species, individuals of different mating types are indistinguishable except by their willingness to mate with each other; some species have hundreds of mating types (M, Pr, F, Pl)

medulla Inner portion of gland or other structure surrounded by cortex (A)

medusa Free-swimming bell-shaped or umbrella-shaped stage (diplophase) in the life cycle of many coelenterates (A-3); jellyfish

medusoid bud Immature form of the medusa when it is still sessile (A-3)

megasporangium Structure in which female meiotic products form; usually produces from one to four megaspores (Pl)

megaspore Haploid spore that develops from megaspore mother cell; develops into the female gametophyte (Pl-4 to Pl-9).

meiosis One or two nuclear divisions in which the number of chromosomes is reduced by half (Pr, F, Pl, A).

membrane tubules Tubular intracellular structures composed of proteins that catalyze respiratory reactions (M)

membranelle Organelle formed by the fusion of rows of adjacent cilia (Pr-18)

merozoite Life cycle stage; agamete produced by multiple mitoses of a trophozoite (Pr-19)

meristem Plant tissue composed of undifferentiated, dividing cells

mesenchyme Gelatinous material containing amoebocytes or other cells (A-1, A-2); unspecialized embryonic tissue that gives rise to circulatory and connective tissue (A)

mesocoel Middle part of the coelom of many deuterostomes (A)

mesoderm Middle layer of body tissue (between the ectoderm and endoderm) formed in gastrulation (A)

mesoglea Gelatinous noncellular layer formed between ectoderm and endoderm (A-2, A-3)

mesosome (1) Midregion of pogonophore body (A-28). (2) Membranous structure associated with DNA segregation in dividing bacteria (M)

messenger RNA (mRNA) RNA produced from DNA-directed polymerization in pieces just large enough to carry the information for the synthesis of one or several proteins

metabolism Sum of enzyme-catalyzed chemical reaction sequences occurring in cells and organisms

metacoel Trunk part of the coelom of an ectoproct (A-16)

metamorphosis An abrupt transition from the immature to an intermediate or adult form (for example, tadpole to frog).

metatroch Middle band of cilia on a trochophore larva (A-19, A-23)

Metazoa Kingdom Animalia; excludes protoctists (protozoa)

microaerophil Aerobic organism requiring an oxygen concentration less than normal (less than 20 percent by volume)

microbe Microscopic organism

microbial mat Matlike community of microorganisms; living precursor of a stromatolite

microcyst Desiccation-resistant sporelike structure that is released from bacterial fruiting bodies and may develop into a new bacterium (M-16)

microfibril Any solid, thin, fibrous proteinaceous structure, generally those in the cytoplasm of eukaryotic cells; some participate in movement (Pr, A)

microfilament See *microfibril*

microgamete Small gamete, typically male (Pr, Pl)

micronucleus Smaller of the two types of nuclei in ciliates (Pr-18); does not synthesize messenger RNA; most are diploid and required for meiosis and autogamy but not for asexual growth and division

micropaleontology Study of fossil microbes and microscopic parts of fossil organisms

micropyle Flower part; the opening in the ovule integument through which the pollen tube grows (Pl-4 to Pl-9)

microsporangium Structure in which male meiotic products form; produces many microspores (Pl)

microspore Haploid spore that develops from microspore mother cell; develops into a male gametophyte (Pl-4 to Pl-9); in a flowering plant (Pl-9), it becomes a pollen grain

microsporophyll Leaflike appendage bearing microsporangia (Pl)

microstrobilus Small male cone (Pl)

microtubule Slender, hollow proteinaceous intracellular structure; most of them are 24 nm in diameter; found in axopods, axonemes, mitotic spindles, undulipodia, haptonemes, nerve cell processes (axons and dendrites), and other intracellular structures; often, their formation can be inhibited by colchicine (Pr, F, A, Pl)

microvillus Minute fingerlike projections of the plasma membrane of a cell (A)

mitochondrion Organelle in which the chemical energy in reduced organic compounds (food molecules) is transferred to ATP molecules by oxygen-requiring respiration (Pr, F, A, Pl)

mitosis Nuclear division in which attached pairs of duplicate chromosomes move to the equatorial plane of the nucleus, separate at their centromeres, and form two separate, identical groups; subsequent division of the cell will thus produce two identical daughter cells (Pr, F, A, Pl)

mitotic apparatus Mitotic spindle

mitotic spindle Microtubular structure formed during mitosis and responsible for the poleward movement of the chromosomes (Pr, F, A, Pl)

mole Quantity of a substance containing a standard number (Avogadro's number, 6.023×10^{23}) of atoms or molecules; quantity of a substance whose mass in grams is equal to the atomic or molecular weight of the substance

molting Shedding of all or part of an outer covering, such as cuticle, skin, or feathers (A)

monocotyledon Member of a subphylum of angiosperms (Pl-9); includes plants having a single cotyledon in the seed

monoecious Hermaphroditic. Having the anthers (male parts) and carpels (female parts) on the same organism either on the same or on different flowers (Pl)

monokaryon Mycelial organism all of whose nuclei are descended from a single parent (F)

monophyletic Of a trait or a group of organisms derived directly from a common ancestor

monoplanetic Of species that produce only one type of zoospore and in which there is only one swarming period (Pr-27)

monopodial Having one main axis of growth (Pl) or pseudopod (Pr)

monosaccharide Simple sugar, such as glucose or ribose

morphogenesis Development of the size, form, or other structural features of an organism

morphology Form and structure, study of form

mother cell Cell that, by mitosis, gives rise to nuclei able to be fertilized and to continue development

mouth cone Protrusible snout that contains the mouth of a kinorhynch (A-11)

mucilage Secreted slime or extracellular cementing material composed in part of nitrogen-containing polysaccharides

mucopolysaccharide Member of a class of polymers composed of nitrogen-substituted simple sugars

mycelium Mass of hyphae constituting the body of a fungus or a funguslike protoctist (Pr-24 to Pr-27, F)

mycorrhiza Association between the hyphae of a fungus and the roots of a plant

myoneme Fibrillar organelle that has a contractile function (Pr-18)

myosin One of the two major proteins of muscle, in which it makes up the thick filaments (A); also found in filaments that participate in the motility of protists (Pr)

myxospore A spore or other desiccation-resistant stage of myxobacteria (M-16)

nematocyst Capsule containing a threadlike stinger used for anchoring, defense, or capturing prey; some, for example, the cnidoblasts of coelenterates (A-3), contain poisonous or paralyzing substances

nebenkörper Dense body at the nucleus surface in some amoebas; composed of bacterium-like obligate symbionts (Pr-3)

nematogen Life-cycle stage of mesozoans (A-5); the ciliated wormlike stage of a mesozoan living in a host cephalopod

nephridiopore Exit pore for wastes secreted by a nephridium (A)

nephridium Excretory organ of many invertebrates (A)

nephrostome Ciliated funnel-shaped inner opening of the nephridium of coelomate animals

nerve ring Circular network of nerve fibers; a component of the central nervous system (A)

niche Role performed by members of a species in a biological community

nitrogen fixation Incorporation of atmospheric nitrogen into organic nitrogen compounds; requires nitrogenase (M)

nitrogenase Enzyme complex containing iron and molybdenum; converts atmospheric nitrogen to organic nitrogen (M-2, M-7, M-9)

node Typically hard or swollen junction of two internodes of a stem or branch; may bear a leaf or leaves (Pl)

notochord Long elastic rod that serves as an internal skeleton in the embryos of chordates; replaced by the vertebral column in most adult chordates (A-32)

nuclear cap Crescent-shaped sac surrounding a third or more of the zoospore nucleus of some chytrids (Pr-26); it contains virtually all the ribosomes of the cell

nuclear envelope Nuclear membrane. Membrane surrounding the nucleus of a cell

nucleocytoplasm That portion of a eukaryotic cell that includes the nucleus and the cytoplasm with its membranes and inclusions but excludes organelles such as plastids and mitochondria (Pr, F, A, Pl)

nucleoid Genophore; DNA-containing structure of prokaryotic cells; not bounded by membrane (M)

nucleolus Structure in the cell nucleus; contains DNA, RNA, protein, precursors of ribosomes (Pr, F, A, Pl)

nucleotide Single unit of nucleic acid; composed of an organic nitrogenous base, deoxyribose or ribose sugar, and phosphate

nucleus Large membrane-bounded organelle that contains most of a cell's genetic information in the form of DNA (Pr, F, Pl, A)

nucleus-associated organelle (NAO) See *spindle pole body*

nummulite Large Cenozoic fossil foraminiferan (Pr-17) having a coin-shaped shell

obligate anaerobe Organism that can survive and grow only in the absence of gaseous oxygen (M)

obligate parasite Organism unable to survive and grow when removed from its host

ocellus Eye or eyespot (Pr, A)

oidium Thin-walled hyphal cell that functions as a spore (F)

oocyst Desiccation-resistant thick-walled structure in which sporozoans are transferred from host to host (Pr-19)

oocyte Cell that eventually develops into an egg or ovum (A)

oogamy Mode of sexual reproduction; anisogamy in which one of the gametes (the egg) is large and nonmotile, while the other gamete (the sperm) is smaller and motile (Pr, A)

oogonium Unicellular female sex organ that contains one or several eggs (Pr)

oosphere Large, naked, nonmotile female gamete (Pr)

opaline Shiny and whitish; resembling amorphous silica (opal)

operculum Lid; covering flap (Pr, F, A, Pl)

organelle ''Little organ''; distinct intracellular structure composed of a complex of macromolecules and small molecules; for example, nuclei, mitochondria, and lysosomes

osculum Excurrent opening of a sponge (A-2)

osmoregulatory Regulating the concentration of dissolved substances in a cell, organ, or organism, despite changes of the concentrations in the surrounding medium

osmotrophic Obtaining nutrients by absorption, direct uptake of molecule-sized food compounds across membranes

ossicle Calcareous structure that, in numbers, forms the endoskeleton of echinoderms (A-29)

ostiole Pore of a fruiting body

ostium Incurrent opening of a sponge (A-2)

ovary Multicellular female reproductive organ (A, Pl)

ovoviviparous Of a mode of sexual reproduction in which eggs of embryos develop inside the maternal body but do not receive nutritive or other metabolic aids from the parent; offspring are released as miniature adults (A, P)

ovule Megasporangium in seed plants; contains specialized tissues and an egg cell, which will become a seed after fertilization (Pl-5 through Pl-9)

ovuliferous scale In conifer cones, plant tissue on which female reproductive structures develop (Pl-7)

ovum Egg (Pr, A, Pl)

palynology Study of pollen and spores

palmate Flat and palm shaped; leaflike structure whose parts radiate in one plane

papilla A small bump

parabasal body Organelle located near kinetosomes; resembles dictyosome and Golgi apparatus in structure and probably in function (Pr-8)

parabasal folds Creases in the parabasal body

paraproct One of a pair of ventrolateral lobes bordering the anus of insects (A-27)

paramylon Carbohydrate composed of glucose units; forms the reserve foodstuff of some protoctists (Pr-5, Pr-6)

paraphysis One of several cellular trichomes that surround the egg in the phaeophyte oogonium (Pr-12) and in mosses (Pl-1)

parapodium Fleshy, segmental appendage of polychete worms (A-23)

parapyle Tubular aperture of the central capsule of radiolarians of the suborder Phaeodorina (Pr-16)

parasexual cycle Production of individuals having more than one parent without meiosis and fertilization (M, Pr, F)

parasite Organism that lives on or in an organism of a different species and obtains nutrients from it

parenchyma Tissue made of unspecialized vegetative cells (Pr, F, Pl)

parthenogenesis Development of an unfertilized egg into an organism (A)

pathogen Organism that causes disease (M, F, Pr, A)

pebrine Disease of silkworm larvae (A-23)

pectin Complex carbohydrate found between plant cells and in their cell walls (Pl)

pedal gland Gland that secretes adhesive for attaching an animal to a substrate (A-9 to A-15); some pedal glands are also called cement glands or adhesive tubes

peduncle Holdfast, base, stalk, or stemlike structure (Pr, A)

pelagic Dwelling in the open ocean

pellicle Thin, typically proteinaceous outer layer of a cell or organism, outside plasma membrane

pelta Swelling at the anterior end of an axostyle (Pr-8)

pen Skeleton or internal horny shell of cephalopods (A-19)

periaxostylar bacteria Endosymbiotic bacteria found around the axostyle (Pr-8)

periaxostylar ring Morphologically distinctive ring of material surrounding the axostyle (Pr-8)

periostracum Proteinaceous outer layer of a shell (A-18, A-19)

perispore Wrinkled outer covering of the spores of some vascular plants (Pl-2 to Pl-4), especially of ferns

peristalsis Successive contraction and relaxation of transverse muscles of tubular organs, resulting in the movement of contained fluid in one direction (A)

peristome A serrated ring around the opening of a moss spore capsule (Pl-1)

perithecium Closed ascocarp with a pore, or ostiole, at the top and a wall of its own (F-2)

peritoneum Mesodermal membrane lining the body cavity (true coelom) and forming the external covering of the visceral organs (A)

peritrichous Bearing flagella uniformly distributed over the body surface (M)

periphysis Short, hairlike growth (Pr, F)

petal Flower part; one of the leaves of a corolla (Pr-9)

petiole Stalk of a leaf (Pl-5 to Pl-9)

pH Scale for measuring the acidity of aqueous solutions; pure water has a pH of 7 (neutral); solutions having a pH greater than 7 are alkaline; less than 7 are acid

phaeoplast Brown plastid, the membrane-bounded photosynthetic structure of brown algal cells (Pr-12)

phagocytosis Ingestion, by a cell, of solid particles by flowing over and engulfing them whole (Pr, A)

pharynx Throat; part of the alimentary tract between the mouth cavity and the esophagus (A)

phloem A plant vascular tissue that transports nutrients; in trees, the inner bark (Pl-2–Pl-9)

phosphorescence Luminescence caused by reemission of absorbed radiation as light that continues after the impinging radiation has ceased

photoautotroph Organism able to produce all nutrient and energy requirements by using inorganic compounds and visible light (M-6, M-7, M-8, Pl)

photoreceptor Specialized accumulation of pigment that reacts to light stimuli (M, Pr, A)

photosynthate Chemical products of photosynthesis (M, Pr, Pl)

photosynthesis Production of organic compounds from carbon dioxide and water by using light energy captured by chlorophyll (M, Pr, Pl)

phototaxis Movement toward a light source

phototropism Growth toward a light source

phragmoplast See *cell plate*

phycobilin One of various blue-green pigments that take part in photosynthesis in cells of cyanobacteria and rhodophytes (M-7, Pr-13)

phycology Study of algae, algology

phylogeny Evolution of a genetically related group of organisms; also, a schematic diagram representing that evolution

phytoplankton Free-floating microscopic photosynthetic organisms (Pr)

pileus Cap or upper part of certain ascocarps and basidiocarps (F-2, F-3); the cap of a mushroom

pilidium Free-swimming larval stage of a nemertine (A-7)

pinacocytes Epithelial cells of sponges (A-2)

pinna Leaflet or primary division of a pinnately compound leaf; pinnae are arranged in two opposite rows along the same axis (Pl)

pinnule Subleaflet or division of a pinna (Pl)

pinocytosis Ingestion, by a cell, of liquid droplets by engulfing them whole (Pr, A)

pioneer plant One of the first plants to colonize an area devoid of vegetation

pit connection Protoplasmic connection joining cells into tissues (Pr-13, F-3)

pith Tissue, usually parenchyma, occupying the center of the stem within the vascular cylinder of a plant (Pl-2 to Pl-9)

placenta Temporary structure formed in part from the inner lining of the uterus and in part from the extraembryonic membranes; the placenta facilitates the exchange of nutrients and wastes between embryo and mother (A-32)

plankton Free-floating microscopic or small aquatic organisms, both photosynthetic and heterotrophic (Pr, A)

planogamete Motile gamete having undulipodia (Pr)

planula Ciliated, free-swimming larva of coelenterates (A-3)

plasmalemma Cytoplasmic membrane; cell membrane; cell envelope; plasma membrane

plasmodium Multinucleate mass of cytoplasm lacking internal cell boundaries (membranes, walls), cells, syncitium, coenocyte (Pr-23, Pr-24)

plasmogamy Fusion of two cells or protoplasts without fusion of nuclei (Pr, F)

plastid Cytoplasmic, photosynthetic pigmented organelle (such as a chloroplast) or its nonphotosynthetic derivative (such as a leucoplast or etioplast) (Pr, Pl)

pleural region Membranous sidewall of an insect thorax (A-27)

podium (1) Food-gathering or locomotory extensions of the cytoplasm of a cell. (2) Tube foot of an echinoderm (A-29)

poikilotherm Organism unable to internally regulate its body temperature (A)

pollen tube Male reproductive structure of plants; nucleated tube formed by germination of the pollen grain; it grows down the style of the pistil and carries the male nuclei to the female nuclei in the ovary (Pl-5 to Pl-9)

polymastigote Motile heterotroph with few (< 16) undulipodia (Pr-8)

polymorphism Morphological or genetic differences between individuals that are members of the same species

polyp Coelenterate body form; cylindrical tube closed and attached at one end, open at the other by a central mouth typically surrounded by tentacles (A-3)

polyphyletic Of a trait or group of organisms derived by convergent evolution from different ancestors

polyploidization The production of a cell whose nucleus contains more than the diploid number of chromosomes (Pr, F, A, Pl)

polysaccharide Carbohydrate composed of many monosaccharide units joined to form long chains; for example, starch, cellulose, glycogen, and paramylon

porocyte Tubular sponge cell through which water moves into the cavity of the sponge (A-2)

porphyrin Nitrogen-containing heterocyclic organic compound; derivatives include chlorophyll and heme

postannular papilla Small bump that probably secretes chitinous tube material or adhesive; part of a pogonophoran (A-28)

posterior Pertaining to the rear end

presoma Barrel-shaped, retractile body region anterior to the trunk of priapulids (A-20)

proboscis Tubular protrusion or prolongation of the head or snout (Pr, A)

prokaryote Cell or organism composed of cells lacking a membrane-bounded nucleus, membrane-bounded organelles, and DNA coated with histone proteins (M)

proloculus Initial test formed by free-swimming foraminifera; precedes mature, multichambered test (Pr-17)

prostheca Stalk (M-14)

protandry Prevention of self-fertilization in hermaphroditic animals by production of sperm and eggs at different times (sperm first)

prothallus Young gametophyte (Pl-1 to Pl-4)

protocoel Anterior part of coelom in freshwater ectoprocts (A-16)

protoconch Embryonic shell of nudibranch mollusks (A-19)

protonema Filamentous thread produced by germinating spore (Pl-1, Pl-4)

protonephridium Rudimentary tubular excretory organ; has a blind inner end, typically occupied by a ciliated flame cell, and an open outer end, or external pore, through which it discharges collected fluid wastes (A)

protoplasmic cylinder Long thin cytoplasm and nucleoid of a spirochete; the part of a spirochete bounded by the plasma membrane (M-3)

protoplast Nucleus and cytoplasm of a cell after cell wall is removed

protosome Anterior part of a pogonophoran (A-28)

protostoma Series of Grade Coelomata; includes all animals in which the blastopore becomes the mouth (A-6 to A-27)

prototroch Most apical girdle or cilia on a trochophore larva (A-19, A-23)

psammophile Organism that lives in sand, especially in the spaces between sand grains (Pr, A)

pseudocoelom Animal body cavity lined by both endoderm and ectoderm (not only by mesoderm) (A-9 to A-15)

pseudopod Temporary cytoplasmic protrusion of an amoeboid cell; used for locomotion or phagocytotic feeding (Pr, A)

psittacosis Disease of birds caused by rickettsia (M-1); can be transmitted to people

psychrophil Organism whose optimum environmental temperature is well below 15°C (M, Pr)

pusule Fluid-filled intracellular sac responsive to pressure changes (Pr-2)

pycnidium Asexual, hollow fruiting body lined inside with conidiophores (F)

pycnotic Darkly staining, as of nongerminative nuclei or those in moribund cells (Pr, A, F, Pl)

pyloric stomach More aboral of the two stomachs of echinoderms (A-29)

pyrenoid Proteinaceous structure inside some plastids; serves as a center of starch formation (Pr, Pl)

pyriform Having the form of a tear or pear

rachis Axis from which the pinnae or leaflets of a frond or a compound leaf arise (Pl-4 through Pl-9)

radial canal (1) Part of the digestive system of medusae (A-3) (2) Tube of the water vascular system that passes into each ray and to the tube feet of echinoderms (A-29) (3) Canals of some sponges (A-2)

radial symmetry Regular arrangement of parts around one longitudinal axis; any plane through this axis will divide the object or organism into similar halves

radula Horny toothed organ of mollusks (A-19); used to rasp food and carry it into the mouth

ray Stiff projection

recombination The appearance, in progeny, of gene combinations that differ from the combinations in the parents

regolith Loose material consisting of soil, sediments, or broken rock overlying solid rock

reservoir Vestibule or holding structure; deep part of the oral structure of some protoctists (Pr-6, Pr-18)

respiration Oxidative breakdown of food molecules and release of energy from them; the terminal electron acceptor is inorganic, it may be oxygen (M, Pr, F, A, Pl) or, in anaerobic organisms, nitrate, sulfate, or nitrite (M)

respiratory membranes Intracellular membranes in which enzymes for oxygen or nitrogen respiration are embedded (for example, in mitochondria and bacteria)

reticulate Netlike or covered with netlike ridges

reticulipod Pseudopod that is part of a network of cross-connected pseudopods (Pr-17)

rhizoid Rootlike structure that anchors and absorbs (Pr, Pl-1)

rhizome Subterranean or creeping stem; differs histologically from a root, it sends out shoots from near its tip (Pl)

rhizomycelium Rhizoidal system extensive enough to resemble mycelium superficially (Pr-26)

rhizoplast Basal rootlet, proteinaceous striated strand connecting the nucleus and a kinetosome; some contractile and calcium sensitive (Pr)

rhodoplast Red plastid, the membrane-bounded photosynthetic structure of red algal cells (Pr-13)

rhombogen Sexually mature life-cycle stage of mesozoans (A-5); attained only when the population is dense

ribosome A spherical organelle composed of protein and ribonucleic acid; the site of protein synthesis

ring canal Ring-shaped tube of the water vascular system of echinoderms (A-29)

RNA Ribonucleic acid; a molecule composed of a linear sequence of nucleotides; can store genetic information; component of ribosomes; takes part in protein synthesis; see Messenger RNA

RNA core Inner, genetic portion of an RNA-type virus

root nodules Spherical growths on the roots of leguminous plants; contain nitrogen-fixing bacteroids

rootstock Subterranean stem; a rhizome (Pl)

rosette Star-shaped structure composed of radiating filaments

Saefftigen's pouch Fluid-filled sac used during mating by an acanthocephalan male (A-12)

samara Dry, one-seeded, indehiscent winged fruit (Pl-9)

saprobe Organism that excretes extracellular digestive enzymes and absorbs dead organic matter (Pr, F)

saprophyte Saprobe

scale Thin, hard surface plate that overlaps with others to coat a surface (Pr, A); small rudimentary or vestigial leaf that protects the bud of a seed plant (Pl)

schistosomiasis See *bilharziasis*

schizogony Multiple mitoses without increase in cell size; gives rise to schizonts (Pr-19)

schizont Organism that is the product of schizogony or that will give rise to more schizonts by schizogony (Pr-19)

Scyphozoa Class of coelenterates (A-3); jellyfish

semen Complex nutrient fluid in which male gametes (sperm) are transferred to a female (A)

seminal receptacle Organ or vessel that receives semen (A)

seminal vesicle Saclike structure containing semen (A-32)

sensorium Entire sensory apparatus (A)

sensory plate Apical mass of nerve tissue in trochophore larvae (A-23)

sepal Flower part; one of the leaves of a calyx (Pl-9)

septum Cross wall in a plasmodium, trichome, or hypha (M, Pr, F)

sessile Attached, not free to move about

seta Bristle

sexual apex Sexually differentiated anterior or top part of a thallus (Pr-12, Pr-13)

sexual reproduction Reproduction leading to individual offspring having more than one parent

sheath External layer or coat of bacteria; typically composed of mucopolysaccharides (M)

sieve tube Vertical cellulose-lined tube made of sugar-conducting cells; component of the phloem of angiosperms (Pl-9)

siphon Tubular organ used for drawing in or ejecting fluids (Pr, A)

skeleton Hardened biogenic scaffolding, structural material often composed of calcium carbonate, silica, or calcium phosphate (Pr, A)

solenocyte Undulipodiated cell at the inner end of a protonephridial tube; distinguished from a flame cell by having only a single long undulipodium (A)

soma The parts of an organism's body that lack genetic continuity, as distinguished from its gametes (eggs, sperm) or reproductive structures (gemma, spores), which have genetic continuity through time

somatic cell Differentiated cell composing the tissues of the soma; any body cell except the germ cells (Pr, A, Pl)

somatic nucleus Nucleus incapable of further division or of fertilization followed by further division (Pr-17, Pr-18, A)

soredium Asexual reproductive structure of lichens; fragment containing both fungal hyphae and algal cells (F-5)

sorocarp Fruiting body (F, Pr, Pl)

sorogenesis Development of fruiting bodies (Pr-22, Pl-4)

sorus Cluster of sporangia or spores (Pl, F); for example, found on the underside of fern fronds (Pl-4)

spectroscope Instrument that measures the quality and quantity of light absorbed by translucent liquids or solids

sphincter Ring-shaped muscle capable of closing a tubular opening by constriction (A)

spicule Slender, typically needle-shaped biogenic crystal; for example, those secreted by sponge cells for skeletal support (A-2)

spindle pole body Nucleus-associated organelle (NAO), granulofibrosal and microtubular material found at the poles of mitotic spindles; centriole (Pr, F, A)

spirillum Spiral or helical filamentous cell (M-4, M-6, M-14)

spongin Horny, sulfur-containing protein that makes up the flexible skeleton of some sponges (A-2); related to the keratin of vertebrate hair and nails

spongocoel Central body cavity of a sponge (A-2)

sporangiophore Branch bearing one or more sporangia (F)

sporangiospore Spore borne within a sporangium (F)

sporangium Hollow unicellular or multicellular structure in which spores are produced and from which they are released (Pr, F, Pl)

spore Small or microscopic propagative unit containing at least one genome and capable of maturation, often desiccation- and heat-resistant (M, Pr, F, Pl)

spore wall Outer tough resistant complex layer of spore; composed of peptidoglycan impregnated with dipicolinic acid (M-2, M-4, M-11) or sporopollenin (Pl)

sporogony Multiple mitoses of a spore or zygote without increase in cell size; produces sporozoites (Pr-19)

sporophore Structure that bears spores (Pr, F, Pl)

sporophyll Leaflike appendage bearing sporangia (Pl)

sporophyte Spore-producing diploid plant (Pl)

sporoplasm Infective body (Pr-19); amoeboid organism within a spore (Pr-20)

sporozoite Life-cycle stage of apicomplexans (Pr-19); motile product of multiple mitoses (sporogony) of zygote or spore; usually infective

spur shoot Branch shaped like a spur (Pl-6)

squamous cell Flattened, platelike epithelial cell (A)

stamen Flower part; male stalk bearing pollen grains (Pl-9)

statoblast Asexually formed resistant internal bud of an ectoproct (A-16); under severe conditions, statoblasts are disseminated and grow into new individuals

statocyst (1) Organ of balance; a vesicle containing granules of sand or some other material that stimulate sensory cells when an animal moves (2) Silicified chrysophyte cell (Pr-4) that overwinters or resists desiccation

sterile Unable to reproduce

sternite Ventral part or shield of a somite of an arthropod (A-27)

sternum Breastbone (A-32)

steroid Member of a class of biogenic organic compounds composed of four carbon rings and attached chemical groups; class includes many hormones, such as testosterone, estrogen, cholesterol, and cycloartanol

stigma (1) Female flower part; receptive surface of carpel upon which pollen germinates (Pl) (2) Eyespot (Pr-10)

stipe Stalk of an organ or organism

stolon Horizontal stalk near the base of a plant or a colonial animal; produces new individuals by budding (A, Pl)

stoma One of many minute openings bordered by guard cells in the epidermis of leaves and stems; route of gas exchange between plant and air (Pl-9)

stone canal Calcareous tube leading from the madreporite to the ring canal of the water vascular system of echinoderms (A-29)

stratigraphy Field of geology that deals with the origin, composition, distribution, and succession of sediments (strata)

stria Stripe, striation

strobilus Cone; modified ovule-bearing leaves or scales grouped together on an axis (Pl)

stromatolite Laminated carbonate or silicate rocks, organosedimentary structures produced by growth, metabolism, trapping, binding, and/or precipitating of sediment by communities of microorganisms, principally cyanobacteria (M-7)

style Female flower part; slender column of tissue that arises from the top of the ovary and down through which the pollen tube grows (Pl-9)

stylet Rigid elongated organ or appendage (A)

substrate (1) Foundation to which an organism is attached; for example, a rock or carapace (2) Compound acted upon by an enzyme

sulcal groove In dinoflagellate hypocone (Pr-2), groove in which longitudinal undulipodium lies

supratidal Normally just above the reach of high tide but inundated during storms

symbiosis Intimate and protracted association between two or more organisms of different species

syncytium Multinucleate mass of cytoplasm lacking internal cell boundaries (membranes, walls) (Pr, A). See *coenocytic, plasmodium*

taproot Stout, tapering main root from which arise smaller, lateral branches (Pl-9)

telotroch Most posterior circlet of cilia on a trochophore larva (A-19, A-23)

tercite Body plate on the dorsal surface of arthropod (A-27)

terminal electron acceptor In a metabolic pathway, the compound finally reduced by the acceptance of electrons and converted to the compound to be excreted

test Shell, hard covering, valve, or theca (Pr, A)

tetrasporangium Structure in which tetraspores are formed (Pr-13, Pr-15)

tetraspore One of a set of four spores, products of meiosis, that will develop into haploid thalli (Pr-13, Pr-15, Pl)

thallus Simple flat leaflike body undifferentiated into organs such as leaves or roots (Pr, F, Pl-1)

theca Test, shell, valve, coat, hard covering, enveloping sheath or case

thermophil Organism whose optimum environmental temperature is well above 30°C

thorax In chordates, the chest, the part of the body that contains the heart and the lungs (A-32); in arthropods, three leg-bearing membranous segments between head and abdomen (A-27)

thylakoid Photosynthetic membrane, saclike membranous lamella that, stacked in numbers, constitutes the grana of chloroplasts

tinsel See *mastigoneme*

tissue Aggregation of similar cells organized into a structural and functional unit; component of organs (Pl, A)

tornaria larva Immature, ciliated form of certain enteropneust hemichordates (A-31)

trachea Air-conducting tube (A-27, A-32); windpipe (A-32)

tracheophyte Vascular plant; plant that contains xylem and phloem (Pl-2 to Pl-9)

trichocyst Organelle underlying the surface of many ciliates (Pr-18) and some mastigotes (Pr-8); capable of sudden discharge to sting prey

trichocyst pore Openings through which contents of a trichocyst are released (Pr)

trichome Filament consisting of a string of connected cells (M, F, Pr, Pl)

trochophore larva Free-swimming, ciliated marine larva (A-19, A-23)

trophi Specialized circlet of cuticle-covered cilia (undulipodia) forming the jaws of rotifers (A-10)

trophozoite Life cycle stage; growing, vegetative stage of Apicomplexa (Pr-19)

true branching Growth habit of a filamentous structure produced by the presence of three growth points on a single cell; see *false branching* (M, Pr, F)

trypanosomiasis Disease caused by trypanosome infection (Pr-8); borne by the tse-tse fly

tube foot Hydraulically controlled foot, part of the water vascular system of echinoderms (A-29)

ultrastructure Detailed structure of cells and organs; structure visible by transmission electron microscopy

undulipodium A cilium or eukaryote ''flagellum''; used primarily for locomotion and feeding (Pr, A, Pl); see the Introduction, Figure 2

uroid Tail-like protuberance

uterus In mammals, the chamber of the female reproductive tract in which the offspring develops and is nourished before birth (A-32)

valve Test, shell, hard covering (Pr, A)

vascular cylinder Column of tissue containing primary xylem and primary phloem (Pl)

vegetative apex Sexually undifferentiated anterior or top part of a thallus

vegetative cell Somatic cell, produced asexually by mitosis (distinguished from germ cells)

vein (1) Rib supporting an insect wing (A-27) (2) Vessel carrying blood from capillaries toward the heart (A-32) (3) Conduit for fluid in a leaf (Pl)

veliger larva Post-trochophoral stage of many mollusks (A-19)

velum A membranous part resembling a veil or curtain (M, A)

ventral On or toward the belly or undersurface; in human beings, toward the front (A)

vermiform Worm-like in form

viscera Collective term for the internal organs (A)

vitellarium Part of the ovary; produces yolk-filled cells that nourish the eggs (A)

viviparous Of a mode of sexual reproduction in which offspring develop inside the maternal parent, from which they receive nutritional and other metabolic aids; offspring are released either immature or as miniature adults (A)

wall See *cell wall*

water vascular system Ambulacral system; set of hydraulic canals derived from the coelom and equipped with tube feet (podia) used for gas exchange, movement, food handling, and sensory reception (A-28)

weed Unwanted plant (not a taxonomic or botanical term, usually Pl-9)

xanthin Member of a class of oxidized isoprenoid derivatives; typically yellow or orange

xylem Plant vascular tissue through which most of the water and minerals are conducted from the root to other parts of the plant; constitutes the wood of trees and shrubs (Pl-2–Pl-9)

zonite Body segment of a kinorynch (A-11)

zoochlorella Green alga living symbiotically within protists or animals (M-8, Pr-15)

zooid Colonial individual that resembles but is not a separate organism (A)

zooplankton Free-floating heterotrophic microorganisms (Pr, A)

zoospore Undulipodiated motile cell capable of germinating into a different developmental stage without being fertilized (Pr)

zooxanthella Yellow alga, often dinoflagellates of the genus *Gymnodinium* (Pr-2), living symbiotically within protists or animals

zygospore Large multinucleated resistant structure (resting spore) that results from the fusion of two gametangia (should be called zygosporangium) (F-1)

zygote Diploid nucleus or cell produced by the fusion of two haploid cells and destined to develop into a new organism (Pr, F, A, Pl)

zygotic meiosis Meiosis that takes place in a zygote immediately after it is formed

INDEX

Page numbers in *italics* refer to an illustration.
For additional genera, see p. 273.

gamete, 11–12
  chlorophyte, 102
  in coelenterate life cycle, *171*
  echiuran, 215
  nucleus, 11
  plasmodiophoran, 128
  sipunculan, 213
gametic meiosis, 11
gametocyst, 88
gametogenesis, 82, 84
gametophyte, 12, 96, 256
gamma-carotene, 80
gamma particle, 133
gamma radiation, 52
gamonts, 110
Gamophyta, 14, 100–101, 102
ganglia, 190, 210, 222, 232
gas, 26, 41, 50, 54, 56
gas exchange, 210, 216, 228
gas gangrene, 32
gas vacuole, 41
Gasteromycetes, 152
gastric gland, 188, *189*
gastrodermis, 170
Gastropoda, 206
Gastrostyla, 112, 113
Gastrotricha, 186–187
  relation to gnathostomulids, 184
  relation to kinorhynchs, 190, 191
  relation to turbellarians, 180
  resemblance to arrow worms, 233
gastrovascular cavity, 170
gastrulation, 229, 230
Gavia, 239
Geastrum, 152
gel, 182, 188
gelatin, 56
Gelidium, 98
Gelliodes, 169
gemmae, 135
gemmule, 169
genera. *See* genus
generative nucleus, 110, *111*
genes, 74, 75
Genicularia, 100
genital pore, *185,* 218
genital wing, *235*
genome, 74
genus, 3, 18, 31, 36
Geococcyx, 239
Geonemertes, 182
Geothallus, 252
Geotrichum, 154
Gephyrocapsa, 80

germ, 25
germinarium, 188
germovitellarium, *189*
germ tube, 135
giant kelp, 69
giant sequoia, 264
giant squid, 206
gill slits, 234, *239*
gills, 118, 152, 214, 216, 230
Ginkgo, 262
Ginkgo biloba, 262, 263
Ginkgophyta, 260, 262–263, 270
girdle, 74, *75, 229*
Glabratella, 110
gland, 185, 228
Glandiceps, 234
glass sponge, 168
gliding bacteria, 62
Globigerina, 110
Gloeocapsa, 42
Glossobalanus, 234
Gluconobacter, 57, 58
glucose
  breakdown of, 26
  fermented by Aphragmabacteria, 30
  fermented by yeasts, 150
  metabolized by bacilli, 50
  metabolized by micrococci, 52
  and Omnibacteria, 56
glucose polymer, 82
Glugea, 118
Glugea stephani, 119
glutamate, 52, 54
glutamic acid, 56
glycerol diether, 48
glycine, 268
glycogen, 62, 72
glycogen body, 73
glycosidic carotenoid, 44
Gnathostoma, 238
Gnathostomula, 184
Gnathostomulida, 184–185
Gnetophyta, 266–267
Gnetum, 266
goats, 220
gold, 28
golden-yellow algae, 14, 94, 95
golden-yellow pigment, 78
golden-yellow plastid, 80
Golfingia, 212
Golgi body
  in *Chlamydomonas, 103*
  in coccolithophorids, 80
  in dinoflagellates, *75*

  in *Euglena, 83*
  in eukaryotes, *9*
  lack of in *Pelomyxa, 72*
  in *Melosira, 95*
  in *Ophiocytium, 91*
gonads
  in *Craspedacusta, 171*
  in echinoderms, 230, *231*
  in hemichordates, 234
  in kinorhynchs, 190
  in mesozoans, 178, *179*
  in nemertines, 182
  in pogonophorans, 228
  in priapulids, 210, *211*
Gonatozygon, 100
Goniotrichum, 98
Gonium, 102
gonoduct, 190, 222
gonopore, *223*
gonorrhea, 57
Gonyaulax, 74
Gonyostomum, 90
Gordius, 198, 199
Gorilla, 239
Gram, Hans Christian, 26
Gram test, 26
Grantia, 168
granules, 55, 101
grasshoppers, 198, 226
Graz Zoological Institute, 166
green algae, 100, 102
  ancestors of green land plants, 249
  beta-carotene in, 44
  Chlorophyta, 102
  Gamophyta, 100
  life cycle in relation to fungi, 14
  as sponge symbionts, 168
green chloroplasts, 82
green hydra, 58
green sulfur bacteria, 24, 40
Greenland, 229
Gregarina, 114
Gregarinasina, 114
ground pine, 254
guanine, 52, 179
gullet, *113*
gut
  of gastrotrichs, 186
  of gnathostomulids, 184, *185*
  of kinorhynchs, 190
  of platyhelminths, 180
  of rotifers, 188
Guttulina, 124
Guttulinopsis, 124

tail fin, *233*
Takakia, 252
tannic acid, 26
tapeworms, 180
tar, 264
Taraxacum, 270
Tardigrada, 218–219, 220
Targionia, 252
tarsiers, 240
tartrate, 56
Tatjanellia, 214
Taxodium, 264
taxon, 3
taxonomy, 3–4, 18
Taxus, 264
tea, 266
teeth, 112, *185, 233*
Telomyxa, 118
Telosporidea, 114
telotroch, 217
Temnogyra, 100
temperature indicators, 233
Temple University, 166
Tenebrio, 224
tentacle
    of coelenterates, 170, *171, 172,* 173
    of ctenophores, 174, *175, 177*
    of ectoprocts, 200, *201*
    of entoprocts, 194, *195*
    of flatworms, 180
    of hemichordates, 234
    oral, of tunicates, *238*
    of phoronids, 202, *203*
    of pogonophorans, 228, *229*
    of rotifers, 188
    of sipunculids, 212, 213
    of starfish, *231*
Terebratulina, 204
tergite, *225*
terminal electron acceptor, 52
termites, 34, 35, 59, *225, 226*
Tertiary Period, 17
test, 76, 94, 95, 110, 214
    dinoflagellate, 74, 188
testes
    of ctenophores, 174
    of gnathostomulids, 184, *185*
    of kinorhynchs, 190
    of platyhelminths, 180
    of salamanders, *237*
tetrad, 52, *151*
Tetrahymena, 112
Tetramyxa, 128

Tetranchyroderma, 186, 187
Tetraphis, 252
Tetrapoda, 238, 239
Tetraspora, 102
tetrasporangium, 98
tetraspore, 98
Textularia, 110
Thalassema, 214
Thalassicola, 104
Thalassiosira, 94, 95
Thalassiosira nordenskjøldii, 94
Thaliacea, 238
Thallochrysis, 78, 79
thallose gametophyte, 252
thallus
    of algae, 184
    of brown algae, 96, *97*
    of bryophytes, 252
    of chlorophytes, 103
    of chytrids, 132
    of lichens, 156, *157*
    of liverworts, 252
    of oomycotes, 135
    of red algae, 98, *99*
    of xanthophytes, 91
Thecamoeba, 76
Thecina, 76
Themiste, 212
Theobroma, 270
Theria, 239
Thermoactinomyces, 60
thermophils, 50
Thermoplasma, 30, 31
Thermoplasma acidophilum, 31
thiaminase, 256
thiamine, 256
thimble jelly, 174
Thiobacillus, 54
Thiobacillus ferrooxidans, 54
Thiobacterium, 54
Thiocapsa, 40
Thiocystis, 40
Thiodictyon, 40
Thiomicrospira, 54
Thiopedia, 40
Thiopneutes, 24, 29, 36–37
Thiosarcina, 40
Thiospira, 54
thiosulfate, 54
Thiothece, 40
Thiovelum, 54
thoracic pleurite, *225*
Thuja, 264

thylakoid, *9*
    in cyanobacteria, 42, *43*
    in haptophytes, 80
    in *Ochromonas,* 79
    in *Rhodomicrobium,* 41
Thyone, 230
tibia, *237*
ticks, 226
tiger salamander, *236*
Tinamus, 239
tinsel, 87
tintinnids, 112
toads, 239
tobacco mosaic virus, 14
toilet plunger, 214
Tokophyra, 112
Tolypella, 102
tongue worms, 220, *221,* 234
tornaria, 234
Tortula, 252
toxins, 250
Toxoplasma, 114
Trachelomonas, 82
tracheophytes, 249, 250
Trachydemus, 190
Tradescantia, 270
transduction, 13
transmission electron microscopy, 4,
    20, *21*
transverse fission, 112, 213
transverse undulipodium, *75*
Trebouxia, 102, 156
Trematoda, 180, 208
Tremella, 152
trepang, 230
Treponema, 34
Treponema pallidum, 34
Triactinomyxon, 118
Triassic Period, 17, 110, 184
Tribonema, 90
Trichechus, 239
Trichida, 126
Trichinella, 196
trichinosis, 196
trichocyst, *75*
trichocyst band, *85*
trichocyst pore, *75*
Trichomitus, 89
Trichomonadidae, 88
Trichomonas, 89
Trichonympha, 89
Trichonympha ampla, 88
Trichophyton, 154